Experimental Fluid Mechanics

R.J. Adrian · M. Gharib · W. Merzkirch
D. Rockwell · J.H. Whitelaw

Springer
Berlin
Heidelberg
New York
Barcelona
Hong Kong
London
Milano
Paris
Singapore
Tokyo

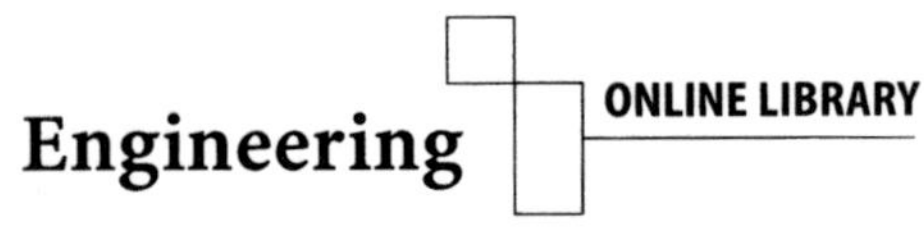

Arnold Frohn · Norbert Roth

Dynamics of Droplets

With 193 figures

Springer

Series Editors

Prof. R.J. Adrian
University of Illinois at Urbana-Champaign
Dept. of Theoretical and Applied Mechanics
216 Talbot Laboratory
104 South Wright Street
Urbana, IL 61801
USA

Prof. M. Gharib
California Institute of Technology
Graduate Aeronautical Laboratories
1200 E. California Blvd.
MC 205-45
Pasadena, CA 91125
USA

Prof. Dr. Wolfgan Merzkirch
Universität Essen
Lehrstuhl für Strömungslehre
Schützenbahn 70
45141 Essen
Germany

Prof. Dr. D. Rockwell
Lehigh University
Dept. of Mechanical Engineering and Mechanics
Packard Lab.
19 Memorial Drive West
Bethlehem, PA 18015-3085
USA

Prof. J.H. Whitelaw
Imperial College
Dept. of Mechanical Engineering
Exhibition Road
London SW7 2BX
UK

Authors

Prof. Dr. rer. nat. Arnold Frohn
Dr.-Ing. Norbert Roth

Universität Stuttgart
Institut für Thermodynamik der Luft- und Raumfahrt
Pfaffenwaldring 31
70550 Stuttgart
Germany

ISBN 3-540-65887-4 Springer Verlag Berlin Heidelberg New York

Cataloging-in-Publication Data applied for
Frohn, Arnold: Dynamics of droplets / Arnold Frohn ; Norbert Roth. – Berlin ; Heidelberg ; New York ; Barcelona ; Hong Kong ; London ; Milano ; Paris ; Singapore ; Tokyo : Springer, 2000
ISBN 3-540-65887-4

Coverdesign: design & production, Heidelberg
Typesetting: Digital data supplied by author
SPIN: 11933380 61/3111 Printed on acid-free paper – 5 4 3 2 –

Preface

The main purpose of this book is the description of dynamic aspects of droplet behavior under various ambient conditions. Engineers and scientists from a large variety of disciplines have become interested in this topic during the last years. Many new applications have been found in different modern technologies as combustion, ink jet printing, spray drying, net shape forming, or microencapsulation and manufacturing. The numerical simulation of dynamic properties of single droplets or droplet systems depends often on information, which must be gained in experiments. The book describes experimental tools, which are appropriate for the investigation of droplets. The book, which has been written for the experimental worker, contains research results from many different scientific fields. A large amount of the presented results is based on experiments performed at Institut für Thermodynamik der Luft- und Raumfahrt, University of Stuttgart (ITLR).

The first chapter, Chapter 1, of this book gives a brief review of the theoretical background, which is essential for the understanding of the behavior of droplets in various environments. A simple approach has been chosen, which provides a short but comprehensive survey of the various mechanical, thermodynamical, physical, and physico-chemical phenomena. The basis for the derivations has been outlined. Ample references have been provided for those, who wish a deeper understanding of the fundamentals. Chapter 2 is devoted to the generation of droplets. The generation of sprays, droplet streams, and single droplets is discussed. In Chapter 3 droplet systems, sprays, linear and planar droplet arrays are discussed. Three-dimensional droplet arrays are considered as a model for the simulation of sprays. Different experimental arrangements for the study of single droplets are considered. The most important measurement techniques for studying the various droplet systems are discussed in Chapter 4, whereas Chapter 5 deals with mechanical interactions of droplets with the surrounding gas, with solid walls, and with other droplets. In Chapter 6 experiments for studying phase transition processes are presented. Experimental setups for the investigation of evaporation or combustion, condensation, and freezing are described and results are presented. Chapter 7 deals with a selected set of practical applications. The material of this chapter may help to avoid the impression that the subject of droplet dynamics is a definitive closed body of knowledge.

The authors are indebted to Professor Dr. Merzkirch for the encouragement to write this book. Contributions of Klaus Anders during the initial planning phase will not be forgotten. Klaus Anders contributed results on evaporating droplets, especially at low Knudsen numbers. Many scientists at ITLR contributed experimental results. Gerd Bauer performed droplet experiments near the critical point, Nils Widdecke and Wolfgang Klenk studied interactions between shock waves and droplets, Guido Funcke interactions between acoustic waves and droplets, Alexander Karl droplet-wall interactions, Thomas Kraut performed wind-tunnel experiments. Frank Herrmann studied the glare points of droplets and developed several electronic devices necessary for various experiments.

Numerical calculations are due to Peter Drtina, Markus Schelke, and Martin Rieber. Markus Schelkle developed a lattice Boltzmann code and Martin Rieber the Navier-Stokes code. These codes were used in Chap. 5 for the simulation of droplet-droplet and droplet-wall interactions. Martin Rieber contributed describing numerical results in Sects. 5.3.5 and 5.4.3. The authors are indebted to Ulrike Schröder and Jutta Schöllhammer for technical assistance received in many experiments described in this book. Harald Hettrich designed devices to examine droplets in shock tubes and under high pressure, the corresponding three-dimensional schematical views in Figs. 5.11 and 6.52 have been drawn by him. Without the help of Peter Fischer it had been impossible to provide so many references. The authors are very grateful to Professor Gouesbet and Professor Gréhan from INSA Rouen for their kind permission to use their computer code for calculations of light scattering by droplets illuminated with a laser beam with Gaussian intensity distribution.

Table of Contents

1. Theory

1.1 Surface Tension and Internal Pressure

The mechanical and thermodynamic behavior of systems composed of two or more phases may be influenced by thin interface regions between any two phases. The interfaces possible can be characterized by the three states of matter solid, liquid and gas. The interface between a liquid and its own vapor is called a surface. If sufficiently mobile the interface will assume an equilibrium shape. Most common examples are drops, meniscusses, liquid sheets surrounded by vapor, air or by another liquid as well as soap bubbles formed by thin films. The most obvious phenomenon associated with the interface is that of surface tension. From a macroscopic point of view the interface between a liquid and its vapor appears sharply defined with a discontinuous change of density and the other thermodynamic properties. Therefore the interface is often considered as if it were a thin, uniformly stretched skin or membrane. Surface tension σ is defined as the force acting per unit length across a line on this fictitious membrane. Under the operation of this force the fluid behaves as if enclosed in an envelope of constant tension. The

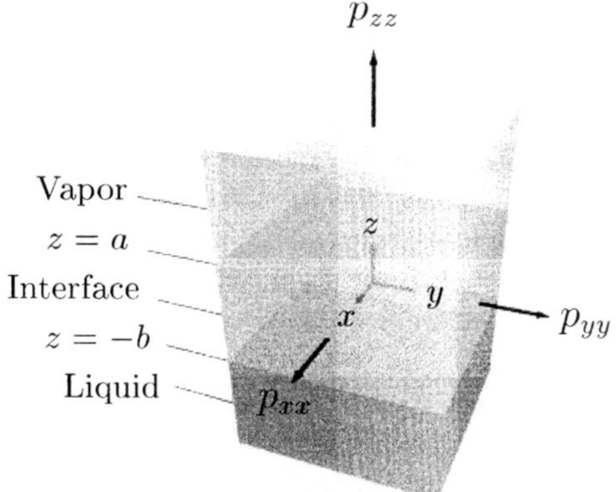

Fig. 1.1. Schematic representation of surface between liquid ($z < -b$) and vapor ($z > +a$). The transition from the density of the liquid to the density of the vapor occurs in the interface zone $-b < z < +a$

physical interface between droplet and surrounding vapor is of course not a geometrical surface of zero thickness but a thin boundary layer or film whose

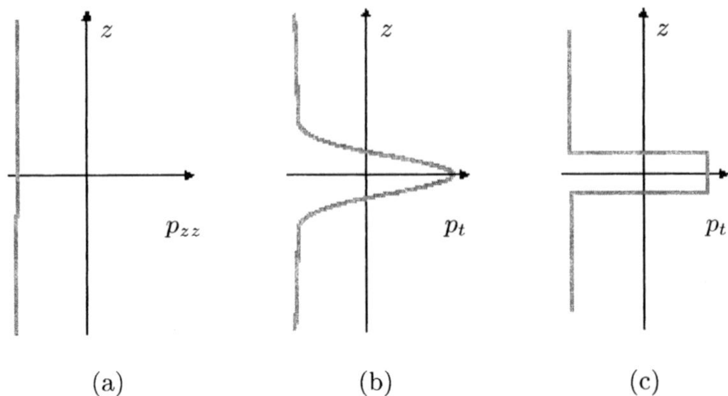

Fig. 1.2. Components of stress tensor across plane interface. The stress normal to the interface (a) must be constant $p_{zz} = -p$. The tangential stress components $p_{xx} = p_{yy} = p_t$ are depicted in (b). In (c) the surface tension is approximated by a thin skin of constant tension

thickness is of the order of a few molecular diameters, as shown schematically in Fig.. 1.1. Thus the density has very steep gradients in this surface layer between liquid and gas. In Fig. 1.2 a plane interface zone between a liquid and its vapor is considered. The z-axis of the Cartesian coordinate system is normal to the plane of the interface. The strong change of density in the interface zone causes an asymmetry in the intermolecular forces. As a consequence the tangential components of the pressure tensor p_{xx} and p_{yy} may differ from the hydrostatic pressure p. It is easy to see, that the component p_{zz} must remain constant in passing from the liquid to the vapor phase. The surface tension is given by the equation

$$\sigma = \int_{-b}^{+a} (p - p_{xx}) dz \ . \tag{1.1}$$

For most practical applications it may be assumed that the interface is a region of zero thickness containing no mass. In this model the density of the system is allowed to jump discontinuously across the interface between two phases whereas the temperature and the tangential velocity are continuous. The velocity normal to the interface is discontinuous when evaporation or condensation occurs. Many scientific articles have been devoted to mathematical descriptions of surface tension [1, 2]. The idea of surface energy is illustrated in Fig. 1.3. The small rectangular element of the interface of a drop with length x and width y is subjected to the surface forces $x\,\sigma$ and $y\,\sigma$. When the line elements x and y are displaced by dy and dx the work

$$dW = \sigma x\, dy + \sigma y\, dx = \sigma(x\, dy + y\, dx) = \sigma\, d(xy) = \sigma\, dA \tag{1.2}$$

will be performed. This formulation shows that the surface tension σ can be interpreted as energy per area

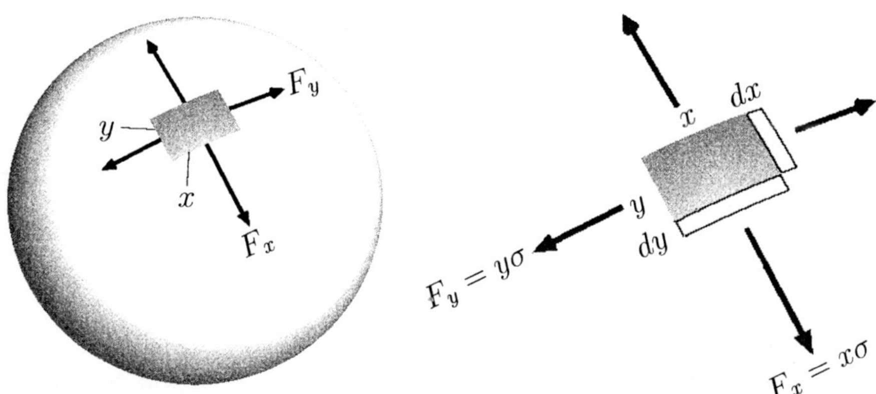

Fig. 1.3. Forces F_x and F_y resulting from surface tension along sides x and y of rectangle with area $A = x \cdot y$ on the surface of a droplet. The work $F_x dy + F_y dx$ is performed when the side elements x and y of the rectangular element are displaced by dx and dy

$$\sigma = \frac{dW}{dA} \; . \tag{1.3}$$

It should be mentioned that the surface tension in the example of Fig. 1.3 cannot be interpreted as force per unit length of the circumference of the entire rectangle.

From the concept of surface tension σ as a contractile skin it follows immediately that there must be a pressure difference Δp between the liquid enclosed in a droplet and the surrounding medium. Figure 1.4 shows the distribution of surface tension along a meridian circle of a spherical droplet. The force resulting from surface tension is $2\pi r\sigma$ and the force resulting from

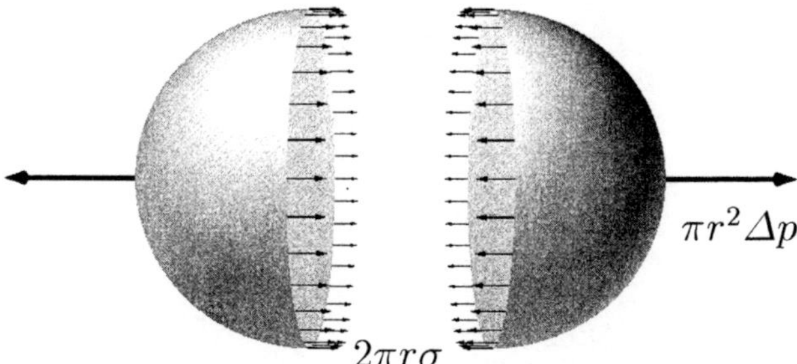

Fig. 1.4. The force resulting from surface tension along a meridian circle of the droplet is $2\pi r\sigma$. This force is in mechanical equilibrium with the force resulting from the pressure difference between the liquid in the droplet and the surrounding vapor $\pi r^2 \Delta p$

pressure $\pi r^2 \Delta p$. For mechanical equilibrium it follows

$$2\pi r\sigma = \pi r^2 \Delta p \tag{1.4}$$

or

$$\Delta p = \frac{2\sigma}{r} \ . \tag{1.5}$$

This relation is of course a special case of the well-known fundamental equation of Young and Laplace

$$\Delta p = \sigma \left(\frac{1}{r_1} + \frac{1}{r_2} \right) \ . \tag{1.6}$$

Anywhere in the bulk of the liquid the pressure tensor will be isotropic. Suppose that the z-direction of a Cartesian coordinate system is normal to the interface. In passing from the liquid to the vapor the tensor component p_{zz} will stay constant. The molecules in the interface will be attracted towards the liquid phase. Thus the liquid tends to minimize its surface area. Macroscopically this phenomenon is observed as surface tension. From a thermodynamic point of view the surface layer is often considered as a third phase with own properties. The energy of the surface of a spherical droplet with radius r is given by

$$E_s = \int_0^V \Delta p \cdot dV = \int_0^r \frac{2\sigma}{r} 4\pi r^2 dr = 4\pi\sigma r^2 \ . \tag{1.7}$$

Values of the surface tension are given for a selection of elements and compounds in Table 1.1. The data are from Ref. [3]. It should be mentioned that the substance of only one phase is specified. This means that the gaseous

Table 1.1. Surface tension σ for various liquids at temperature t

Substance	σ N/m	t °C
Helium	$.12 \cdot 10^{-3}$	-269
Helium	$.353 \cdot 10^{-3}$	-271.5
Hydrogen	$2.31 \cdot 10^{-3}$	-255
Nitrogen	$6.6 \cdot 10^{-3}$	-183
Nitrogen	$10.53 \cdot 10^{-3}$	-203
Oxygen	$13.2 \cdot 10^{-3}$	-183
n-Hexane	$18.43 \cdot 10^{-3}$	20
Isopropanol	$21.7 \cdot 10^{-3}$	20
n-Octane	$21.8 \cdot 10^{-3}$	20
Ethanol	$22.75 \cdot 10^{-3}$	20
Acetone	$23.70 \cdot 10^{-3}$	20
Sulfuric acid (98.5%)	$55.1 \cdot 10^{-3}$	20
Water	$72.75 \cdot 10^{-3}$	20

phase is assumed to consist of the vapor of the liquid with the saturation pressure corresponding to the temperature of the liquid. This true surface tension is usually measured when the substance in the liquid phase is very volatile. The surface tension of common liquids like water are often measured against the ambient air. For medical applications the surface tension of biological liquids is important. The surface tension of human blood is for instance in between 55.5 and $61.2 \cdot 10^{-3}\,\mathrm{N/m}$ and for saliva approximately $53 \cdot 10^{-3}\,\mathrm{N/m}$ [4, 5].

The interfacial layer depicted schematically in Fig. 1.1 has the finite thickness $\delta = a - (-b)$. The magnitude of δ has the order of molecular dimensions. Assuming $\delta = 10^{-9}\,\mathrm{m}$ and $\sigma = 20 \cdot 10^{-3}\,\mathrm{N/m}$ one finds for the average tangential stress $p_t = \sigma/\delta \approx 2 \cdot 10^{7}\,\mathrm{N/m^2}$.

Better knowledge of thermodynamic and mechanical properties of liquid metals is essential for accurate description of present-day liquid metal processing operation. For the understanding of metallurgical processes the knowledge of the surface tension of liquid metals is important. The surface tension is an essential property in melting and refining operations, in casting, brazing and sintering [3, 6-8]. Accurate measurements of the surface tension of liquid metals are not easy since impurities on the surface of liquid metals may have large influence on the surface tension. Most solute elements are highly surface active. Experimental values for the surface tension of liquid metals should therefore be considered with caution. Some of the problems have been discussed in the book of Iida and Guthrie [9]. These authors discuss also briefly the surface tension of binary mixtures of liquid metals. Values of the surface tension of molten metals are presented in Table 1.2. The source for these data is Ref. [9].

Table 1.2. Surface tension σ of pure liquid metals at their melting temperature

Metal	σ N/m
Caesium	$70 \cdot 10^{-3}$
Platinum	$195 \cdot 10^{-3}$
Gold	$197 \cdot 10^{-3}$
Mercury	$498 \cdot 10^{-3}$
Aluminium	$914 \cdot 10^{-3}$
Silver	$966 \cdot 10^{-3}$
Titanium	$1650 \cdot 10^{-3}$
Chromium	$1700 \cdot 10^{-3}$
Nickel	$1778 \cdot 10^{-3}$
Iron	$1872 \cdot 10^{-3}$
Vanadium	$1950 \cdot 10^{-3}$
Tungsten	$2500 \cdot 10^{-3}$

Table 1.3. Interfacial tension σ_i at temperature $t = 20°C$ for a selection of liquids against water

Second liquid	σ_i N/m
Butanol	$1.8 \cdot 10^{-3}$
Ethyl acetate	$6.8 \cdot 10^{-3}$
n-Heptane	$50.2 \cdot 10^{-3}$
n-Octane	$50.8 \cdot 10^{-3}$
Mercury	$375 \cdot 10^{-3}$

The concept of surface tension can be extended to the interface between two liquids. Instead of the term surface tension the term interfacial tension is used. The symbol σ_i has been chosen for the interfacial tension. The physical dimensions of both quantities σ and σ_i are the same. Values for the interfacial tension between water and a few other liquids are shown in Table 1.3. The data are from Refs. [3, 8]. The interfacial tension between two liquids which are fully miscible is of course zero. When an oil drop is put on a water surface a liquid lens is formed.

At the critical temperature T_c of a substance the difference between liquid and gas phase disappears. This means that the surface tension cannot exist, or $\sigma = 0$ for $T = T_c$. The linear relation

$$\sigma = \alpha(T_c - T_x - T) \tag{1.8}$$

has been suggested for the temperature dependence in the vicinity of the critical point. In this equation the coefficient α is characteristic for the liquid, and the constant T_x is often assumed to have the value 6°C . Other

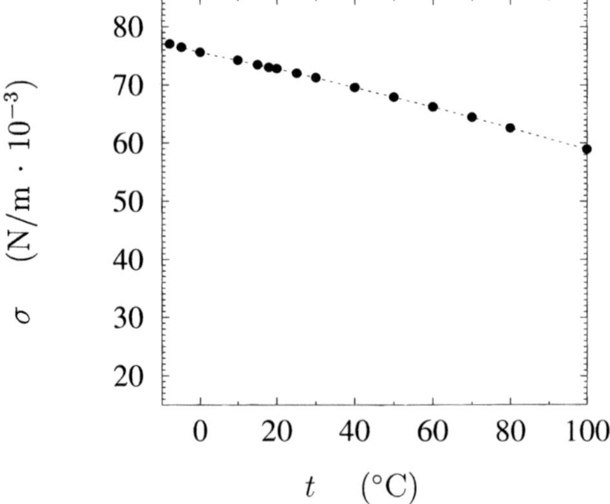

Fig. 1.5. Surface tension σ of water against air as a function of temperature

relations for the temperature dependence of σ may be found in the literature [7, 8]. Figure 1.5 shows the surface tension for water in the temperature range between −8°C and 100°C. The data are from Ref. [3].

The surface tension in mixtures depends of course on the concentrations of the components. Experimental data are available for many binary mixtures. Some results from Ref. [3] for binary aqueous mixtures are shown in Figs. 1.6 and 1.7. The surface of water-ethanol mixtures is shown in Fig. 1.6 as a function of the ethanol concentration. The values of surface tension of the pure liquids, ethanol and water, differ appreciably for this system. It can be seen that the surface tension of the mixture shows a marked decrease, when small amounts of ethanol are added to pure water. Figure 1.7 shows the surface tension for the system NaCl - water. In this case the surface tension rises with the concentration of NaCl [3].

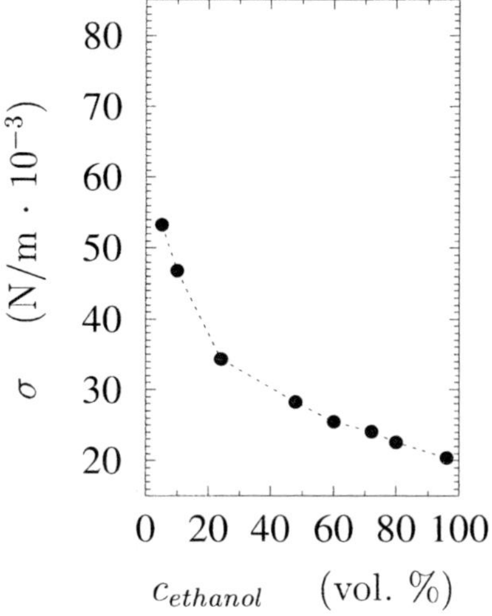

Fig. 1.6. Surface tension σ of binary water-ethanol mixture against air as a function of ethanol concentration at $t = 50°C$.

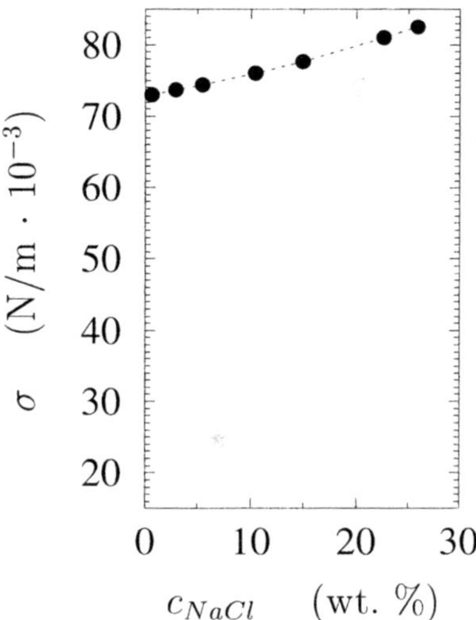

Fig. 1.7. Surface tension σ of water-NaCl solution against air as a function of NaCl-concentration at $t = 20°C$.

The dynamic behavior of the surface tension and the aging of fresh surfaces have been studied by different authors for pure liquids and solutions [7, 8, 10]. The surface tension of liquids can change due to dissolution of air in the liquid, due to chemical reactions or evaporation of one component. Another reason for a temporal change of surface tension of mixtures is the concentration difference existing in equilibrium between the surface of a droplet and the inner liquid. When a droplet is created from the bulk liquid this concentration difference must be achieved by diffusion. The thermodynamic explanation for the accumulation of one component at the surface of the droplet will be given in Sect. 1.9.4. The value of the surface tension is affected when the droplet radius becomes comparable with molecular dimensions. This effect is important for nucleation theories. The surface tension of extremely small droplets has been addressed in the literature [7, 11-13]. Effects of electrical charges on liquid surfaces are discussed for example in the book by Adamson [7]. Some properties of droplets carrying electrical charges will be described in Sect. 1.3.

1.2 Liquid-Liquid and Solid-Liquid Interfaces

So far surfaces between liquids and their vapor or a foreign gas have been discussed. For certain applications interfaces between two liquids or between a liquid and a solid are important. Liquid lenses on liquid surfaces are very common. An example for such a lens consisting of oil on water is shown schematically in Fig. 1.8. Indicated are the forces resulting from the interface tension and acting per unit length along the three-phase line. Mechanical equilibrium between these forces is possible only if

$$\frac{\sin\theta_1}{\sigma_{12}} = \frac{\sin\theta_2}{\sigma_{23}} = \frac{\sin\theta_3}{\sigma_{13}} \ . \tag{1.9}$$

This equation follows, when the law of sines is applied to the triangle formed by the interface forces.

The shape of a liquid droplet placed on a solid surface is shown schematically in Fig. 1.9. The contact angle between the liquid (2) and the solid (3) is θ. It is assumed, that Young's equation holds. This equation, which has the form

$$\sigma_{13} = \sigma_{23} + \sigma_{12}\cos\theta \ , \tag{1.10}$$

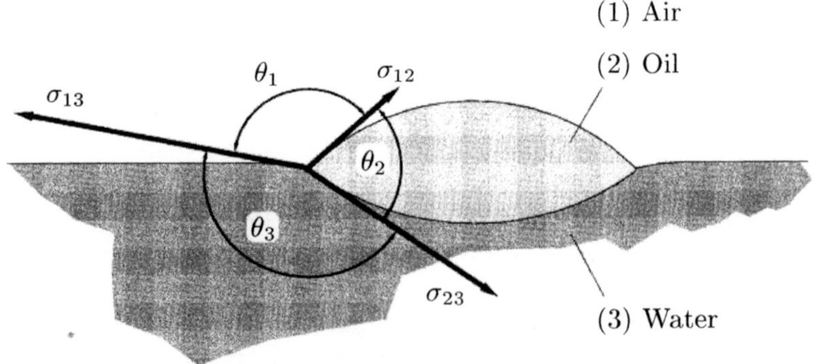

Fig. 1.8. Liquid lens on liquid surface

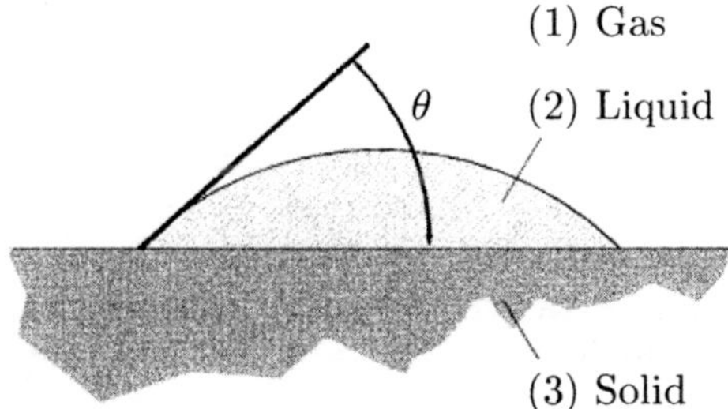

Fig. 1.9. Liquid droplet on solid surface with contact angle θ. For $\theta \rightarrow 0°$ complete wetting of the solid occurs, whereas for $\theta \rightarrow 180°$ the contact between the two phases and liquid would disappear

is often derived by considering the mechanical equilibrium between the horizontal components of the forces acting per unit length along the 3-phase line. A critical discussion of this equation with objections against it are given in Ref. [8]. Contact angles existing at the boundary line between the three phases vapor, liquid, and solid depend on the physical and chemical properties of each of these.

In various practical applications the terms wetting and nonwetting are used [14, 15]. Wetting means that the contact angle θ between a liquid and a solid is so close to zero, that spreading of the liquid occurs. The surface of a solid is said to be completely wetted if the contact angle is zero. Nonwetting is often defined by $\theta > 90°$. A contact angle of 180° would mean, that there is no adhesion between the liquid and the solid.

When wetting occurs the spreading coefficient defined by

$$S_{23} = \sigma_{13} - \sigma_{23} - \sigma_{12} \tag{1.11}$$

is important. For technical applications wetting is often achieved by adding a surfactant to the liquid [7].

There are many situations where good contact between a solid surface and a liquid is important. Examples include paint sprays, ink, solder, lubricants and detergents. In other situations repelling of a liquid is desired. This means that the contact angle θ should be large. Practical applications are water-repellent or waterproof fabric. More examples will be mentioned in Sect. 7.32.

The concept of the critical surface tension has been proposed as a measure of the wettability of solids. The critical surface tension of a solid is defined as the surface tension, which a liquid has that just wets the solid. It has been observed that the cosine of the advancing contact angle θ for various series of liquids is a function of the surface tension of the liquid. The empirical equation

$$\cos\theta = 1 + b(\sigma_{crit} - \sigma_{liq}) \tag{1.12}$$

has been proposed for describing the relation between contact angle θ, critical surface tension of the solid σ_{crit}, and surface tension of the liquid σ_{liq}, with b as a constant. Different aspects of equilibrium, kinematic, thermodynamic, and dynamic phenomena of interfacial transport phenomena are described in the literature [16, 15].

1.3 Charged Droplets

Natural sprays of water and other fluids consist of droplets which carry electrical charges. Charging mechanisms of liquid droplets and electrical properties of charged droplets are very complex subjects. The main reason for droplet charges are the existence of charge carriers and ions within the liquid before atomization. The droplets formed by atomization of the liquid carry a random excess of either positive or negative charges. Hydrodynamic processes during the disruption of liquid into a spray influence the droplet size distribution as well as the charge distribution over the droplet population. At high levels of charge carried by droplet systems sparking may occur. This is apparent from atmospheric lightning. Bailey points out that during water jet washing of very large oil cargo tanks space potentials of several thousand volts may occur [17]. In the presence of an external electrostatic field during the atomization process the charging may be strongly enhanced. The different charging mechanisms include induction charging, corona charging, diffusion charging, and field charging [17]. Smoluchowsky suggested a random distribution of charges which can be expressed by a Gaussian distribution.

In technical applications it is often advantageous to charge the droplets of a spray, since the trajectories of charged droplets can be influenced with the purpose to distribute the liquid onto a target. This applies for example to ink-jet printers as well as to paint and crop spraying.

Surface charges of a droplet cause an outward electrostatic pressure acting in opposition to the inner pressure, which arises from surface tension. Forces resulting from surface tension stabilize the droplet shape, whereas the electrostatic forces resulting from the charges tend to disintegrate the droplet. With increasing charges a droplet becomes instable. This limit is reached, when the effect of surface tension forces is compensated by the electrostatic forces. The droplet can be considered as a spherical electrical capacitor of capacity $C = 4\pi\epsilon_0 r$, where $\epsilon_0 = 8.842 \cdot 10^{-12}\,\mathrm{Coulomb\,Volt^{-1}\,m^{-1}}$ is the permittivity of vacuum. The electrostatic energy corresponding to the charge Q is $Q^2/(2C) = Q^2/(8\pi\epsilon_0 r)$, whereas according to Eq. 1.7 the surface energy is $4\pi\sigma r^2$. The total potential energy of the droplet is therefore

$$W_{tot} = \frac{Q^2}{8\pi\epsilon_0 r} + 4\pi\sigma r^2 \; . \tag{1.13}$$

The total energy W_{tot} has a minimum at the so-called Rayleigh radius

$$r = r_{Ray} = \sqrt[3]{\frac{Q^2}{64\pi^2\epsilon_0\sigma}} \; . \tag{1.14}$$

The charge in the so-called Rayleigh limit is

$$Q_{Ray} = 8\pi\sqrt{\epsilon_0\sigma r^3} \; , \tag{1.15}$$

whereas the total energy in the Rayleigh limit is

$$W_{ges,Ray} = 12\pi\sigma r^2 \ . \tag{1.16}$$

The mobility b of a charged particle moving with velocity v in an electric field E is defined by the equation

$$b = \frac{v}{E} \ . \tag{1.17}$$

When Stokes law is valid one has $QE = 6\pi\eta r v$ and it follows for the mobility

$$b = \frac{Q}{6\pi\eta r} \ . \tag{1.18}$$

The maximum mobility is obtained in the Rayleigh limit. From Eqs. 1.15 and 1.18 it follows

$$b_{Ray} = \frac{4\sqrt{\epsilon_0 \sigma r}}{3\eta} \ . \tag{1.19}$$

For the acceleration and manipulation of charged droplets in electrostatic fields the charge-to-mass ratio Q/m is an important parameter. In the Rayleigh limit one has

$$\left(\frac{Q}{m}\right)_{Ray} = \frac{6\sqrt{\epsilon_0 \sigma}}{\varrho r^3} \ . \tag{1.20}$$

The Rayleigh limit may be reached for example in the laboratory by controlled evaporation of charged droplets. According to Ref. [17] charge is not lost during evaporation until the Rayleigh limit is reached and ejection of highly charged very tiny droplets occurs. The levitation of single droplets in electrostatic fields will be described in Sect. 3.4.4. Soot control by charging fuel droplets has been studied in Ref. [18]. Some aspects of electrical phenomena at interfaces are touched on in Ref. [19], whereas the electrical state of atmospheric clouds has been described in Ref. [20]. Charge-carrying particles from boiling were studied by Pounder [21, 22].

Induction charging of droplets has been studied in a laboratory experiment by Lord Kelvin using the famous water-dropper apparatus depicted schematically in Fig. 1.10. Water drips simultaneously from the two nozzles N_1 and N_2. The droplets formed after disintegration of the laminar water jets will carry small electrical charges. Details of the disintegration of liquid jets will be discussed in Sect. 1.7. A random difference between the charges of the two droplet streams will result in different charges of the two isolated collector cups C_1 and C_2. As a consequence the electrical potential of the two ring electrodes R_1 and R_2 will differ as shown in the figure. Due to induction negatively charged droplets will break away when a ring electrode has a positive potential with respect to the nozzle. The feed-back effect due to cross connecting of the collectors and ring electrodes leads to a rapid increase of the potential difference between the collector cups C_1 and C_2. Bailey reports that a voltage between 10 and 20 kV is obtained within a few minutes [17, 23, 24].

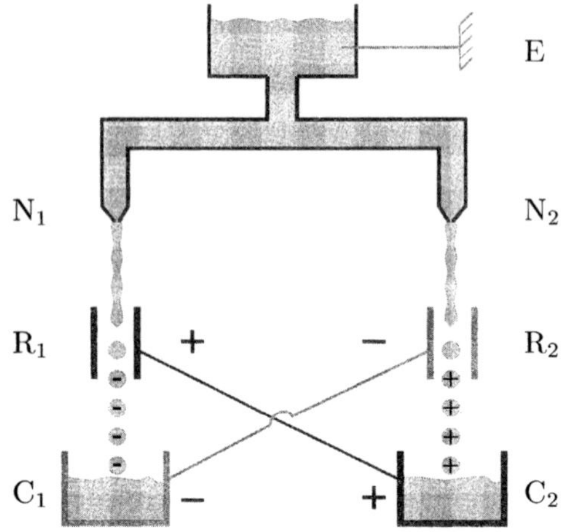

Fig. 1.10. Schematic of Lord Kelvin's water dropper apparatus with earthed water reservoir E, nozzles N_1, N_2, ring electrodes R_1, R_2, and electrically isolated collector cups C_1, C_2. A negative charge in C_1 will cause the production of positively charged droplets from N_2

1.4 Small-Amplitude Droplet Oscillations

The analysis of droplet oscillations is of considerable interest for a wide variety of phenomena both in nature and industry. Examples are the behavior of raindrops or the fluids in biological cells. The restoring force maintaining mechanical equilibrium is a result of surface tension. Consider small perturbations of a liquid drop held together by surface tension at the free boundary. The discussion will be restricted to the case when the effect of the external density can be neglected [25]. An expression for the oscillation frequency of inviscid droplets has been presented by Lamb [26]. The normal-mode analysis of inviscid liquid droplets has been described in detail in the published literature [25, 27, 28]. The velocity potential of the liquid within the droplet satisfies the equation

$$\Delta\psi = 0 \; . \tag{1.21}$$

In spherical coordinates the surface of the droplet is given by an equation of the form $r = r(\theta, \varphi)$. For the undisturbed surface of the droplet this equation is $r = R$, where R is the radius of the undisturbed droplet. For small deviations from the spherical surface the configuration is described by

$$r = R + \zeta(\vartheta, \varphi, t) \qquad \text{with} \quad \zeta \ll R \; . \tag{1.22}$$

The pressure difference across a curved surface is given by Eq. 1.6, which here takes the form

$$p - p_0 = \sigma \left(\frac{1}{R_1} + \frac{1}{R_2} \right) \; , \tag{1.23}$$

where R_1 and R_2 are the principal radii of curvature, p is the pressure in the droplet, and p_0 is the constant outer pressure. Equation 1.23 must be satisfied on the deformed boundary of the droplet. Following Landau and Lifschitz for a slightly disturbed droplet with $r = R + \zeta$ the sum of the reciprocals of the two principal radii of curvature can be expressed as

$$\frac{1}{R_1} + \frac{1}{R_2} = \frac{2}{R} - \frac{2\zeta}{R^2} - \frac{1}{R^2}\left\{\frac{1}{\sin^2\theta}\frac{\partial^2\zeta}{\partial\varphi^2} + \frac{1}{\sin\theta}\frac{\partial}{\partial\theta}\left(\sin\theta\frac{\partial\zeta}{\partial\theta}\right)\right\} . \quad (1.24)$$

With

$$\frac{\partial\zeta}{\partial t} = v_r = \frac{\partial\psi}{\partial r} \quad (1.25)$$

and taking into account that p_0 and R are constants, one obtains the equation

$$\varrho\frac{\partial^2\psi}{\partial t^2} = \frac{\sigma}{R^2}\left\{2\frac{\partial\psi}{\partial r} + \frac{\partial}{\partial r}\left[\frac{1}{\sin^2\theta}\frac{\partial^2\psi}{\partial\varphi^2} + \frac{1}{\sin\theta}\frac{\partial}{\partial\theta}\left(\sin\theta\frac{\partial\psi}{\partial\theta}\right)\right]\right\} . \quad (1.26)$$

It is assumed that the solution has the form of a stationary wave

$$\psi(r,\theta,\varphi,t) = \Psi(r,\theta,\varphi)\cdot e^{-i\omega t} . \quad (1.27)$$

It can be shown that $\Psi(r,\theta,\varphi)$ satisfies Laplace´s equation, $\Delta\Psi = 0$. This means that $\Psi(r,\theta,\varphi)$ can be expressed by Legendre polynomials $P_l(cos\theta)$ or spherical harmonics $Y_e^m(\theta,\varphi)$, with $l = 0, 1, 2, \ldots$. For the oscillation frequency it follows

$$\omega^2 = \frac{l(l-1)(l+2)}{R^3\varrho_i}\sigma . \quad (1.28)$$

In this equation σ is the surface tension of the droplet liquid, ϱ_i the density of the droplet liquid, R the radius of the undisturbed droplet and l describes the mode of vibration. Equation 1.28 defines the frequencies of oscillation of the normal modes in the absence of viscosity. The mode $l = 0$ describes movement, mode $l = 1$ expansion of the droplet. For the first vibration mode one has therefore $l = 2$. The various resonance modes are uncoupled [27, 28]. An arbitrary disturbance can be expressed by a superposition of the basic modes. For the lowest mode with $l = 2$ it follows

$$\omega_{min} = \sqrt{\frac{8\sigma}{R^3\varrho_i}} . \quad (1.29)$$

When the density of the ambient medium is not zero the frequency is given by the expression

$$\omega^2 = \frac{l(l+1)(l-1)(l+2)}{R^3[(l+1)\varrho_i + l\varrho_a]}\sigma , \quad (1.30)$$

where ϱ_a is the density of the ambient medium. Droplet shapes for the oscillation modes $l = 2$ to $l = 7$ with rotational symmetry are depicted qualitatively in Fig. 1.11.

Frequency oscillations of droplets have been studied by many authors in both theory and practice [25-30]. The oscillations of a fluid droplet immersed in another fluid have been investigated theoretically by Miller and Scriven [31]. Small-amplitude oscillations of droplets immersed in a host liquid have

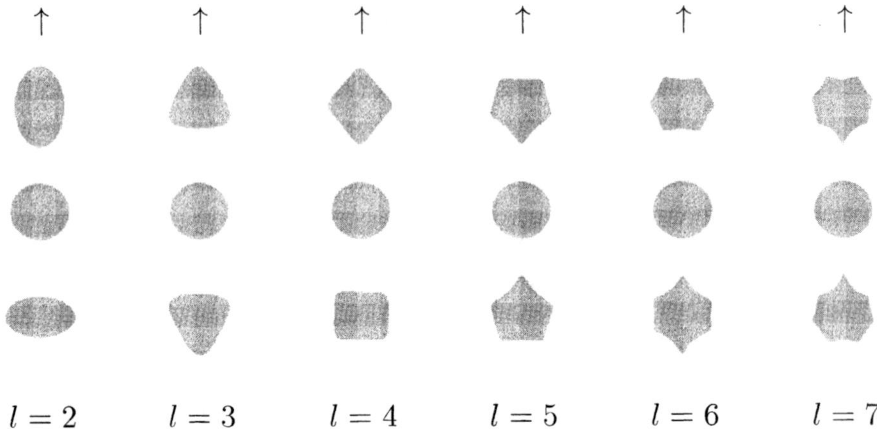

Fig. 1.11. Droplet shapes for oscillation modes $l = 2$ to $l = 7$

been investigated experimentally by Trinh et al. using acoustic levitation [29]. A numerical method for studying liquid droplet behavior is given by Foote [32]. Other investigators have studied nonlinear [33, 34] and viscous [35-37] effects. Rayleigh has studied oscillations of droplets carrying electrical charges [38]. He has shown that the angular frequency for a droplet carrying the electrical charge Q is given by the expression

$$\omega^2 = \frac{l(l-1)}{R^3 \varrho_i} \left((l+2)\sigma - \frac{Q^2}{16\pi^2 \epsilon_0 R^3} \right) . \tag{1.31}$$

An oscillation of the mode $l = 2$ will grow for $\omega^2 < 0$ or $Q^2 > 64\pi^2 \epsilon_0 \sigma R^3$. This result is in agreement with Eq. 1.15.

1.5 Internal Circulation

Available experimental and theoretical investigations reveal that the shear stress at the interface between moving droplets and the surrounding gas can become large enough to induce internal circulation. Fluid motion within droplets enhances the rate of mixing, as well as evaporation and combustion. Internal circulation can also be generated by gradients of surface tension, which result from temperature or concentration variations at the interface between the liquid phase and the surrounding gas. This thermocapillary convection is known as Marangoni convection. Many papers dealing with different aspects of internal circulation have been published in the last years [39-45]. Other authors presented experimental results for the internal circulation obtained with flow tagging velocimetry [46] or investigated the internal circulation and shape of droplets in wind tunnels [47]

1.6 Instability of Droplets

Droplets moving through air or another gas may become dynamically unstable and eventually break up. Distortions of free-falling droplets have been studied extensively by many authors both theoretically and experimentally. The type of distortion caused by aerodynamic forces depends on the flow field around the droplet. When the viscosity of the liquid is low the deformation is determined primarily by aerodynamic forces and forces resulting from surface tension. The balance of these two forces requires

$$2\pi r\sigma = C_D \frac{\varrho_a v^2}{2} \pi r^2 \ , \tag{1.32}$$

where v represents the relative velocity between droplet and the surrounding air, ϱ_a the density of the air, and C_D the drag coefficient. Introducing the Weber number

$$We = \frac{2r\varrho v^2}{\sigma} \tag{1.33}$$

one obtains the relation

$$We = \frac{8}{C_D} \tag{1.34}$$

as condition for breakup of a droplet. For a given droplet diameter the critical relative velocity can be determined from Eq. 1.32. When the relative velocity is given, the maximum diameter of a stable droplet follows. The influence of viscosity and turbulence on droplet breakup are discussed in the literature [48]. Droplet breakup due to shock waves will be described in Sect. 5.2.5.

1.7 Instability of Jets

The behavior of fluids under the influence of surface tension has been studied by many investigators since pioneering work in the last century [49-54]. The axisymmetric flow of liquids issuing from tubes with circular cross section into another immiscible fluid is of considerable importance for different technical applications. Under the operation of this force the fluid behaves as if enclosed in an envelope of constant tension, and the recurrent form of the jet is due to vibrations of the fluid column about the circular figure of equilibrium. A liquid jet subjected to small disturbances may break up into small droplets. The disturbances can be in form of fluctuations in the liquid supply system as well as fluctuations in liquid properties. The stability and the disintegration of liquid jets has been of interest since the work of Rayleigh, who presented a linear stability analysis of an inviscid jet of infinite length.

For jets of Newtonian fluids a small contraction or expansion of the diameter has been observed [55]. This diameter change occurs over a distance of a few orifice diameters. The behavior of viscoelastic fluids has been described by different investigators in the literature [56-59]. Due to surface tension the cylindrical column of liquid is unstable to necking in and becomes a stream of alternate smaller and larger droplets, which are periodically spaced along the axis of the jet. The smaller droplets are called satellite droplets. The theory of an inviscid jet with radius r_{jet} in a vacuum has been given in a classical paper by Rayleigh [50]. In this analysis it is assumed that the surface of the jet is subjected to a spectrum of small disturbances. By comparing the surface energy of the disturbed jet with that of the undisturbed jet it can be shown that the liquid column is unstable to axisymmetrical disturbances with wavelengths

$$\lambda > \lambda_{min} = 2\pi r_{jet} \; . \tag{1.35}$$

All disturbances on the jet with wavelengths greater than λ_{min} will grow. Rayleigh has shown that the disturbance with the wavelength

$$\lambda = \lambda_{opt} = 9.02\, r_{jet} \tag{1.36}$$

has the fastest growth rate. The volume $V_{jet} = \pi r_{jet}^2 \lambda_{opt}$ of a liquid column of length λ_{opt} must be equal to the volume $V = 4/3\, \pi r^3$ of the droplet formed with radius r. Hence

$$\lambda_{opt} \pi r_{jet}^2 = \frac{4}{3} \pi r^3 \; . \tag{1.37}$$

It follows

$$r = 1.89\, r_{jet} \; . \tag{1.38}$$

This result is important for the so-called droplet stream generator. This device will be described in Sect. 2.3.

The stability of a viscous jet in a vacuum has been studied by Rayleigh and by Weber [53, 60]. For viscous fluids λ_{min} remains the same as in the nonviscous case. For the optimum wavelength one obtains for viscous fluids the result

$$\lambda_{opt} = \sqrt{8}\, \pi r_{jet} \sqrt{1 + \frac{3\, \eta_{liq}}{\sqrt{2\, \varrho_{liq} r_{jet} \sigma}}} \; . \tag{1.39}$$

This equation can written in the form

$$\lambda_{opt} = \sqrt{8}\, \pi\, r_{jet} \sqrt{1 + 3\, Oh} \; , \tag{1.40}$$

where Oh is the Ohnesorge number.

1.8 Relaxation Phenomena

There are many natural and man-made flows consisting of liquid droplets suspended in a gas. Typical examples are fuel sprays, medical sprays or rain drops. When these flows involve changes of the gas velocity and the gas temperature, the droplet-gas interaction through viscous drag and heat transfer will produce corresponding changes in the droplets. The droplets cannot follow the changes in the gas velocity and temperature without time lag. This means that typical relaxation processes are observed.

1.8.1 Relaxation of velocity

Constant gas velocity. In order to understand the mechanical behavior of droplet-gas systems, the motion of a droplet relative to the carrier gas is described by introducing the forces acting on the particle. Effects due to acceleration, rotation, rarefaction, internal circulation in the droplet, evaporation or condensation, deformation, surfactants, contamination of the surface and 'history'-effects will not be considered. These effects have been discussed to some extent in the literature [61-64]. The equation of motion for a droplet of mass $m = 4\pi\,\varrho\, r^3/3$ is written in the form

$$m\frac{dv}{dt} = F_{ext} + F_{gas} \ . \tag{1.41}$$

Here F_{ext} represents external forces and F_{gas} fluid forces. Examples for external forces are forces due to gravity or electrical fields. Assuming that the combined effects of all fluid forces acting on the droplet can be taken into account by means of the drag coefficient C_D, one has

$$F_{gas} = C_D \pi r^2 \frac{\varrho_{gas}}{2} \left| v - v_{gas} \right| (v_{gas} - v) \ , \tag{1.42}$$

where ϱ_{gas} is the density of the surrounding carrier gas. External forces are neglected in the following. When Stokes law is valid one has

$$m\frac{dv}{dt} = 6\pi\eta_{gas} r (v_{gas} - v) \ . \tag{1.43}$$

The solution of this linear differential equation depends on the characteristic time

$$\tau_v = \frac{2r^2 \varrho}{9\eta_{gas}} \ . \tag{1.44}$$

Values of τ_v for a water droplet in air are represented in the diagram of Fig. 1.12 for a wide range of droplet radii. It should be mentioned, that the characteristic time is proportional to the square of the droplet radius. For a droplet with the velocity $v = v_0 = 0$ at $t = 0$ in a gas flow with velocity v_{gas} the solution of Eq. 1.43 can be written in the form

$$v = v_{gas}(1 - e^{(-t/\tau_v)}) \ . \tag{1.45}$$

According to this equation the droplet does not attain the gas velocity v_{gas} in a finite time. But from a practical point of view the difference between v and v_{gas} will be negligible after a time of approximately $5\tau_v$ as can be seen

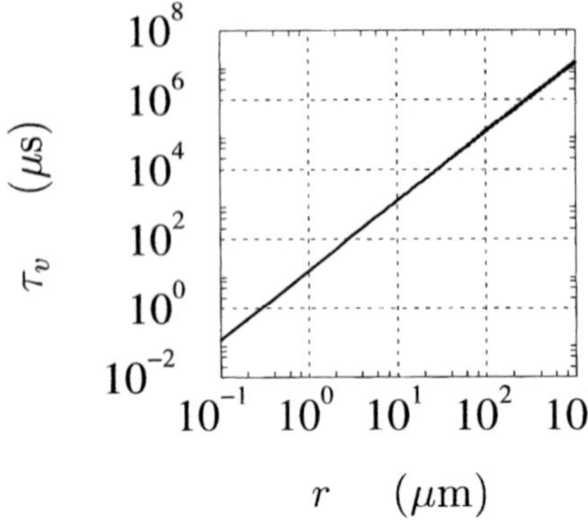

Fig. 1.12. Characteristic time τ_v as a function of droplet radius for a water droplet in air at the temperature 20°C

in Fig. 1.13. If droplets or particles are used as tracers for measurements of the gas velocity the slip between gas velocity and tracer velocity is essential. Tracers play an essential role in laser Doppler velocimetry (LDV) or particle image velocimetry (PIV). These measurement techniques will be described briefly in Chap. 4.

In many practical applications a droplet is injected with the initial velocity v_0 into still air ($v_{gas} = 0$) or another gas. This case with $v_{gas} = 0$ can easily be derived from Eq. 1.45 by a simple transformation of the coordinate system. One obtains

$$v = v_0 e^{(-t/\tau_v)} \ . \tag{1.46}$$

The distance the droplet travels in this case before it comes to rest is readily obtained as

$$s = \int_0^\infty v dt = v_0 \int_0^\infty e^{(-t/\tau_v)} dt = v_0 \tau_v \ . \tag{1.47}$$

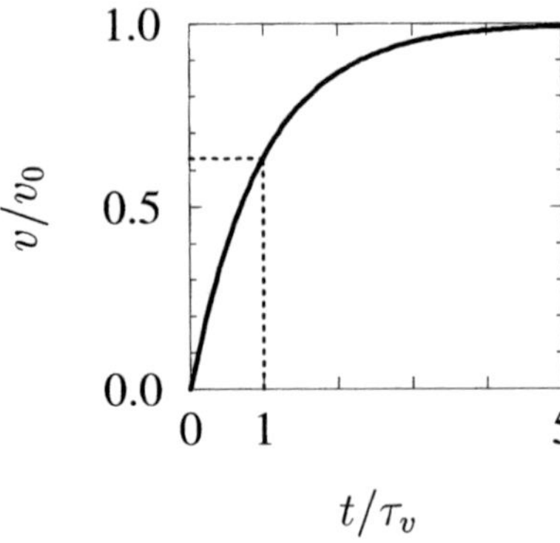

Fig. 1.13. Dimensionless representation of droplet velocity versus time. After the time $t = \tau_v$ the droplet has reached approximately 63 % of the gas velocity. For $t \approx 5\tau_v$ the droplet velocity is practically indistinguishable from the gas velocity

The quantity s is called stop distance of the droplet. It is the distance within which a particle comes to rest, when projected with velocity v_0 into still air. For another mathematical form for the drag coefficient Eq. 1.43 would be different. Nevertheless the relaxation time τ_v would be a characteristic measure for the duration of the relaxation process.

Oscillating gas velocity. For certain practical applications the dynamic behavior of droplets in standing acoustic waves is of interest. A few examples will be given in Sect. 5.2.2. The velocity field in a standing wave can be described by the relation

$$v_{gas}(x,t) = v_{gas,0}\sin(kx)\sin(\omega t) \ , \tag{1.48}$$

where k is the wave number and ω the angular frequency of the acoustic wave. For the sound velocity c the relation $\omega = kc$ holds. The differential equation for the droplet velocity v is

$$\tau_v \dot{v} + v = v_{gas,0}\sin(kx)\sin(\omega t) \ . \tag{1.49}$$

Some characteristics of the particle motion can be discussed when it is assumed that the droplet remains in the velocity loop, i.e. $\sin(kx) = 1$. With this assumption it follows

$$\tau_v \dot{v} + v = v_{gas,0}\sin(\omega t) \ . \tag{1.50}$$

The solution of the homogeneous equation $\tau_v \dot{v} + v = 0$ decays for $t \gg \tau_v$. It can therefore be assumed, that the solution for $t \gg \tau_v$ can be written in the form

$$v = a\cos(\omega t) + b\sin(\omega t) \ , \tag{1.51}$$

where a and b are constants. It follows

$$\frac{v}{v_{gas,0}} = \frac{\sin(\omega t - \epsilon)}{\sqrt{1+(\omega\tau_v)^2}} \tag{1.52}$$

with

$$\tan\epsilon = \tau_v \omega \ . \tag{1.53}$$

The dynamic droplet behavior depends on the dimensionless parameter $\omega\tau_v$. For $\omega\tau_v \to 0$ the droplet follows the acoustic field without phase shift. For $\omega\tau_v \gg 1$ one has the phase shift $\epsilon \to \pi/2$, whereas the amplitude of the droplet oscillation tends to zero.

An approximate solution of Eq. 1.49 can be found by iteration. For $\tau_v = 0$ one obtains in a first step from Eq. 1.49 the approximation

$$v_1 = v_{gas,0}\sin(kx)\sin(\omega t) \ , \tag{1.54}$$

which does not predict a phase shift of the oscillating droplet. Therefore it is assumed in a second step, that the droplet velocity can be approximated by

$$\frac{v_2}{v_{gas,0}} = \frac{1}{\sqrt{1+(\omega\tau_v)^2}}\sin(kx)\sin(\omega t - \epsilon) \ , \tag{1.55}$$

where the phase shift ϵ is given by Eq. 1.53. This approximation is used to determine $\dot{v}_2$, which is used to replace $\dot{v}$ in Eq. 1.49. The resulting equation is now solved for v. After averaging with respect to time one obtains for the drift velocity $\bar{v}$ of the droplet the result

$$\frac{\bar{v}}{v_{gas,0}} = -\frac{1}{8}\frac{v_{gas,0}}{c}\sin(2\epsilon)\sin(2kx) \ . \tag{1.56}$$

This approximate solution reveals that the phase shift is essential for the drift velocity. The interaction between a particle and an acoustic field has been studied by many investigators [65].

1.8.2 Relaxation of temperature

The relaxation of the droplet temperature may be described in a similar way as the relaxation of the droplet velocity. Denoting the droplet temperature by T, the temperature of the ambient gas by T_{gas}, the specific heat of the droplet by c_{liq} and the heat transfer coefficient by α one has for the heat balance of the droplet the relation

$$\frac{4}{3}\pi r^3 \varrho c_{liq}\frac{dT}{dt} = 4\pi r^2 \alpha (T_{gas} - T) \ . \tag{1.57}$$

The heat transfer coefficient α is usually expressed as a function of the Nusselt number

$$Nu = \frac{2r\alpha}{\lambda_{gas}}. \tag{1.58}$$

Here λ_{gas} is the thermal conductivity of the surrounding gas. Frequently the empirical relation

$$Nu = 2 + 0.6 Pr^{1/3} Re^{1/2}. \tag{1.59}$$

is used [66]. For pure heat conduction one has $Re = 0$ and hence

$$Nu = 2 \ . \tag{1.60}$$

For this special case the differential equation Eq. 1.57 has the same form as Eq. 1.43 and the solution of Eq. 1.57 is easily obtained as

$$T - T_{gas} = (T_0 - T_{gas})e^{-t/\tau_T} \tag{1.61}$$

where T_0 is the initial temperature of the droplet and

$$\tau_T = \frac{r^2 \varrho\, c_{liq}}{3\lambda} \tag{1.62}$$

a characteristic time for the relaxation of the temperature. For the ratio of τ_T and τ_v from Eq. 1.44 one obtains the relation

$$\frac{\tau_T}{\tau_v} = \frac{3}{2}\frac{\eta_{gas} c_p}{\lambda_{gas}}\frac{c_{liq}}{c_p} = \frac{3}{2} Pr \frac{c_{liq}}{c_p} \ , \tag{1.63}$$

where Pr is the Prandtl number of the gas. Since c_{liq}/c_p is of order one and the Prandtl number of air is approximately 0.7, it follows that τ_v and τ_T are approximately equal.

1.9 Thermodynamics

1.9.1 Multiphase Systems

The states of thermodynamic equilibrium of a single-component substance can be represented by a surface in the three-dimensional p, V, T phase diagram shown in Fig. 1.14. In the regions labeled 'solid', 'liquid' and 'vapor' the pressure and the temperature can be varied independently, which means that the system has two degrees of freedom. In the regions 'liquid, solid', 'vapor, liquid' and 'vapor, solid' the system consists of a mixture of two phases. The projections of these regions on the p, T plane are curves, not surfaces. In these regions the system has only one degree of freedom. Droplet liquid in thermodynamic equilibrium with its vapor exists in the region 'vapor, liquid'. The critical point represents the upper limit of the pressure for the possible existence of a liquid in equilibrium with its vapor. Above the critical point a fluid state of uniform density in the whole exists and no distinction between

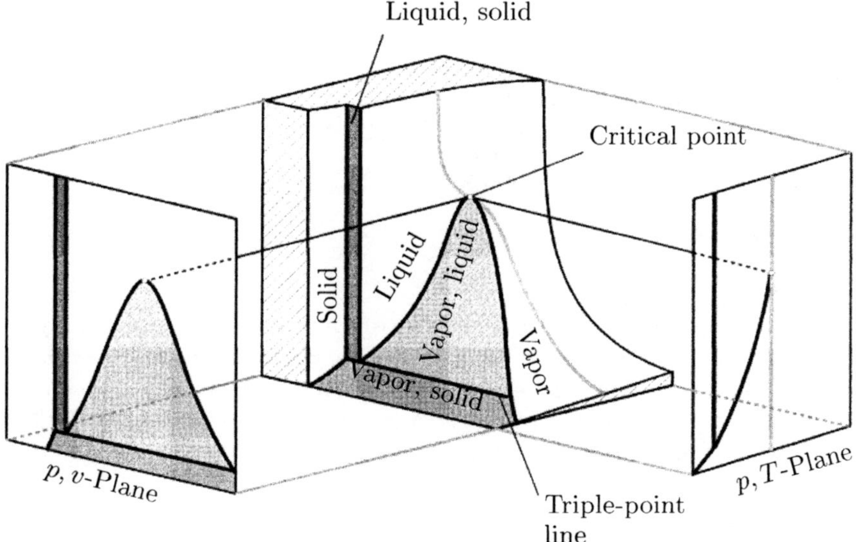

Fig. 1.14. The p, v, T surface for a single-component substance with projections of the two-phase regions on the p, v plane and on the p, T plane. The curves in the p, T phase diagram are the vapor pressure curve, the sublimation curve and the melting curve. The intersection of these three curves is the triple point. The vapor pressure curve is terminated by the triple point and the critical point. The p, v, T surface is shown for a substance which contracts upon freezing. It should be mentioned that there are some substances, like water, which expand upon freezing

liquid and vapor can be made. Values of the critical pressure and the critical temperature are shown in Table 1.4 for some substances.

Table 1.4. Values of critical pressure p_{crit} and critical temperature T_{crit} for different substances

Substance	p_{crit} Mpa	T_{crit} °C
Methane	4.64	-82.1
Ethane	4.88	32.2
Propane	4.26	96.8
n-Pentane	3.37	196.6
n-Hexane	3.03	234.2
n-Heptane	2.74	267.1
n-Octane	2.51	296
n-Decane	2.11	344.4
n-Dodecane	1.81	386
Methanol	7.95	240
Ethanol	6.38	243
n-Propyl alcohol	5.17	263.6
Isopropyl alcohol	4.76	235
Acetone	4.76	235.5
Acetylene	6.24	35.5
Helium	.229	-267.9
Hydrogen	1.3	-239.9
Nitrogen	3.39	-147
Oxygen	5.08	-118.4
Carbon dioxide	7.39	31
Water	22.12	374.1

The lower pressure limit for the existence of a liquid is given by the triple-point line. The states on the triple-point line contain all three phases. Thermodynamic equilibrium between a liquid phase and its vapor at constant pressure and constant temperature is determined by the condition that Gibbs free energy G of the system has a minimum. Since G is an extensive quantity is follows

$$G = m'g'(p,T) + m''g''(p,T) \;, \tag{1.64}$$

where g' and g'' represent the specific Gibbs free energy and m' and m'' the mass of the liquid and the vapor. For passage of a small quantity of material from one phase to the other at constant pressure and temperature one has $m' + m'' =$ constant or $dm' = -dm''$. It is easy to see that G would decrease, when $g' \neq g''$ and a minimum value of G would be achieved if the substance transformed into one phase. Hence a necessary condition for equilibrium is

$$g'(p,T) = g''(p,T) \;. \tag{1.65}$$

This relation represents the vapor pressure curve and in general the phase-equilibrium line. Differentiating this relation yields $dg' = dg''$ or

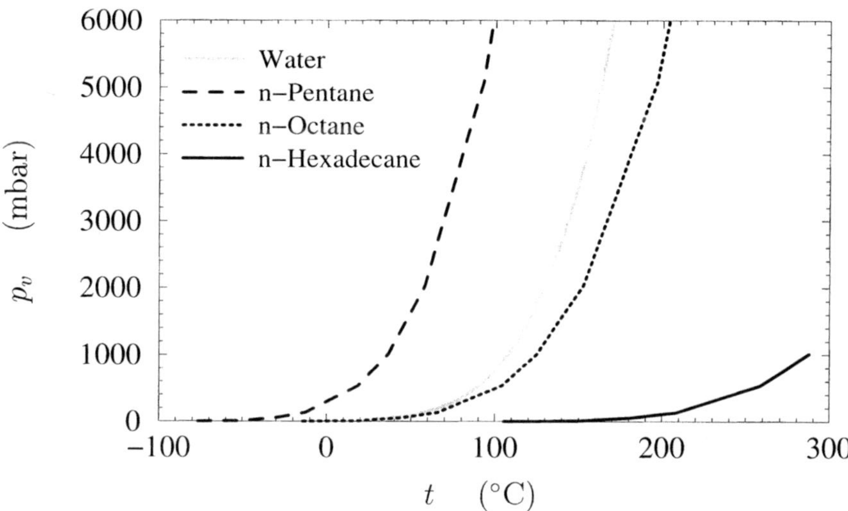

Fig. 1.15. Vapor pressure curves for n-pentane, n-octane, and n-hexadecane in comparison with water

$$-s'dT + v'dp = -s''dT + v''dp \ , \tag{1.66}$$

where s is the specific entropy. Rearranging the terms gives the Clausius-Clapeyron equation in the general form

$$\frac{dp}{dT} = \frac{s'' - s'}{v'' - v'} \ . \tag{1.67}$$

Introducing $s'' - s' = r_h(T)/T$ where $r_h(T)$ is the latent heat of vaporization one obtains

$$\frac{dp}{dT} = \frac{r_h(T)}{T(v'' - v')} \ . \tag{1.68}$$

It should be mentioned that the phase transitions liquid-solid and vapor-solid can be treated analogously.

An approximate equation for the vapor pressure of a liquid can easily be derived from Eq. 1.68 by assuming, that the density of the vapor is much smaller than the density of the liquid $v' \ll v''$ and that the vapor can be treated as an ideal gas with $pv'' = RT$. It follows

$$\frac{dp}{dT} = \frac{r_h(T)}{RT^2} \ . \tag{1.69}$$

This equation can be immediately integrated for $r_h =$ constant to give

$$p = p_0 \, e^{-\frac{r_h}{R}\left(\frac{1}{T} - \frac{1}{T_0}\right)} \ , \tag{1.70}$$

where $p = p_0$ for $T = T_0$. In Fig. 1.15 the vapor pressure curves of different substances are represented, the data used is from Ref. [3].

1.9.2 Kelvin-Helmholtz Equation

From Eq. 1.7 it follows that a spherical droplet possesses the surface energy $E_s = 4\pi\sigma r^2$. When vapor transforms into droplet liquid the droplet surface increases and the energy increases also. When the mass dm condenses on the droplet, then the area increase will be

$$dA = \frac{2}{\varrho}\frac{dm}{r} \ . \tag{1.71}$$

This corresponds to an increase of surface energy

$$d(\sigma A) = \frac{2\sigma}{r}\frac{dm}{\varrho} = p_r dV \ . \tag{1.72}$$

The Gibbs free energy must be a homogeneous function in n', n'' and A. Using Euler's theorem is follows

$$G = \left(\frac{\partial G}{\partial n'}\right) n' + \left(\frac{\partial G}{\partial n''}\right) n'' + \left(\frac{\partial G}{\partial A}\right) A = \mu' n' + \mu'' n'' + \sigma A \ , \tag{1.73}$$

where the chemical potentials of the liquid and vapor μ' and μ'', and the thermodynamic relation $\sigma = (\partial G/\partial A)_{p,T,n',n''})$ have been introduced This is equivalent to Eq.1.64, when G does not depend on A. For the system considered here one has $\mu' = Mg'$ and $\mu'' = Mg''$, where M is the molar mass. A system consisting of a liquid droplet in equilibrium with its vapor must possess a minimum of the Gibbs free energy. A virtual displacement of this system is considered in which dn' mole of vapor condense. In this case one has

$$dn'' = -dn' \ \text{ and } \ dA = \frac{2}{r} V'_m dn' \ , \tag{1.74}$$

where $dA = (dA/dr)(dr/dV)(dV/dn')dn'$. This means that the infinitesimal displacements of the system dn', dn'' and dA are not independent. From the condition $G = G_{min}$ it follows

$$\delta G = 0 = \mu' dn' + \mu'' dn'' + \sigma dA = (\mu' - \mu'') dn' + \frac{2\sigma V'_m}{r} dn' = 0 \tag{1.75}$$

or

$$\mu'' - \mu' = \frac{2\sigma}{r} V'_m \ . \tag{1.76}$$

For constant temperature it follows from thermodynamics

$$\int (V''_m - V'_m) dp = \mu'' - \mu' \ . \tag{1.77}$$

Assuming that $V''_m \gg V'_m$ and that the equation $pV''_m = \mathcal{R}T$ holds for the vapor one obtains the relation

$$\ln p = \frac{2\sigma V'_m}{\mathcal{R}Tr} + \ln p_\infty(T) \ . \tag{1.78}$$

The quantity

$$r^* = \frac{2\sigma V'_m}{\mathcal{R}T} \tag{1.79}$$

must have the dimension of a length. Equation 1.78 can now be written in the form

$$\frac{p}{p_\infty} = e^{r^*/r} \ . \tag{1.80}$$

This relation is known as Kelvin-Helmholtz equation. For water at the temperature $T = 300\,\mathrm{K}$ one has $r^* \approx .001\,\mu\mathrm{m} = 1\,\mathrm{nm}$. The ratio r^*/r has very small values for droplets, which are larger than a micron. Thus one may use Eq. 1.80 in the form

$$\frac{p - p_\infty}{p_\infty} = \frac{r^*}{r} \ . \tag{1.81}$$

It is easy to see that $p_\infty(T)$ represents the vapor pressure for $r \to \infty$, or for a plane surface. A revision of the Kelvin-Helmholtz equation for very small droplets has been presented in Ref. [67].

Internal boiling can occur in a droplet, when the saturation pressure belonging to the hottest temperature in the droplet becomes larger than $p_\infty + 2\sigma/r$.

1.9.3 Thermodynamics of the Interface

In Fig. 1.16 a two-phase system consisting of liquid phase with the volume V' and vapor phase with the volume V'' is shown schematically. For simplicity a plane interphase between the two phases is considered. The boundary is assumed a mathematical plane of zero thickness at $z = 0$. It should be emphasized that the choice of the position of this dividing surface determines the magnitude of the volumes V' and V''. A different choice of the position of the dividing surface leads to different values of V' and V''. Independent of the position of the dividing surface is of course the relation

$$V = V' + V''. \tag{1.82}$$

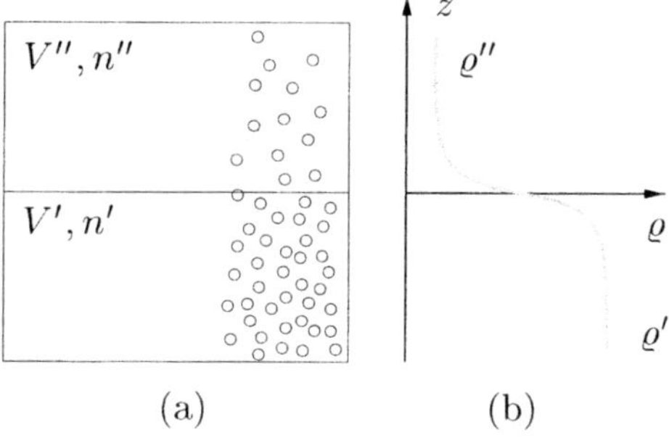

Fig. 1.16. Two-phase system consisting of n' moles of liquid and n'' moles of vapor (a). In the interlayer region (b) the density changes from $\varrho = \varrho'$ to $\varrho = \varrho''$. Note that well below the critical point $\varrho'' \ll \varrho'$

Now it is assumed that the position of the dividing surface has been chosen. The number of moles in the liquid phase n_i' is of course proportional to V', and the number of moles in the vapor phase n_i'' is proportional to V''. This means that the values of n_i' and n_i'' depend on the choice of the dividing surface. It would therefore in general be wrong to write for the mole number of component i the equation $n_i = n_i' + n_i''$. Instead one has

$$n_i = n_i' + n_i'' + n_i^A \tag{1.83}$$

and for Helmholtz free energy of the system

$$F = F' + F'' + F^A . \tag{1.84}$$

Here the so-called excess quantities n_i^A and F^A have been introduced. These quantities depend on the choice of the dividing surface. The first law of thermodynamics is written in the usual form

$$dU = TdS - pdV + \sigma dA + \sum_{i=1}^{i=k} \mu_i dn_i \ , \tag{1.85}$$

where U is the internal energy of the system, T the temperature, p the thermodynamic pressure, σ the surface tension, A the area of the interface, k the number of species and μ_i the chemical potential of species i. Introducing Helmholtz free energy $F = U - TS$ one obtains

$$dF = d(U - TS) = -SdT - pdV + \sigma dA + \sum_{i=1}^{i=k} \mu_i dn_i \ . \tag{1.86}$$

For isothermal and incompressible fluids with $dT = 0$ and $dV = 0$ this relation reduces to

$$dF = \sigma dA + \sum_{i=1}^{i=k} \mu_i dn_i \ . \tag{1.87}$$

If the Helmholtz free energy F is known as a function of V, T, A and n_i the surface tension can be determined from the relation

$$\sigma = \left(\frac{\partial F}{\partial A}\right)_{T,V,n_i} \tag{1.88}$$

and the thermal equation of state follows from

$$p = -\left(\frac{\partial F}{\partial V}\right)_{T,A,n_i} \ . \tag{1.89}$$

Helmholtz free energy F must be a homogeneous function of first degree in V, n_i and A. From Euler's theorem it follows

$$F = V\left(\frac{\partial F}{\partial V}\right)_{T,A,n_i} + A\left(\frac{\partial F}{\partial A}\right)_{T,V,n_i} + \sum_{i=1}^{i=k} n_i \left(\frac{\partial F}{\partial n_i}\right)_{T,V,n_{i\neq j}} \tag{1.90}$$

or

$$F = -pV + \sigma A + \sum_{i=1}^{i=k} \mu_i n_i \ . \tag{1.91}$$

Differentiating Eq. 1.91 and combining with Eq. 1.86 one obtains the Gibbs-Duhem equation

$$-V dp + A d\sigma + S dT + \sum_{i=1}^{i=k} n_i d\mu_i = 0 \ . \tag{1.92}$$

Next Helmholtz free energy is considered for the liquid phase and for the vapor. One has

$$F' = \sum_{i=1}^{i=k} \mu_i n_i' - pV' \tag{1.93}$$

and

$$F'' = \sum_{i=1}^{i=k} \mu_i n_i'' - pV'' \ . \tag{1.94}$$

Combining these two equations and Eq. 1.91 with Eqs. 1.84 and 1.83 one obtains

$$F - F' - F'' = F^A = A\sigma + \sum_{i=1}^{i=k} \mu_i n_i^A \ . \tag{1.95}$$

Next the first law is considered for both phases. One has

$$dF' = -pdV' - S'dT + \sum_{i=1}^{i=k} \mu_i dn_i' \tag{1.96}$$

and

$$dF'' = -pdV'' - S''dT + \sum_{i=1}^{i=k} \mu_i dn_i'' \ . \tag{1.97}$$

Combining these two Equations with Eq. 1.86 and using $dF - dF' - dF'' = dF^A$ one finds

$$dF^A = -S^A dT + \sigma dA + \sum_{i=1}^{i=k} \mu_i dn_i^A \ , \tag{1.98}$$

since according to Eq. 1.83 one has $dn_i - dn_i' - dn_i'' = dn_i^A$. Equation 1.98 shows that the interface can be treated thermodynamically as a third phase, which possesses the entropy $S^A = S - S' - S''$ and consists of n_i^A moles of component i.

1.9.4 Gibbs Isotherm

From a molecular point of view interfacial and surface tension may be explained by considering the intermolecular forces acting on a molecule in the bulk of the droplet liquid as shown in Fig. 1.17. In the interior of the liquid the forces acting on a molecule will compensate, the resulting force is zero, (1) and (3). Molecules at the droplet surface will experience a resulting force directed towards the droplet center (2). Macroscopically the effect of this force is interpreted as surface tension. When these considerations are extended to solutions, it will be seen, that a component, whose molecules experience smaller intermolecular forces (3), will tend to accumulate in the droplet surface. From Eq. 1.7 it is clear that the surface energy of the droplet $E_s = 4\pi\sigma r_s^2$ decreases when the concentration of the component with the smaller σ is higher in the droplet surface. These concentration variations have an effect on the entropy of mixing. Thermodynamic equilibrium requires that Helmholtz free energy has a minimum. From Eq. 1.98 it follows

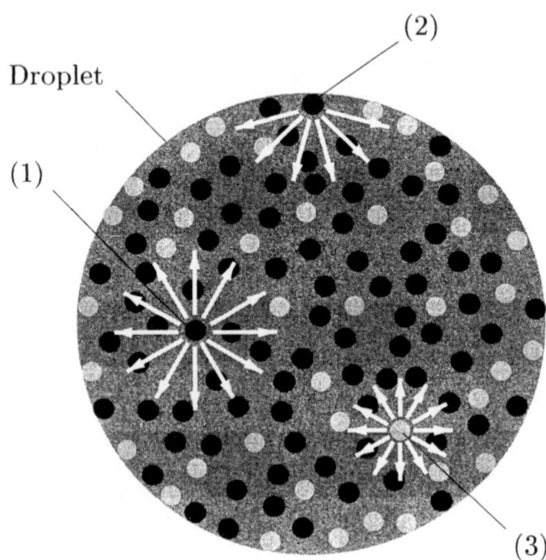

Fig. 1.17. Schematic representation of intermolecular forces acting on the molecules of binary mixture in the interior of a droplet (1) and (3) and at the surface (2)

$$dF^A = Ad\sigma + \sum_{i=1}^{i=k} n_i^A d\mu_i = 0 \ . \tag{1.99}$$

Applying this relation to a binary system with $k = 2$ one obtains

$$Ad\sigma = -n_1^A d\mu_1 - n_2^A d\mu_2 \ . \tag{1.100}$$

Depending on the choice of the dividing surface the excess quantities n_1^A and n_2^A are positive, zero, or negative. A special position of the dividing surface is $n_1^A = 0$. For this case one has

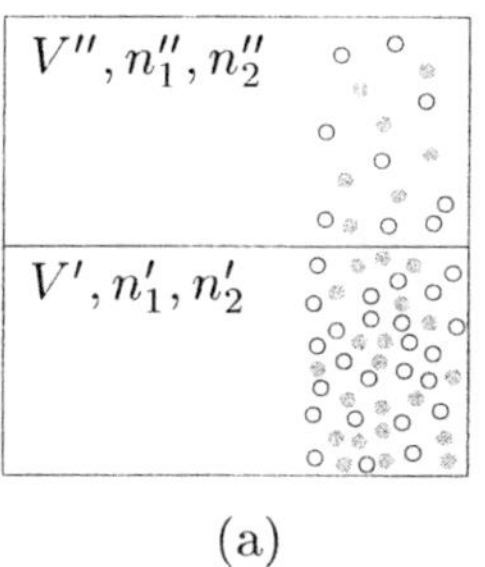

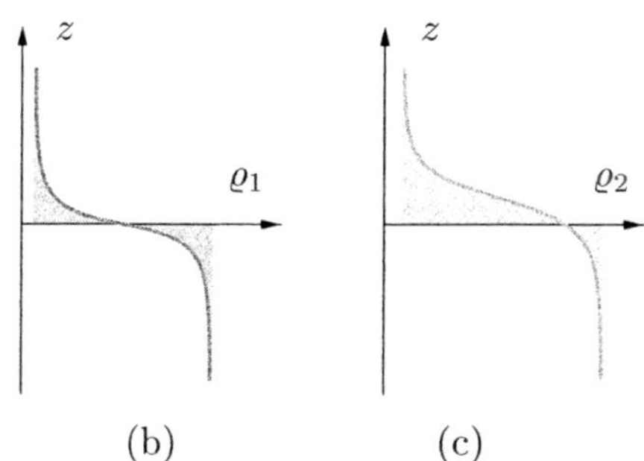

(a) (b) (c)

Fig. 1.18. Binary two-phase system containing liquid and vapor of the components 1 and 2 with interface (a). Concentration profiles for both components (b), (c). The excess mole number n_1^A is zero, when the shaded areas for $z > 0$ and $z < 0$ are equal (b). For this choice of the dividing surface n_2^A will in general not be zero (c)

$$-\frac{d\sigma}{d\mu_2} = \frac{n_2^A}{A} \equiv \Gamma \, , \tag{1.101}$$

where Γ represents the excess mole number of component 2 per area of the interface. This Equation is known as Gibbs equation for adsorption [7, 68].

1.10 Phase Transition Processes

Phase transition processes with droplets have been discussed in the literature from different viewpoints such as meteorology, thermodynamics, heat and mass transfer, drying and combustion [20, 11, 69-72].

The evaporation of droplets in a spray can cause significant changes of the size distribution. It will be assumed in the following that the rate of loss of liquid by evaporation depends on properties of the droplet surface, the vapor pressure of the liquid and the vapor concentration in the environment of the droplet. Additional important effects may be produced by droplet motion. A burning fuel droplet in an oxidizing atmosphere represents an important application.

1.11 Evaporation of a Single Droplet

A single droplet with radius r_s and temperature T is brought into an environment with the temperature T_∞ and the mass fraction $Y_{vap\infty}$ of the vapor of the droplet material. The evaporation process of the droplet depends on these parameters. Two examples for different initial boundary conditions will be explained qualitatively in the following. If the temperatures of droplet and ambiance are the same initially and the mass fraction $Y_{vap\infty}$ corresponds to the vapor pressure for the given droplet temperature, neither evaporation of the droplet liquid nor condensation of the surrounding vapor on the droplet will occur. If $Y_{vap\infty}$ is increased supersaturation is obtained and vapor will begin to condense on the droplet. Vice versa, if this mass fraction is decreased an evaporation process will begin and a mass flux $\dot{m}_{vap}$ of droplet material will leave the droplet. The latent heat of vaporization comes from the inner energy of the droplet liquid. The resulting heat flux $\dot{Q}_{liq}$ within the droplet liquid to the droplet surface causes a heat flux $\dot{Q}_{gas}$ in the surrounding gas to the droplet surface. As a consequence the temperature of the droplet will decrease and a temperature gradient will be established within the droplet and in the surrounding gas as shown schematically in Fig. 1.19(a). Due to heat conduction the temperature gradients within the droplet will decrease. Figure 1.19(b) shows a situation with much higher ambient temperature. It is assumed, that the ambient temperature is higher than the boiling temperature of the droplet liquid. This results in a heat flux $\dot{Q}_{gas}$ towards the droplet, which delivers the latent heat for evaporation and heats the droplet until a homogeneous droplet temperature is obtained, which is close to, but below the boiling temperature of the droplet liquid. The situation during droplet heating is shown qualitatively in Fig. 1.19(b).

In the following analysis it will be assumed that steady conditions prevail. This model implies, that the temperature and concentration profiles can be determined with constant droplet radius. The neglect of surface regression requires that the density of the vapor is much smaller than the density of the liquid. The vaporizing droplet consists of a pure liquid, whereas the surrounding atmosphere is described as a mixture of ideal gases. For a single fuel droplet under steady-state conditions in a nonconvecting atmosphere the continuity equation of the vapor of the droplet material and the energy equation of the surrounding gas can be formulated as

$$r^2 \varrho_{gas} v_{gas} \frac{dY_{vap}}{dr} = \frac{d}{dr}\left(r^2 D \varrho_{gas} \frac{dY_{vap}}{dr}\right) + r^2 \dot{\omega}_{vap} \tag{1.102}$$

and

$$r^2 \varrho_{gas} v_{gas} \frac{dc_p T}{dr} = \frac{d}{dr}\left(\frac{\lambda_{gas}}{c_p} r^2 \frac{dc_p T}{dr}\right) + r^2 \dot{Q} \; . \tag{1.103}$$

In these equations r represents the radial coordinate, ϱ_{gas} the density and v_{gas} the radial velocity of the gas surrounding the droplet, Y_{vap} the mass fraction of the vapor, D the diffusion constant, $\dot{\omega}_{vap}$ the vapor production rate, $\dot{Q}$ the heat production rate, c_p the specific heat of the gas, and λ_{gas} the heat conductivity of the ambiance. For the mass flux $\dot{m}_{vap}$ of the evaporated droplet liquid the equation

$$\dot{m}_{vap} = 4\pi \varrho_{gas,s} D_s r_s \ln(1 + B) \tag{1.104}$$

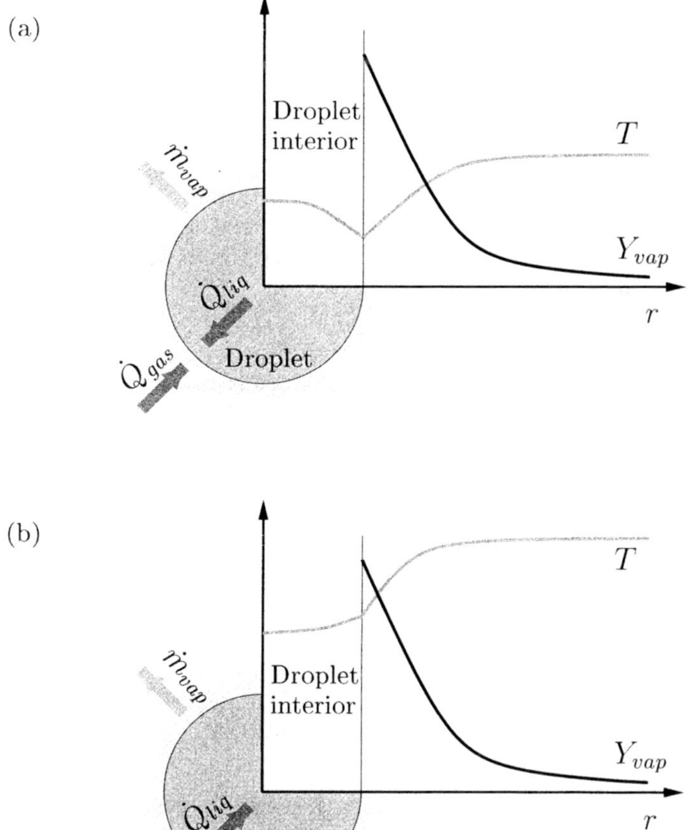

Fig. 1.19. Schematic representation of evaporating droplets indicating the mass flux of the vapor $\dot{m}_{vap}$ and the heat fluxes $\dot{Q}_{gas}$ in the gas phase and $\dot{Q}_{liq}$ in the liquid droplet. In addition the radial profiles of the temperature T and of the vapor mass fraction Y_{vap} are shown qualitatively. The temperature profile of case (a) is observed, when ambient and droplet temperature are equal initially. In case (b) the temperature of the ambiance is much higher than the initial droplet temperature

is obtained, where r_s is the droplet radius, $\varrho_{gas,s}$ and D_s are the density and the diffusion coefficient of the surrounding gas at the droplet surface. The so-called Spalding transfer number B is given by

$$B \equiv \frac{Y_{vap,s} - Y_{vap,\infty}}{1 - Y_{vap,s}} . \tag{1.105}$$

For the calculation of the fuel mass fraction at the surface $Y_{vap,s}$ of the droplet the energy equation and the Clausius-Clapeyron equation must be considered. For the Lewis number it is assumed $Le = 1$ with $D_s = \alpha_s = \lambda_s/(\varrho_{gas,s}c_p)$, where α_s is the thermal diffusivity. The details of this analysis have been described in the literature [71, 73]. From Eq. 1.104 the so-called d^2-law for evaporation follows in the form

$$r_s^2 = r_{s,0}^2 - \beta_v t , \tag{1.106}$$

where $r_{s,0}$ is the initial droplet radius and β_v the evaporation coefficient

$$\beta_v = \frac{2\varrho_{gas,s}\alpha_s}{\varrho_{liq}} ln(1 + B) \tag{1.107}$$

with ϱ_{liq} as density of the droplet liquid. From these equations one obtains for the lifetime of the droplet

$$t_v = \frac{r_{s,0}^2}{\beta_v} = \frac{r_{s,0}^2 \varrho_{liq}}{2\varrho_{gas,s}\alpha_s \ln(1 + B)} . \tag{1.108}$$

When the vapor concentration at the droplet surface is very low, due to low volatility of the liquid or low temperatures, the evaporation process can be described as so-called diffusion-controlled evaporation.

1.11.1 Diffusion-controlled evaporation of a single droplet

The evaporation of droplets at low temperatures has been described by Maxwell and Fuchs [74, 75]. In this case it is assumed that the droplet is surrounded by an atmosphere, which has approximately the same temperature as the droplet liquid. It is also assumed that the temperature is low in comparison with the boiling temperature of the droplet liquid. Under the described conditions the vaporization process is dominated by diffusion processes in the vapor phase. In this special case with $v_{gas} = 0$ and $\dot{\omega}_{vap} = 0$ the vapor species continuity equation Eq. 1.102 can be integrated. The result can be written in the form

$$4\pi r^2 D \frac{dc}{dr} = -\dot{m} , \tag{1.109}$$

where $c = dm/dV$ is the concentration of the droplet vapor and $\dot{m}$ the evaporation rate. Upon solving for dc/dr and integrating one finds

$$c(r) - c_\infty = \frac{\dot{m}}{4\pi D r} , \tag{1.110}$$

where c_∞ is the vapor concentration for $r \to \infty$. Introducing the droplet radius r_s and the vapor concentration $c(r_s) = c_s$ one obtains for the rate of evaporation the expression

$$\dot{m} = 4\pi D r_s (c_0 - c_\infty) . \tag{1.111}$$

The concentration c can be expressed by the partial pressure of the vapor, when it is assumed that the vapor can be described as ideal gas. It follows

$$\dot{m} = 4\pi r_s D \frac{M}{\mathcal{R}T}(p_0 - p_\infty) . \tag{1.112}$$

Froessling has extended this equation to describe a droplet moving through a gas. He obtained the empirical equation

$$\dot{m} = 4\pi D r_s \frac{M}{\mathcal{R}T}(p_0 - p_\infty)(1 + 0.276 Sc^{1/3} \cdot Re^{1/2}) , \tag{1.113}$$

where Sc is the Schmidt number and Re the Reynolds number.

It is easy to show, that the d^2-law follows from Eq. 1.112. Using $\dot{m} = dm/dt = \varrho \cdot 4\pi r^2 dr/dt$ the rate of evaporation can be expressed by the radius. From Eq. 1.112 it follows

$$-\frac{dr_s^2}{dt} = \frac{2DM}{\varrho \mathcal{R}T}(p_0 - p_\infty) . \tag{1.114}$$

By integration one obtains

$$r_{s,0}^2 - r_s^2 = \frac{2DM}{\varrho \mathcal{R}T}(p_0 - p_\infty)t , \tag{1.115}$$

where $r_{s,0}$ is the droplet radius at $t = 0$. Many studies of evaporation and burning rates have revealed that the so-called d^2-law under certain conditions may yield reasonable results for the gasification rate of droplets. The restrictions of this model will be discussed in Sect. 1.11.2.

When vapor diffuses away from the droplet, the ambient air has to diffuse in the other direction to keep the pressure constant. The resulting mean motion has been taken into account by Stefan [75]. In experimental investigations of droplet evaporation Stefan flow is often minimized by the use of relatively nonvolatile species.

Evaporation at larger Knudsen numbers. The discussion of evaporation rates in Sect. 1.11.1 is based on equations of the continuum theory, which are valid for small values of the Knudsen number. Rarefaction effects depend on the magnitude of the Knudsen number Kn, which can be expressed as the ratio of the mean free path of a gas l and a characteristic macroscopic length L. One has

$$Kn = \frac{l}{L} . \tag{1.116}$$

In the case of an evaporating droplet the diameter may be chosen as characteristic macroscopic length, hence $L = 2r_s$ or $Kn = l/(2r_s)$. In air under normal conditions the mean path is approximately $l_0 = 0.1\,\mu$m. The mean free path is a function of density, which can be expressed as $l = l_0 \varrho_0/\varrho$. It is usually assumed that the equations of the continuum theory are valid, when $Kn < 0.01$. In the limiting case of free molecule flow with $Kn \to \infty$ the description of the evaporation rate becomes rather simple. It is assumed that the vapor pressure at the droplet surface is equal to the saturation pressure. The mass flux rate follows from the effusion equation of the kinetic theory of gases. When the pressure and temperature of the surrounding vapor are p_∞ and T_∞ one obtains the net evaporation rate

$$\dot{m}_{FM} = \frac{\alpha_e p_s}{\sqrt{2\pi R T_s}} - \frac{\alpha_c p_\infty}{\sqrt{2\pi R T_\infty}} \,. \tag{1.117}$$

In this equation, which is known as Hertz-Knudsen-Langmuir or HKL equation, α_e and α_c represent the evaporation and condensation coefficient.

It has been suggested to combine Eq. 1.112 and Eq. 1.117 by assuming that the evaporation is diffusion-controlled in the surrounding medium for $r_s > \Delta$ and controlled by effusion for $r_s < \Delta$, where Δ is a distance of the order of the mean free path. The result can be written in the form

$$\dot{m} = \frac{\dfrac{4\pi r_s D M}{RT}(p_0 - p_\infty)}{\dfrac{r_s}{r_s + \Delta} + \dfrac{D}{\alpha r_s^2 \sqrt{RT/2\pi}}} \,. \tag{1.118}$$

This equation can be considered as an interpolation between diffusion-controlled evaporation ($\Delta \to 0$) and kinetic-controlled evaporation ($\Delta \to \infty$). In the derivation of this equation it has been assumed, that $\alpha_e = \alpha_c = \alpha$. A similar expression has been derived by Monchick and Reiss by introducing a nonequilibrium distribution function of the molecular velocities [76]. A rigorous kinetic description of the evaporation problem must be based on the Boltzmann equation of the kinetic theory of gases. In this treatment the droplet surface is the source of vapor, the vapor phase is described with the Boltzmann equation. An approximate analytical solution of this problem has been given by Shankar using the Maxwell moment method in combination with two-stream Maxwellian distribution functions [77]. In other publications the problem of evaporation and condensation has been treated using a linearized version of the collision term in the Boltzmann equation [78-80]. Sitarski and Nowakowski solved the Boltzmann equation by expanding the distribution function in a series of Hermite polynomials around a Maxwellian distribution [81].

1.11.2 Combustion of a single droplet

An initially cold fuel droplet placed in a hot stagnant environment will heat up as described above. The heat conducted to the droplet is used to raise the temperature of the droplet interior and to vaporize liquid. As a consequence fuel vapor at its saturation pressure will exist at the surface of the liquid droplet. Since the vapor pressure is lower in the droplet environment fuel vapor will be transported in radial direction by diffusion. In an oxidizing environment ignition of the droplet can be achieved. The resulting spherically symmetric temperature and concentration profiles shown in Fig. 1.20 imply that transport of heat and mass occur only in radial direction. In theoretical

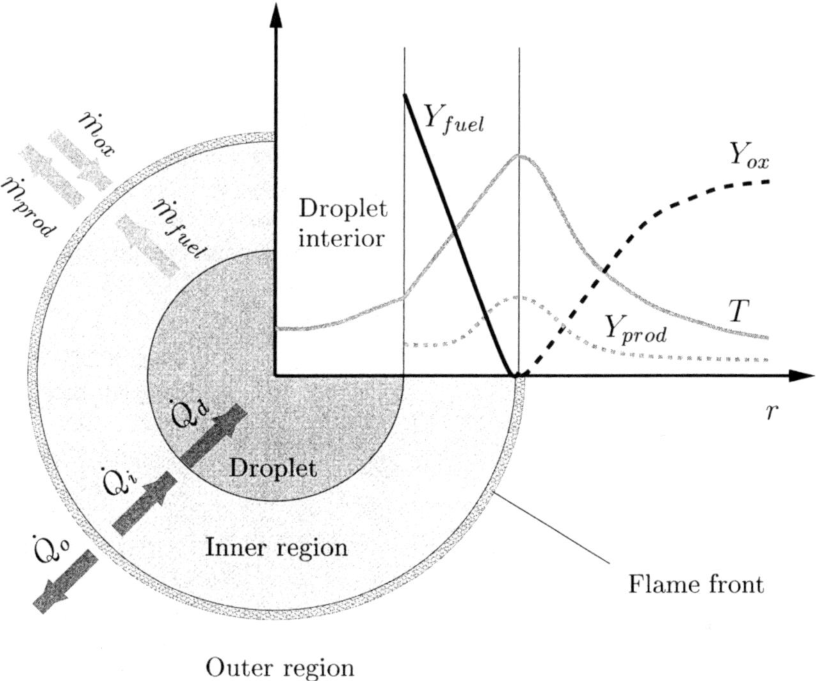

Fig. 1.20. Schematic representation of burning droplet with flame front. Temperature, mass and heat fluxes as well as mass fractions of the fuel, oxidizer and transient products are shown qualitatively as a function of radial distance

studies of burning fuel droplets it is usually assumed that reacting rates are very fast and the flame zone can be described as an initially thin flame sheet which surrounds the droplet. Fuel diffuses from the droplet surface to the flame front, while the oxygen diffuses from the surrounding atmosphere to the flame front. The concentrations of fuel and oxygen are zero at the flame front due to the reaction. The reaction products diffuse to the surroundings. Part of the energy released by the reactions serves to heat the fuel in the droplet, the remaining energy is transported to the surrounding atmosphere.

In analytical studies of burning droplets it is often assumed that the temperature of the droplet surface is close to the boiling temperature, whereas the temperature distribution within the droplet is uniform and only a few degrees below the boiling temperature. With a number of assumptions, such as neglecting thermal radiation and the radial motion of the droplet surface, different analyses have been presented. It can be shown that the problem of a burning single fuel droplet leads to an equation which has the same form as Eq. 1.104, but the transfer number B is given by a different expression [71]. This means, that the burning process can be described by the d^2-law under the mentioned assumptions.

When heat conduction within the droplet is taken into account, the square of the droplet radius does not decrease linearly with time any more. In Fig. 1.21 results obtained with this so called conduction-limit model are shown. In the end phase the droplet temperature is approximately uniform and independent of time. Then the burning behavior is nearly linear, as described by the d^2-law. However, in the initial phase of droplet burning, where droplet heating occurs, deviations from the d^2-law are found, as can be seen from Fig. 1.21.

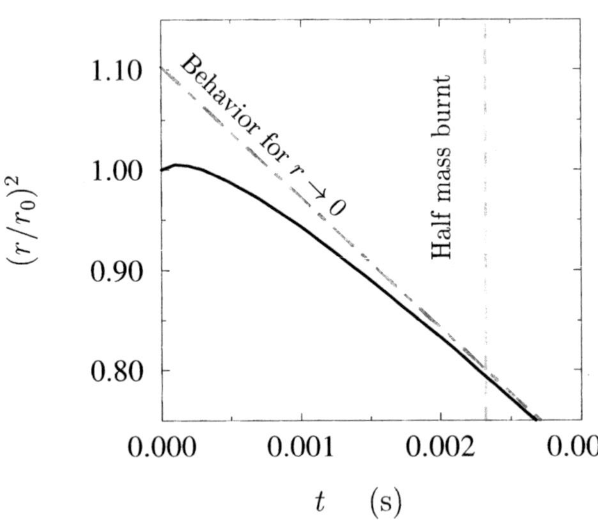

Fig. 1.21. Square of droplet radius versus time for burning ethanol droplet with the initial radius $r_0 = 30\,\mu\text{m}$ and initial temperature $T_0 = 293\,\text{K}$. Shown is the initial phase calculated with the conduction limit model. In addition the burning behavior in the end phase is shown. The radius, which the droplet has, when half of the mass is burnt is indicated

The results discussed so far must be modified when the burning droplet moves through the oxidizing atmosphere [71]. For the Nusselt number the expression

$$Nu_{r_s} = 0.39 \frac{\ln(1+B)}{B} Re_{r_s}^{1/2} \tag{1.119}$$

has often been used, where the Reynolds number Re_{r_s} is based on the droplet radius.

There exists an extended literature on different aspects of droplet burning. Droplets above the critical pressure, and effects of internal circulation on droplet vaporization have been discussed. Other topics covered in the literature include multicomponent fuel droplets, water/oil emulsions and coal/oil mixtures, micro-explosions and spray effects.

The effects of turbulence on vaporization rates have been studied by Birouk et al. [82]. In these experimental investigations the authors used a chamber, which allowed to generate homogeneous and isotropic turbulence with zero mean velocity. It has been found that the turbulent velocity fluctuations increase the vaporization rate. Results are presented for n-alcanes at normal pressure and temperature. A few aspects of internal circulation have been described in Sect. 1.5.

1.11.3 Evaporation of multicomponent droplets

Droplets in technical sprays are mostly multicomponent with components of different physical and chemical properties. In the analysis of the evaporation of multicomponent droplets mass and energy transport processes in the surroundings and in the interior of the droplet must be taken into account. Additional important phenomena for multicomponent evaporation are unsteady-state behavior, variation of temperature and physical properties, as well as Stefan flow. The general conservation equations for multicomponent systems are discussed in Ref. [71].

Ravindran and Davis studied diffusion-controlled evaporation of micron and submicron droplets theoretically and experimentally neglecting Stefan flow, the Kelvin effect and the influence of the Knudsen number [83]. Assuming ideal mixing and ideal liquid solution behavior inside the droplet an expression for the evaporating species is obtained. Experimental results for dioctyl phthalate and dibutyl phthalate agree with the simple theoretical analysis. An analysis of diffusion-controlled concentration profiles in the presence of an inert gas outside an evaporating binary droplet has been presented by Kalkkinen et al. [84].

Interest in combustion of technical sprays in industrial burners, Diesel engines, or gas turbine combustors has led to numerous studies of the evaporation of multicomponent droplets with convective heat and mass transfer. Law et al. presented a theoretical model for the vaporization of an alcohol droplet in humid air. Theoretical and experimental investigation of alcohol is accompanied by simultaneous condensation of water vapor on the droplet surface. The gasification rate of alcohol is enhanced by the released condensation heat [85]. Spark ignition of monodisperse multicomponent fuel sprays has been studied by Lee et al. [86].

1.11.4 Evaporation through films

Surface films can be formed on liquid surfaces by substances which are insoluble in the liquid phase but can spread over the surface. When the surface of a droplet is covered with a film the evaporation rate of the droplet liquid will be reduced, whereas the equilibrium vapor pressure is not affected. In order to obtain a mathematical description of this phenomenon the evaporation rate is expressed in the form $\dot{m} = (c_0 - c_\infty)/R_{vap}$, where the concentration difference $(c_0 - c_\infty)$ is interpreted as driving force for the evaporation and R_{vap} as evaporation resistance. In the presence of a film it is assumed, that R_{total} can be regarded as being the sum of two resistances in series

$$R_{total} = R_{vap} + R_{film} \ . \tag{1.120}$$

The evaporation rate with a film is

$$\dot{m}_{film} = \frac{c_0 - c_\infty}{R_{total}} = \frac{c_0 - c_\infty}{R_{vap} + R_{film}} \ . \tag{1.121}$$

Using the relation $\dot{m} = (c_0 - c_\infty)/R_{vap}$ it follows

$$R_{film} = R_{vap} \frac{\dot{m} - \dot{m}_{film}}{\dot{m}_{film}} \ . \tag{1.122}$$

The evaporation resistances of monolayers have been studied in many investigations [87, 88]. An important application is the retardation of water evaporation in open storages by covering the water surface with monolayers of long-chain alcohols. A large spreading rate of the monolayer is desirable in this case. It has been attempted to find theoretical relations between the evaporation resistance and properties of the monolayer. In the accessible area theory it is assumed that permeation is only possible through sufficiently large gaps and holes in the monolayer. The total area of the holes is called the accessible area [88].

1.11.5 Evaporation and combustion in droplet arrays

Many studies of droplet evaporation and combustion of fuel droplets have been conducted with single droplets. These investigations were motivated by the expectation of a better basic understanding of fuel sprays. Droplets in technical sprays are, however, surrounded by neighbors which influence the evaporation and combustion characteristics. In order to study these interactions, theoretical and experimental investigations have been performed on simple multi-droplet configurations. Brzustowski et al. investigated theoretically the interaction of two burning fuel droplets of arbitrary size [89]. Labowski introduced a burning rate correction factor η which represents the ratio of the actual burning rate and the burning rate of an isolated droplet [90]. Values for the burning rate correction factor were determined numerically for droplet arrays consisting of two, three, and four droplets of equal size. For the four droplet array with a droplet spacing of five radii $\eta \approx 0.6$ is obtained, which means that the droplets in the array burn approximately 40 % slower than isolated droplets. The presented results reveal that η decreases, when the number of droplets in the array increases or the droplet distance decreases. It could be shown that the d^2-law is only approximately correct for describing the history of interacting droplets. In an experimental investigation Xiong et al. studied the influence of different important parameters on the vaporization and combustion of two interacting fuel droplets [91, 92]. The influence of neighboring parallel droplet streams on the burning rate of droplet in these streams are presented in Sect. 6.4.1. More complicated systems like sprays can be described with group combustion models [71].

1.12 Interaction with Light

In this section the interaction between light and single spherical particles will be considered. Different aspects of elastically scattered light will be described. The elastically scattered light contains information on droplet properties such as velocity, size, and refractive index. The refractive index depends on the density and therefore on the temperature, or composition of the droplet liquid. Therefore elastically scattered light is detected and evaluated in many optical measurement techniques, in order to obtain droplet properties nonintrusively. The influence of the droplet on the incident light can be characterized by the extinction cross section C_{ext}. The light absorbed and scattered by the droplet is characterized with the absorption and scattering cross sections C_{abs} and C_{sca} respectively. It holds $C_{ext} = C_{sca} + C_{abs}$. The corresponding dimensionless scattering efficiencies are obtained, when the cross sections are divided by the geometrical cross section πr^2 of the droplet. One obtains $Q_{ext} = Q_{sca} + Q_{abs}$. A detailed description of cross sections and scattering efficiencies may be found for example in the textbooks by Bohren and Huffman [93] or van de Hulst [94]. Scattered light is observed in all directions around the droplet. The scattering angle θ is defined between the forward direction of the incident light and the direction of observation. The scattered light intensity distribution is determined by the complex refractive index of the droplet liquid $N = n + ik$ and the droplet size in comparison with the wavelength of the incident light. The complex refractive index is composed of the real part n, defined by Snell's law of refraction, and the imaginary part k, which is a measure for absorption of light. In addition the scattered light is influenced by the intensity distribution and the form of the wave fronts of the incident light, which are in most cases treated as homogeneous plane waves. Examples for a Gaussian intensity distribution will be given in Sect. 1.12.5. For droplets of arbitrary size and refractive index, which are illuminated by homogeneous plane waves, the scattered light can be described with the so-called Mie theory. Some calculations based on this theory will be presented in Sect. 1.12.4. Depending on droplet diameter d and wavelength λ different approximations may be applied [95]. For droplet radii $d \leq \lambda/10$ a homogeneous isotropic intensity distribution of the scattered light is found, which can be described with the so-called Rayleigh approximation. For diameters up to the wavelength the so-called Rayleigh-Gans-Debye approximation may be suitable. For very large droplets the intensity distribution of the scattered light can to some extent and for some scattering angles be described with the approximation of geometrical optics.

1.12.1 Geometrical approximation

Geometrical optics can be applied to droplets, when the droplet size is very large in comparison with the wavelength of the light. Foundations of geometrical optics are described in textbooks of optics [96, 97]. The intensity distribution of scattered light can be calculated with the approach of geometrical optics, however, not for all scattering angles. Geometrical optics does for instance not account for diffraction; deviations are therefore found in the close forward direction. In focal points and on focal lines infinitely large intensities are predicted, which are of course not realistic [94]. However, basic phenomena of light scattering, like glare points, rainbow phenomena, or the regular fringe pattern in the forward direction can be explained by geometrical optics by tracing the paths of light rays inside and outside of the droplet. Intensities in the far-field of the droplet are obtained by applying Huygens' principle on light rays. In Fig. 1.22 the path of a representative incident light ray through a droplet of radius r is shown. The plane of drawing shows an intersection through the center of the droplet. The incident ray has the distance a from the center. This location of the incident ray can be characterized by the angle τ between the incident ray and the tangential plane to the droplet at the point where the ray hits the droplet, with $\cos\tau = a/r$. At the intersection of the ray with the droplet surface part of the light is reflected, the remaining part is refracted into the droplet. All reflected and refracted rays are in the plane of drawing, which is perpendicular to the tangential planes. The reflected ray is called ray of order $p = 0$. The angle with respect to the tangential plane is τ again. The scattering angle of the ray with order $p = 0$ is then $\theta_0 = 2\tau = 2\cos^{-1}(a/r)$. The angle τ' between the refracted ray and the tangential plane is determined by Snells' law of refraction $n = \sin\alpha/\sin\alpha'$ or $n = \cos\tau/\cos\tau'$ with $\tau = 90° - \alpha$ and $\tau' = 90° - \alpha'$. The refracted ray travels through the droplet until it intersects the droplet surface, where part of the light is refracted and leaves the droplet as ray of order one, i.e. $p = 1$. The remaining part is reflected and traverses the droplet again until the next intersection with the droplet surface occurs, where part of the light leaves the droplet, and so on. The order number p of the rays leaving the droplet increases by one after each path through the droplet. For each incident ray only a limited number of intersections have to be taken into account for calculating the scattered light intensity with sufficient accuracy,

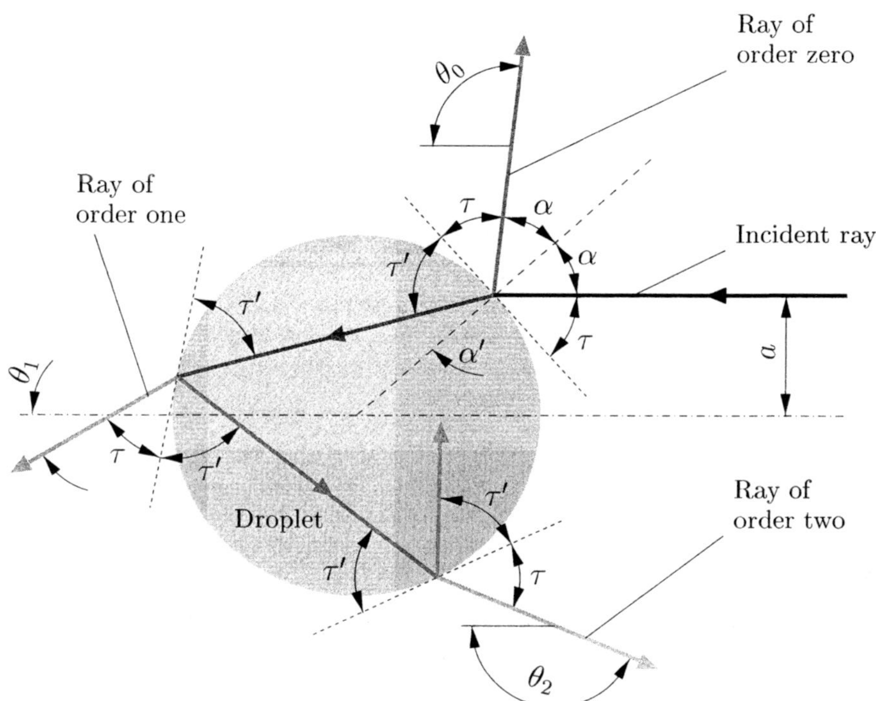

Fig. 1.22. Representative ray incident on spherical droplet. In the example shown the refractive index of the droplet is $n = 1.49$ and the location of the incident ray is given by $a/r = 2/3$. Shown are three interactions with the droplet surface

because the intensity decreases rapidly with increasing order number p. Inspection of Fig. 1.22 reveals that the angles τ and τ' are the same for all interactions at the surface of the droplet. For the scattering angle θ_p of the ray of order p leaving the droplet one has therefore

$$\theta_p = 2(\tau - p\tau') = 2\left(\cos^{-1}\frac{a}{r} - p\cos^{-1}\frac{a}{rn}\right) . \tag{1.123}$$

For determining the intensity distribution of the scattered light the distance a has to be varied from $a = -r$ to $a = r$ to take all rays into account. The intensity for the scattering angle θ is obtained from the superposition of all rays leaving the droplet with this angle taking into account the optical path length and the phase shift between the rays. With numerical calculations the optical path of the rays inside the droplet can be determined even if the refractive index of the droplet changes with the radius coordinate [98]. As mentioned above, phenomena like glare points can be explained on the basis of geometrical optics.

1.12.2 Glare Points

On a droplet, which is illuminated for instance by sunlight, one or more bright spots, the so-called glare points, can be observed with the naked eye. The lens of the eye images the droplet on the retina. With a technical lens the droplet with the glare points can be imaged on a screen or on the film of a camera. A setup for observing the glare points with illumination of the droplet by monochromatic and parallel light of a laser is shown schematically in Fig. 1.23. The droplet is observed at the angle θ with respect to the direction of the incident light. Light scattered in the region $\theta \pm \Delta\theta/2$ is collected by the lens and contributes to the image on the film in the camera. With this method the near-field of the scattered light is observed, when the object plane, which contains the droplet center, is imaged.

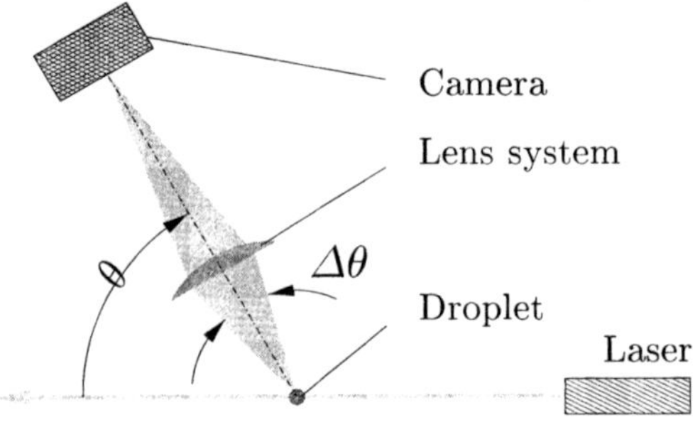

Fig. 1.23. Schematic top view of arrangement to image a droplet illuminated by a laser beam

The far-field of the scattered light is obtained without a lens, when the distance between the screen or the CCD chip of a camera and the droplet of radius r is larger than $4r^2/\lambda$. In the far-field, interference between rays scattered at the same angle can be observed. A closer look at the droplet is shown in Fig. 1.24, which represents an intersection through the droplet center. Shown are rays, which contribute to the two brightest glare points observed in the forward hemisphere. One glare point is due to rays of order zero, which are reflected directly at the droplet surface. All reflected rays collected by the lens are within beam A. The other glare point comes from rays of order one, which are refracted and have passed the droplet once. All refracted rays collected by the lens are within beam B. Parallactic effects can be neglected since the distance between lens and droplet is large in comparison with the droplet size. The object plane is imaged by a camera or the naked eye. The observed extension of the two glare points is indicated by two short bold solid lines in the object plane. The brightness and the distance between the glare points depends on the direction of observation θ. The distance of the glare points is a function of droplet size. Using these effects the droplet size and the refractive index of the droplet can be determined, as will be described briefly in Chap. 4. Glare points coming from rays of higher order number p possess naturally a smaller intensity. A photograph of a Plexiglas

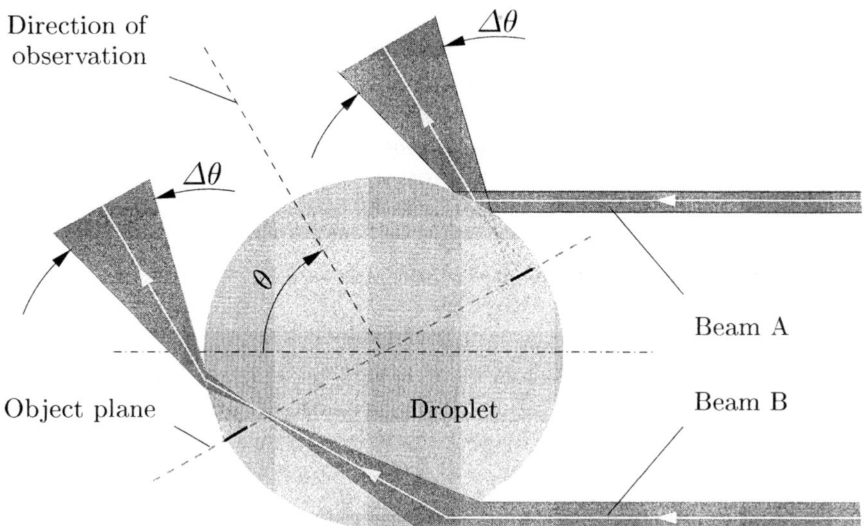

Fig. 1.24. Rays contributing to glare points. Beam A contributes to the glare point due to reflected light, whereas beam B contributes to the glare point, which is due to refracted light

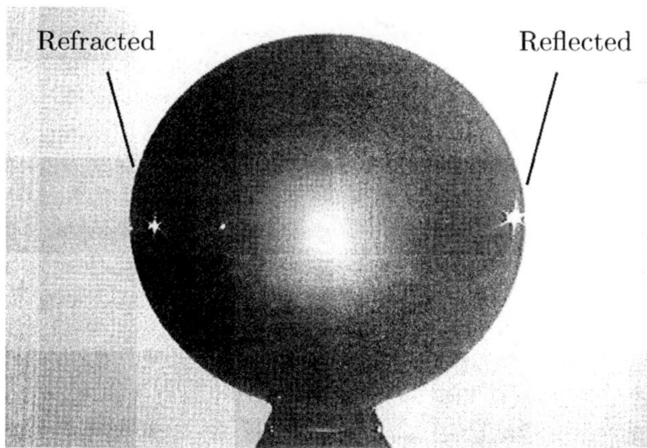

Fig. 1.25. Photograph of a Plexiglas sphere, which is is illuminated by a laser beam arranged as shown in Fig. 1.24. Two bright glare points can be recognized. One of them is due to reflected light, the other one is due to refracted light. An additional illumination with diffuse white light has been used to observe the contour of the sphere. The refractive index of the sphere is $n \approx 1.49$

sphere showing two bright glare points is presented in Fig. 1.25. The photo has been taken with a camera in the configuration sketched in Fig. 1.23. The two small star-shaped spots are the glare points produced by reflected and refracted light rays of order $p = 0$ and $p = 1$. The smooth large white spot in the middle of the sphere is caused by the diffuse light used for the illumination of the Plexiglas sphere. Without this additional illumination only the glare points would be visible. A description of glare points on the basis of the Mie theory has been given by van de Hulst and Wang [99]. Other important scattering phenomena are the rainbows. The brightest rainbow can be observed in the backward hemisphere. An explanation based on geometrical optics will be given in the next section.

1.12.3 Rainbow

The rainbow is well known as a bright and colorful phenomenon in nature. It can be observed for example, when sunlight interacts with rain or a spray near water falls. Rainbows observed in nature are the product of light scattered by many droplets [100]. For experimental or measurement purposes the light scattered by a single droplet can be studied. The intensity maxima observed in the backward hemisphere are responsible for the first rainbow. The brightest maximum is therefore called rainbow maximum. A German monk, called Theodoric of Freiberg, was first reported to connect rainbow phenomena with the path of light rays within a droplet; whereas it was René Descartes, who gave in the year 1637 a physical explanation of rainbow phenomena based on ray optics [101]. The intensity of the first rainbow is essentially determined by rays of order $p = 2$. According to Descartes' theory of rainbows the angular position θ_{rg} of the first rainbow is given by the minimum scattering angle of the rays of order two. In Fig. 1.26 this ray with minimum deflection or minimum scattering angle is shown. In addition neighboring incident rays are shown with larger scattering angles. In the vicinity of the ray with minimal deflection the density of the rays is higher, which results in a higher intensity near the rainbow. It is found, that minima of the scattering angle are obtained only for rays with order numbers $p \geq 2$. The rainbows from rays with higher order numbers, however, have fainter intensities. The second rainbow, which consists essentially of rays of order three can be observed in nature under favorable conditions. The region in the sky between the first and second rainbow appears darker. This region is called Alexander's dark space according to Alexander of Aphrodisias, who lived approximately two hundred years before Christ. Scattering angles of rays of order p can be calculated with Eq. 1.123. According to the theory of Descartes the angular position θ_{rg} of the different rainbows are given by the relation

$$\theta_{rg} = 2\left\{p\cos^{-1}\left(\frac{1}{n}\sqrt{1-\frac{n^2-1}{p^2-1}}\right] - \sin^{-1}\left[\sqrt{\frac{n^2-1}{p^2-1}}\right)\right\}, \quad (1.124)$$

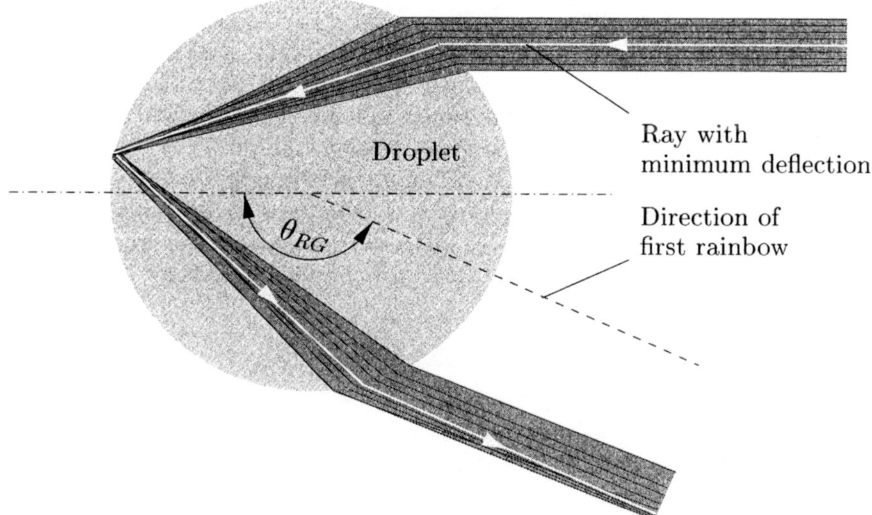

Fig. 1.26. Schematic view of rays contributing to the light near the first rainbow

where n is the refractive index of the droplet liquid and p is the order number of the scattered rays contributing to the rainbow. In nature the first and second rainbow can be observed. The position of the first and the second rainbow is obtained for $p = 2$ and for $p = 3$. Descartes' theory is based on ray paths and therefore independent of droplet size. Detailed measurements of the angular position of the rainbow show that for smaller droplets deviations from Descartes' theory occur. The angular position of the rainbow maximum increases with decreasing droplet size. However, for droplets in the range of a millimeter or larger the theory of Descartes describes the rainbow position with sufficient accuracy in most cases.

In 1838 George B. Airy developed an extension of Descartes' theory, which accounts for the effect of the droplet radius [102]. The rainbow position is according to Airy's theory

$$\theta_{ra} = \theta_{rg} + \frac{1.0845}{\sqrt{(n^2-1)/(p^2-1)}} \left[\lambda^2 \, \frac{\sqrt{1-(n^2-1)/(p^2-1)}}{64\, r^2} \right]^{1/3} , \quad (1.125)$$

where θ_{rg} represents the rainbow position according to Descartes' theory, which is given by Eq. 1.124. The theory of Airy includes Huygens' principle and curved wave fronts of the light scattered in the direction of the rainbows. [94, 103].

Airy's theory predicts not only one intensity maximum in the rainbow region but several maxima with decreasing intensity, called supernumerary rainbow maxima. These maxima can be explained by the interference of rays of order two, which leave the droplet at the same scattering angle, but have different optical paths. This corresponds with observations of the far-field of the scattered light. The far-field of the scattered light can be obtained on a screen as described in the next section, or by using a lens with the screen in its focal plane [104]. For this purpose the distance between lens and droplet must be sufficiently large. Using this observation technique a ripple structure is found around the main rainbow maximum and the supernumerary maxima. These ripples are caused by interference of rays of order zero and order two [100]. An apparently exact intensity distribution is obtained with the theory of Mie. When the Mie theory is used, different definitions of the rainbow positions may be appropriate [105].

1.12.4 Mie scattering

The far-field intensity distribution of the light scattered by a droplet can be made visible on a screen. A suitable optical setup is shown schematically in Fig. 1.27. The screen has been curved to obtain an equidistant scale for the

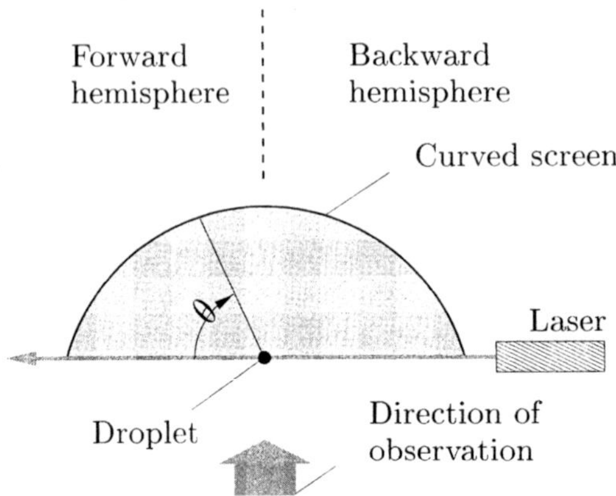

Fig. 1.27. Schematic view of arrangement to visualize the scattered light from droplets on a curved screen. The laser beam illuminates a droplet, which scatters the light into the forward and backward hemisphere

scattering angle θ on photographs of the screen. The scattering angle θ is defined by the direction of the incident light and the direction of the line from the droplet to the point observed on the screen, as shown in Fig. 1.27 and described above. The sides of the scattering angle θ define the scattering plane, which contains the center of the droplet and coincides with the plane of drawing in Fig. 1.27. The incident light is polarized perpendicular to the scattering plane. The observed intensity distribution is not homogeneous. A photo of the screen is shown in Fig. 1.28, where the laser light of wavelength $\lambda = 514.5\,\text{nm}$ comes from the right hand side. A monodisperse stream of droplets crosses the laser beam approximately in the center of the picture, where the bright spot can be seen. It is assumed, that the droplets are illuminated by plane waves. Each droplet produces the same scattering pattern on the screen, since all the droplets have the same size. Above the bright spot

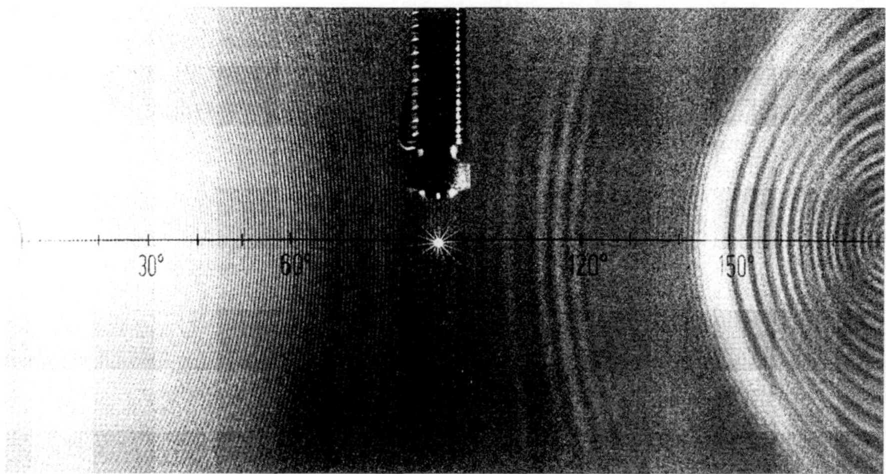

Fig. 1.28. Scattering pattern from iso-propanol droplets with the diameter $d = 46\,\mu$m. The horizontal scale, approximately in the middle of the picture, indicates the scattering angle θ, as described in the text

part of the droplet stream generator producing the monodisperse stream is visible (see Sect. 2.3). The scale on the screen indicates the scattering angle θ. For small scattering angles, that means in the close forward direction, diffraction phenomena are dominant. For larger scattering angles regular fringes are observed. It will be shown in Sect. 4.6.3, that the droplet size can be obtained from the distance of these fringes. At $\theta \approx 90°$ the scattering intensity is very low. In the backward hemisphere, that means for $\theta > 90°$, several maxima of the scattered light are observed. The brightest maximum represents the main maximum of the first rainbow. The maxima found for larger scattering angles are the so-called supernumerary rainbow maxima, mentioned above. From the angular positions of the rainbow maxima the droplet size and the refractive index of the droplet liquid can be determined, as will be described in Sects. 4.6.6 and 4.7.4 in more detail.

For droplets, which are small in comparison with the wavelength, the intensity distribution can be determined with Rayleigh's theory. For very large droplets rules of geometrical optics combined with Huygens' principle give to some extent good results [93]. For the intermediate size range a formal solution of the Maxwell equations has been developed with the boundary conditions of a homogeneous spherical droplet of given size and refractive index, which is illuminated with plane waves. An essential contribution for the solution of this problem has been given by Mie [106]. The most common term is therefore Mie theory, although other scientists, for instance Lorenz and Debye, contributed to the solution [107]. Although the solution of this scattering problem is rather old, systematic numerical calculations can be performed only since large digital computers are available. The diagrams of

Figs. 1.29 and 1.30 have been obtained using a computer code given in the appendix of Ref. [93].

The incident light was in this example a homogeneous plane wave with linear polarization. The imaginary part k of the complex refractive index $N = n + ik$ becomes zero, when absorption is neglected. For practical applications three parameters are usual, namely droplet radius r, wavelength of the incident light λ, and refractive index N. But for the calculations according to the Mie theory only two independent parameters have to be considered. These parameters are the complex refractive index N and the Mie parameter $\alpha = 2\pi r/\lambda$, which represents the ratio of the droplet circumference and the wavelength. In Fig. 1.29(a) the scattered light intensity for the scattering plane perpendicular to the polarization of the incident light is shown as a function of the scattering angle θ. In the backward hemisphere the main maximum and the supernumerary maxima of the first rainbow can be recognized. In the forward hemisphere oscillations of the intensity are found, which correspond to the regular fringes shown in Fig. 1.28. The intensity variations comprise several orders of magnitude. This can be seen in the diagram of Fig. 1.29(b). When the incident light is polarized parallel to the scattering plane the results presented in Fig. 1.30(a) are obtained. An obvious difference in comparison with Fig. 1.29(a) can be seen in the backward hemisphere. There is no main maximum of the first rainbow. This means, that the light of a rainbow is polarized in nature, where the sun illuminates the droplets [101]. To show the increase of the scattered light intensity for smaller scattering angles results have been plotted in logarithmic scale in Figs. 1.29(b) and 1.30(b).

In the presented calculations the absorption within the droplet has been neglected. The intensity of a light ray with wavelength λ traveling the distance l through an absorbing medium is given by the law of Lambert-Beer-Bouguer

$$I = I_0 e^{-\mu l}. \tag{1.126}$$

The equation

$$\mu = \frac{4\pi k}{\lambda} \tag{1.127}$$

describes the relation between the coefficient of absorption μ and the imaginary part of the refractive index k. The influence of absorption is shown in Fig. 1.31 for the same values of droplet diameter and index of refraction as in the example shown above. These calculations have been performed for three different values of the imaginary part k of the complex refractive index N. The incident light was polarized perpendicular to the scattering plane. The result calculated with $k = 0$ is of course the same as in Fig. 1.29(a). For most scattering angles the mean scattered light intensity decreases with increasing absorption. The reflected light at the droplet surface and the diffracted light are not influenced by absorption. However, the light intensity refracted into

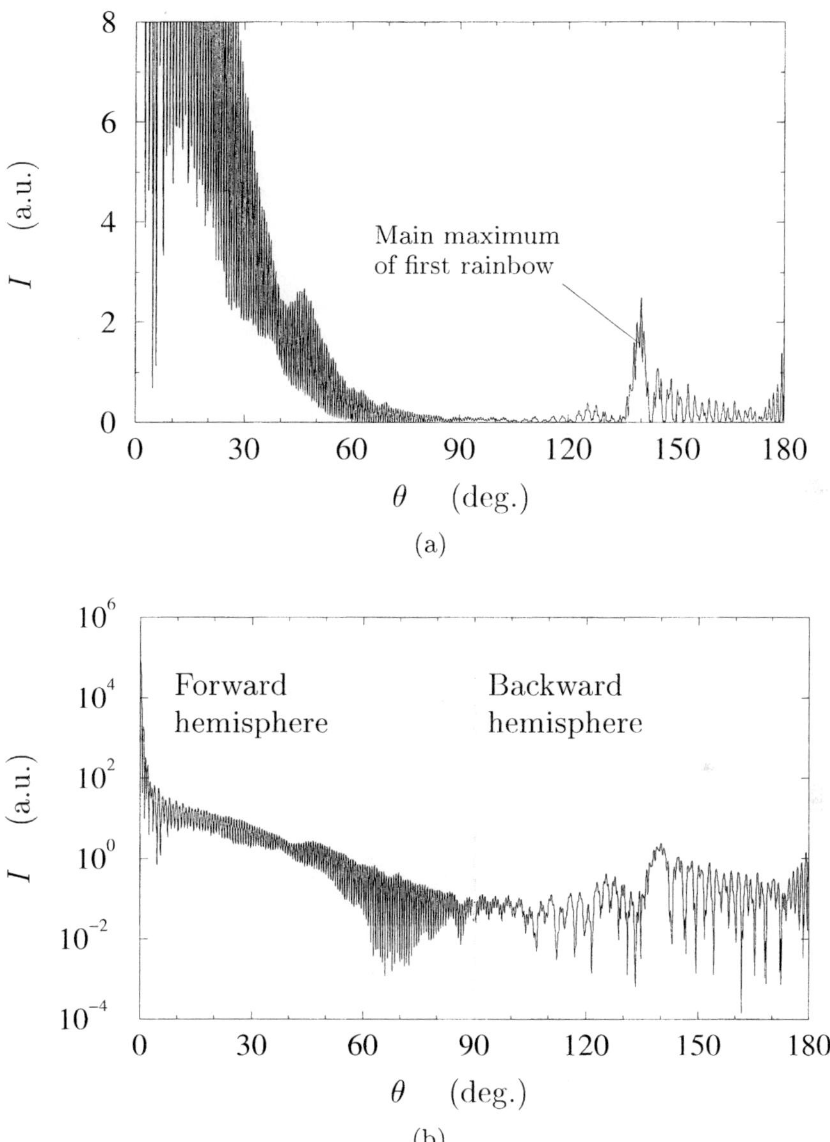

Fig. 1.29. Linear (a) and logarithmic (b) plot of intensity of scattered light as a function of scattering angle θ for a droplet with radius $r = 25\,\mu\text{m}$ and refractive index $n = 1.333$. The results are for incident light with wavelength $\lambda = 514.5\,\text{nm}$ polarized perpendicular to the scattering plane

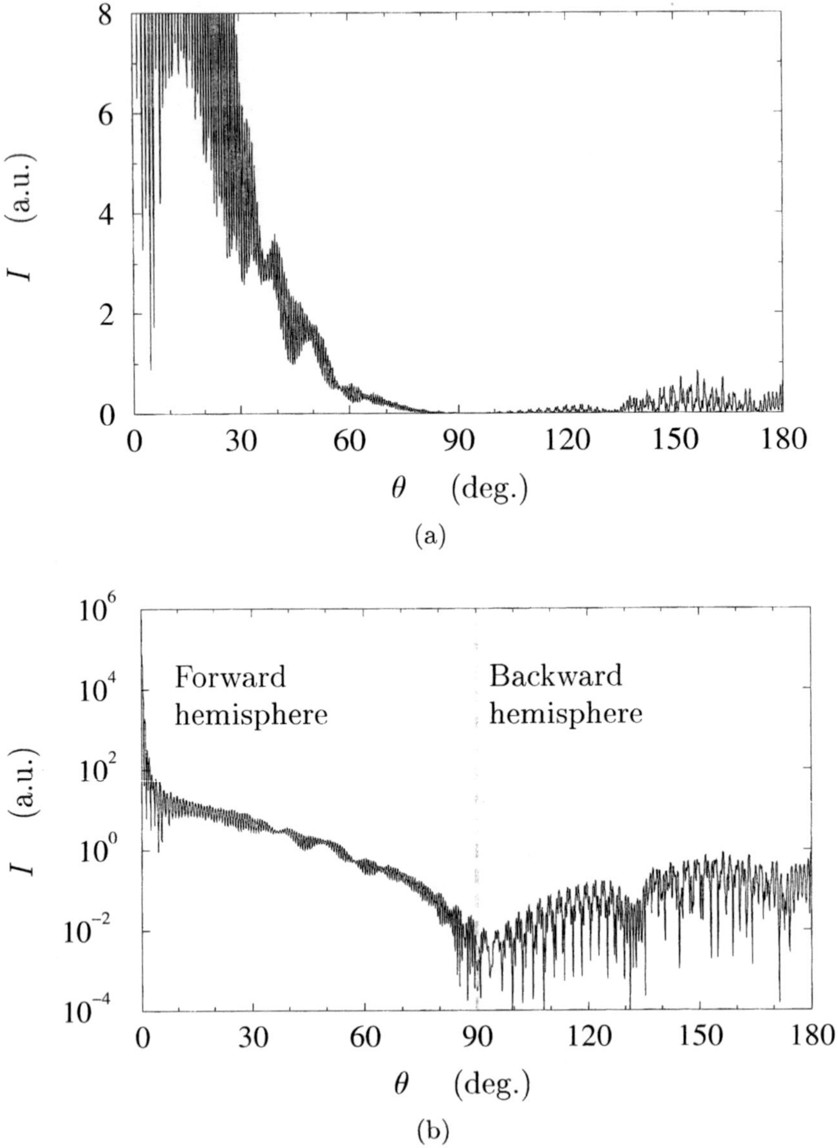

Fig. 1.30. Linear (a) and logarithmic (b) plot of intensity of scattered light as a function of scattering angle θ for a droplet with radius $r = 25\,\mu\mathrm{m}$ and refractive index $n = 1.333$. The results are for incident light with wavelength $\lambda = 514.5\,\mathrm{nm}$ polarized parallel to the scattering plane

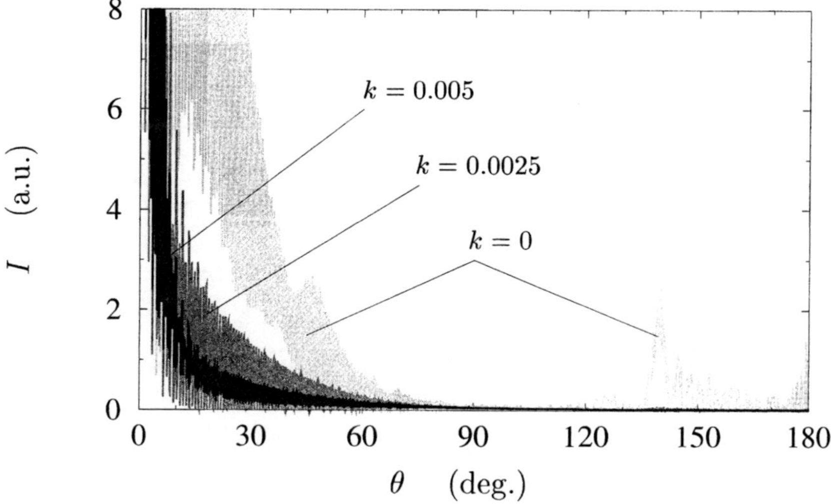

Fig. 1.31. Intensity of scattered light as a function of scattering angle for a droplet with radius $r = 25\,\mu$m and refractive index $N = 1.333 + \mathrm{i}k$. Shown are results for incident light with wavelength $\lambda = 514.5$ nm polarized perpendicular to the scattering plane. Results for three different values of the imaginary part k of the complex refractive index are shown

the droplet is attenuated by absorption. Due to different path lengths and due to the dominant diffraction phenomena in the close forward region, the light scattered backwards is influenced more.

The effect of absorption may be estimated from the numerical results shown in Fig. 1.32. In the diagram the intensity of the scattered light for the two scattering angles $\theta = 30.14°$ and $\theta = 139.44°$ is presented as a function of the imaginary part k of the complex refractive index. The intensities are normalized with the intensity I_{max} obtained without absorption for the same angle. The angle chosen in the forward hemisphere corresponds to an intensity maximum of the regular fringe system. The angle chosen in the backward hemisphere corresponds to the absolute maximum in the region of the first rainbow. It has been found, that the angular position of these maxima are practically independent of absorption as long as the maxima are not damped out. Since light rays of order two, which pass the droplet twice, contribute mainly to the rainbow, the decrease of the intensity is higher in the rainbow region than in the forward direction, as can be seen from Fig. 1.32. The intensity found behind a plane layer of liquid, described by equation 1.126, would result in a straight line in Fig. 1.32, due to the logarithmic scale. The results for a droplet, however, show deviations from a straight line. The scattered light is not only determined by refracted light, which is influenced by absorption, but also by directly reflected and diffracted light.

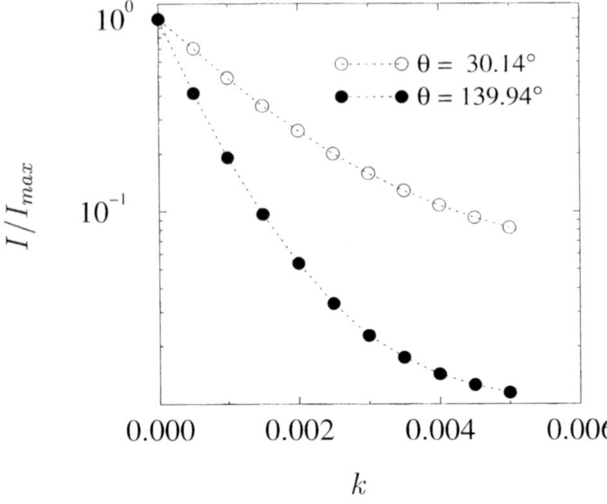

Fig. 1.32. Intensity ratio I/I_{max} as a function of the imaginary part k of the complex refractive index $N = n + ik$. The shown results for $\theta = 30.14°$ in the forward hemisphere and for $\theta = 139.94°$ in the backward hemisphere

The influence of the intensity distribution of the light used for the illumination of the droplet will be discussed in the following.

1.12.5 Influence of Gaussian Intensity Distribution

Laser beams applied in modern nonintrusive measurement systems have usually Gaussian intensity distributions. Normally the laser is operated in the normal or TEM_{00} mode. For this mode the radially symmetrical Gaussian light intensity distribution can be expressed by

$$I(r_b, z) = I_0(z)\,\mathrm{e}^{-2(r_b/w(z))^2} , \tag{1.128}$$

where r_b and z are cylindrical coordinates. The axis of the laser beam coincides with the z-axis. For $r_b(z) = w(z)$ the intensity has decreased to

$$I(w(z), z) = \frac{I_0(z)}{\mathrm{e}^2} . \tag{1.129}$$

With increasing z the radius $w(z)$ of the laser beam increases according to

$$w(z) = \sqrt{w_0^2 + \left(\frac{\lambda z}{\pi w_0}\right)^2} , \tag{1.130}$$

where w_0 is the radius of the laser beam at $z = 0$

The example of Fig. 1.33 gives an impression of the waist form of the laser beam. The calculations were performed for the beam waist radius $w_0 = 125\,\mu\mathrm{m}$ and the wavelength $\lambda = 514.5\,\mathrm{nm}$. The wave front in the beam waist is a plane wave with a Gaussian intensity distribution [108]. The light intensity scattered by a droplet located in the laser beam can be calculated

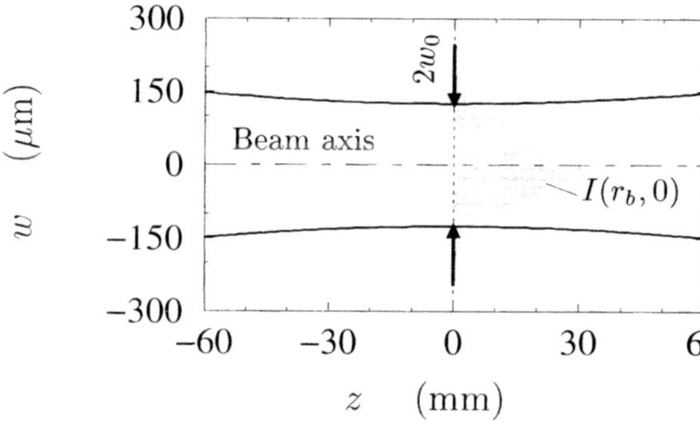

Fig. 1.33. Shape of laser beam along z-axis. In addition the profile of the intensity distribution $I(r_b, 0)$ in the beam waist is shown

with the so called Generalized Lorenz-Mie theory (GLMT), which has been developed by Gouesbet and Gréhan at INSA in Rouen [109-111]. With a computer code based on GLMT several calculations have been performed. A few examples are shown in the following to show the influence of a Gaussian intensity distribution[1]. The system of coordinates for indicating the position of the droplet in the laser beam is given in Fig. 1.34. The axis of the laser coincides with the z-axis, the location of the beam waist is $z = 0$. The incident light is polarized parallel to the x-axis, and the scattering plane is the y, z-plane. In Figs. 1.35 and 1.36 the influence of the beam waist w_0 on the scattered-light intensity is shown for a nonabsorbing droplet with the radius $r = 25\,\mu$m and the refractive index $n = 1.39$ located in the origin of the coordinate system. Results for the forward hemisphere are shown in Fig. 1.35, whereas Fig. 1.36 shows results for the backward hemisphere with the rainbow region. The ratio r/w_0 has to be sufficiently large for a significant scattered light intensity and sufficient contrast. It should be mentioned, that

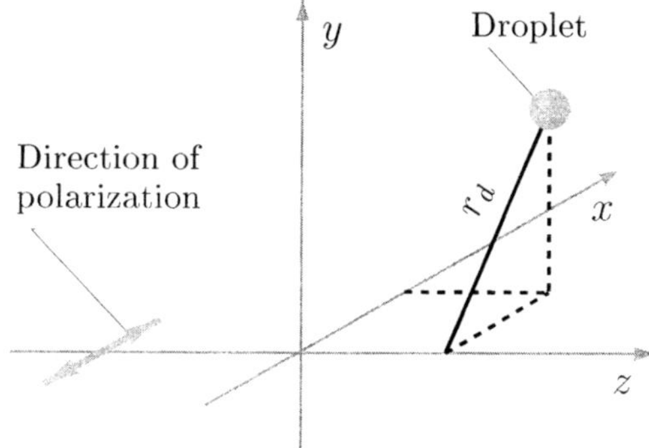

Fig. 1.34. Coordinate system for describing the position of the droplet in the laser beam. The direction of illumination is the positive z-direction. The light is polarized in x-direction. The distance of the droplet from the z-axis is indicated with r_d

[1] The code for the calculations in Figs. 1.35-1.39 has been used with kind permission of Professor Gérard Gouesbet and Professor Gérard Gréhan.

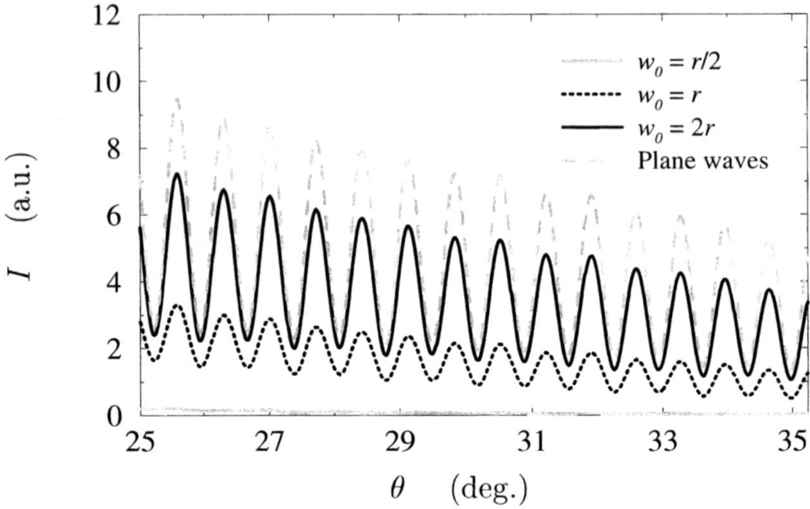

Fig. 1.35. Intensity of light scattered in the forward hemisphere for different values of the beam waist radius w_0

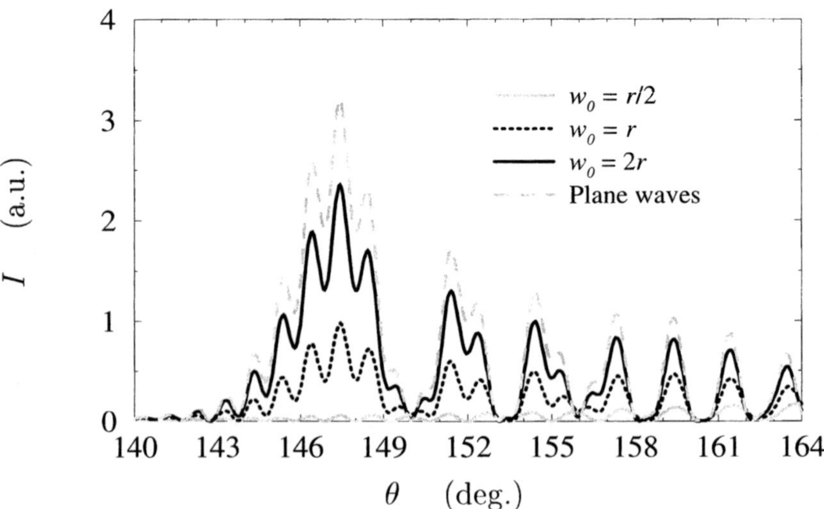

Fig. 1.36. Intensity of light scattered in the backward hemisphere around the first rainbow for different values of the beam waist radius w_0

both intensity and contrast have an essential influence on the accuracy of measurements. With the exception of some special individual applications the beam waist radius should be sufficiently large in comparison with the droplet radius.

Not only the beam waist radius has an influence on the scattered light intensity, but also the position of the droplet in the laser beam. As the position of the droplet is arbitrary only a few examples will be discussed here. The scattering angle is restricted here to an interval in the forward hemisphere with $\theta \approx 30° \pm 5°$. The influence of a displacement in y-direction has been analyzed for two different z-positions. Droplet radius and refractive index are the same as in Figs. 1.35 and 1.36. The beam waist radius is $w_0 = 125\,\mu\text{m}$, which is five times the droplet radius. In Fig. 1.37 the scattered light intensity is shown for three different values of y at $z = 0$. A logarithmic scale has been

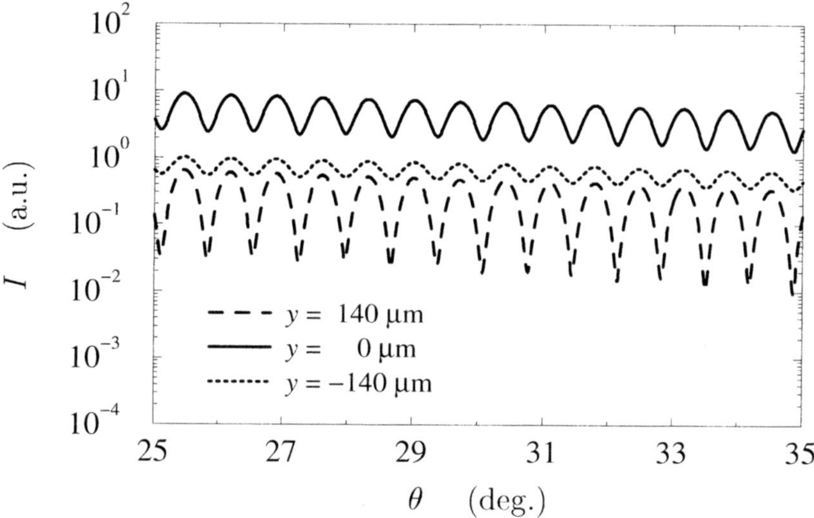

Fig. 1.37. Intensity of scattered light in the forward hemisphere for different values of y within the beam waist at $z = 0$

used since the scattered light intensity in the center position is very large in comparison with the positions at the border of the laser beam. It can clearly be seen, that the intensity distribution of the scattered light shows no symmetrical behavior with respect to the center position, when the y-position is varied. The amplitudes of the oscillations differ between the side positions $y = \pm 140\,\mu\text{m}$ as shown in the figure. The mean value I_m and the standard deviation σ have been calculated for the intensity distribution of the scattered light in the interval $\theta \approx 30° \pm 5°$. In Fig. 1.38 the mean intensity I_m is shown as a function of y-position for $z = 0$ and $z = 50\,\text{mm}$. The values of I_m, which are approximately symmetrical with respect to the center position

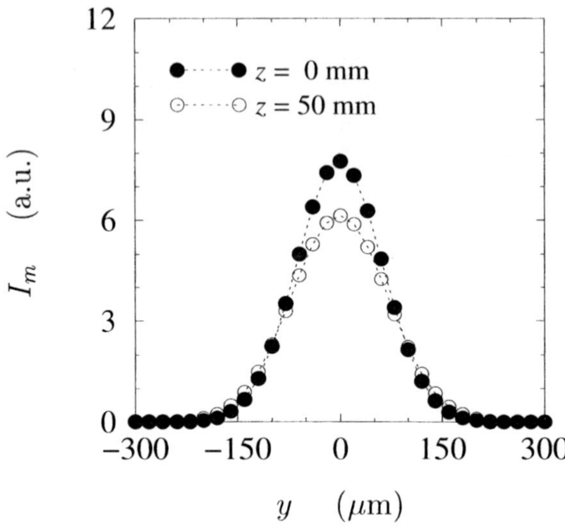

Fig. 1.38. Mean intensity of scattered light I_m as a function of the droplet's y-position for two different z-positions. The calculations of the mean value have been performed for the window $\theta \approx 30° \pm 5°$

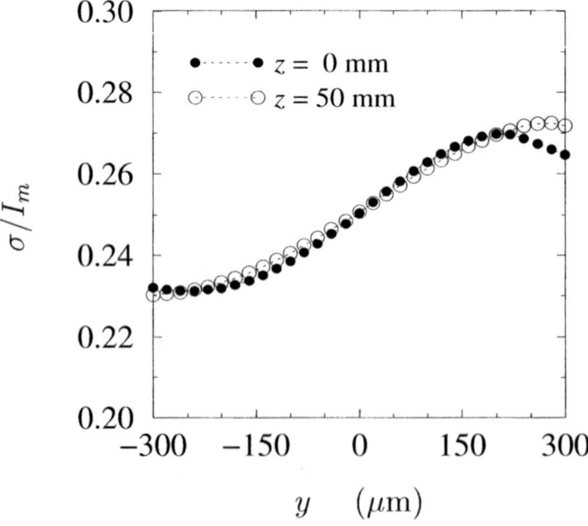

Fig. 1.39. Ratio of standard deviation and mean scattered light intensity σ/I_m as a function of y-position for two different values of z-position. For the calculations of the mean value I_m and the standard deviation σ the window $\theta \approx 30° \pm 5°$ has been used

at $y = 0$, follow, as might be expected, approximately the Gaussian intensity distribution of the laser beam. For the beam waist with $z = 0$ the maximum value is higher than the maximum value for $z = 50\,\mathrm{mm}$. The values of the mean intensity I_m are larger in the outer region of the beam as the beam radius at $z = 50\,\mathrm{mm}$ is larger than the beam waist radius.

The amplitudes of the regular fringes in the forward direction may be characterized by the ratio of the standard deviation and the mean intensity σ/I_m. For higher values of σ/I_m the fringes can be better distinguished from the background, due to the higher contrast. The ratio σ/I_m as a function of the y-position is shown in Fig. 1.39 for two different values of the z-position of the droplet. The asymmetrical behavior of this ratio is evident. It should be noticed, that at the y-position of the maximum ratio the fringes of the scattered light have the highest contrast, but a very low intensity. Which value of the y-position is best for experimental investigations of the fringes depends not only on the contrast of the fringes, but also on the sensitivity of the detector.

The calculations shown above are focused on the determination of the fringe spacing in the forward hemisphere. The examples show, however, that the influence of the Gaussian intensity distribution has to be taken into account. For other experimental needs corresponding calculations have to be performed to simulate and optimize the experiment. It should be mentioned again, that the classical Mie theory is limited to spherical and homogeneous droplets. A review of elastic light scattering theories for nonspherical and nonhomogeneous particles may be found in the literature [112].

1.12.6 Radiation Pressure

A light ray conveys energy as well as momentum. During the interaction with a droplet, light is scattered and absorbed, as discussed in the previous sections. The change of momentum of the radiation results in a force F_{pr} acting on the droplet [93, 94], which can be used to compensate the weight and to levitate and stabilize droplets within a laser beam [113]. An expression for the radiation force acting on a transparent particle based on Maxwell's equations has been given by Debye [114]. The forces on a droplet in a laser beam with Gaussian intensity distribution have been determined by Roosen [115] using an approach based on geometrical optics. Radiation pressure forces on a droplet arbitrarily located in a laser beam can be calculated using GLMT [116].

The force, which acts on the droplet, can be decomposed into two components, one parallel, the other perpendicular to the axis of the laser beam. For transparent droplets the force perpendicular to the axis is always directed towards the center of the laser beam and vanishes, when the droplet is on the axis. In this case the droplet is stabilized on the axis. For fully reflecting spheres this component is directed away from the axis; the position on the axis is therefore instable. In Fig. 1.40 the radiation pressure cross section C_{pr},

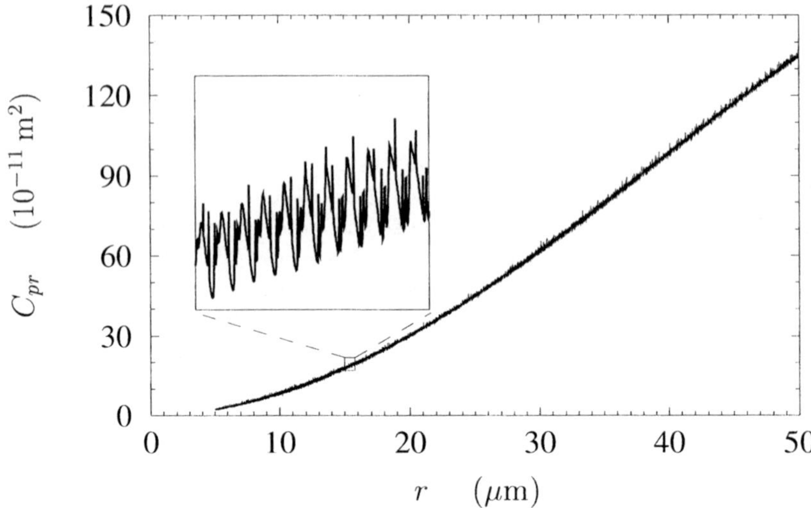

Fig. 1.40. Radiation pressure cross section C_{pr} as a function of droplet radius r. The enlarged view reveals the oscillatory behavior of C_{pr}. These calculations were performed for the beam waist radius $w_0 = 100\,\mu\mathrm{m}$, refractive index $n = 1.33$, and wavelength of incident light $\lambda = 514.5\,\mathrm{nm}$

which is a measure for the force acting on a droplet parallel to the beam axis of the laser beam, is presented as a function of the droplet radius. The calculations have been performed using a computer code based on GLMT. The force perpendicular to the axis is zero, as the droplet is located on the axis in the beam waist. The mean of the force along the beam axis increases, when the droplet radius increases. The enlarged view in Fig. 1.40 reveals oscillations of this force. These oscillations are caused by partial wave resonances, which will be described in the next section.

Radiation pressure forces are important for optical levitation techniques, which will be described in Sect. 3.4.9, whereas experimental setups and results of measurements on optically levitated droplets will be described in Sects. 6.2.2 and 6.3.1.

1.12.7 Partial Wave Resonances

When the droplet radius is varied, peaks are observed in the intensity of the scattered light. These so-called partial wave resonances depend on the shape of the droplet, on the wavelength of the incident light, and on the refractive index of the droplet liquid. The resonances are also called morphology dependent resonances as they depend on droplet properties. For homogeneous spherical droplets the resonances are determined by the Mie parameter α, which describes the droplet size, and by the refractive index. Resonances can

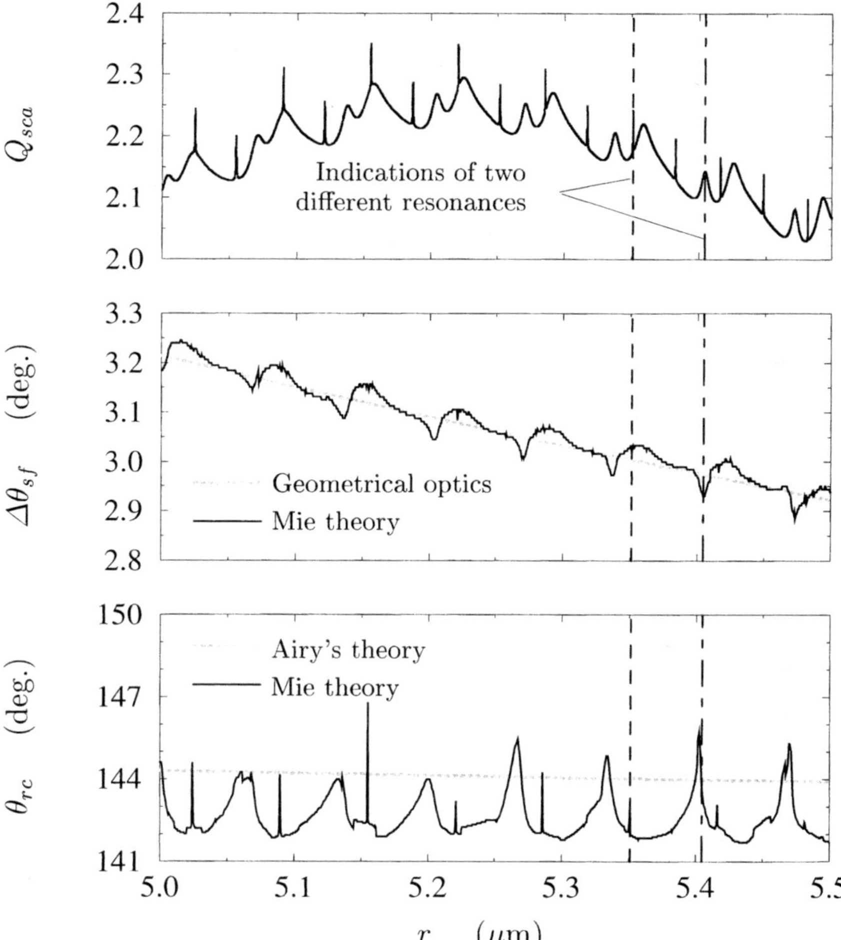

Fig. 1.41. Scattering efficiency Q_{sca}, fringe spacing $\Delta\theta_{sf}$ and rainbow position θ_{rc} as a function of droplet radius r. For the numerical calculations of the scattering efficiency the radius was varied in steps of $\Delta r = 0.0125\,\text{nm}$, for the fringe spacing $\Delta\theta_{sf}$ and the rainbow position θ_{rc} in steps of $\Delta r = 0.5\,\text{nm}$. The calculations were performed for the refractive index $n = 1.333$ and the wavelength $\lambda = 514.5\,\text{nm}$. In each diagram the position of two different resonances is indicated by two vertical lines

be observed for instance in the scattered light intensity as ripples or peaks, when the Mie parameter or the droplet radius respectively is varied. These characteristic effects can be described with the scattering efficiency Q_{sca}. The top diagram of Fig. 1.41 shows Q_{sca} as a function of droplet radius r for incident light of wavelength $\lambda = 514.5\,\mathrm{nm}$ and the refractive index $n = 1.333$. These calculations have been performed using the Mie theory. The range of the Mie parameter shown is approximately $61 \leq \alpha \leq 67$. The structure of the resonances is very complicated. Ripples and sharp peaks are observed. Two different resonances are indicated by vertical lines. Some resonance peaks are so small, that they are not resolved in the diagrams, even when the small increment of the droplet radius $\Delta r = 0.0125\,\mathrm{nm}$ is used. Detailed descriptions of the resonances and their classification are given in the literature [95, 117, 118].

Resonance peaks are not only observed in the scattering efficiency Q_{sca}. The diagrams of Fig. 1.41 reveal, that other different properties of the scattered light are influenced by partial wave resonances.

The fringe spacing $\Delta\theta_{sf}$ between the regular fringes of the light scattered in the forward hemisphere is shown as a function of droplet radius in the center diagram of Fig. 1.41. Part of the resonance structure can clearly be seen in the presented results. The droplet size can be determined from $\Delta\theta_{sf}$, as will be described in Sect. 4.6.3. The results without resonances are based on geometrical optics and are obtained with Eq. 4.4. Also size measurements with phase Doppler systems, which will be described briefly in Sect. 4.6.2, are influenced to some extent by the resonance structure.

The angular position of the first rainbow can be used to determine the refractive index of droplets, as will be described in Sect. 4.7.4. In the diagram at the bottom of Fig. 1.41 the angular position of the first rainbow θ_{rc} is shown as a function of droplet radius. The position θ_{rc}, which is derived from Mie calculations using a special correlation technique described in Ref. [105], shows a resonance structure. These results are compared with the rainbow position obtained with Eq. 1.125, which is based on Airy's theory and does not account for the resonances.

As can be seen from the enlarged view of Fig. 1.40 in the preceding section the radiation pressure characterized by C_{pr} varies according to the resonances. An optically levitated evaporating or condensing droplet will oscillate in its vertical position due to these resonances. If it is desired this can be avoided by varying the laser power appropriately.

The examples presented reveal, that partial wave resonances are found in many scattering phenomena and may influence the performance of measurement techniques. However, partial wave resonances found in the elastically and nonelastically scattered light can be used to determine droplet parameters like size and refractive index very precisely, as described by different authors [119-124]. One technique to determine the rate of size change with high accuracy using partial wave resonances will be described in more detail in Sect. 4.10.

2. Droplet Generation

2.1 Introduction

Droplets can be created in principle by condensation of vapors or by disintegration of bulk liquid. Droplet formation by homogeneous condensation of a vapor may occur under highly supersaturated conditions, whereas inhomogeneous condensation is observed at lower supersaturation in the presence of very small particles serving as condensation nuclei. Droplets generated by condensation play an important role in many natural and technical processes. The disintegration of bulk liquid can be observed during the splashing of raindrops on solid surfaces or the breakup of larger raindrops into smaller droplets. The transformation of bulk liquid into sprays is of enormous importance for numerous practical applications. There exists a vast literature on the properties of sprays and their production [48, 70, 71].

The lower limit of the droplet diameter is in many cases a few micrometer. Typical size spectra of technical sprays range from approximately 10 to 100 μm. Raindrops may be as large as 5 to 6 mm. Very small droplets exist in tobacco smoke.

The transformation of bulk liquid into fine droplets is associated with an enormous increase of surface. If liquid with the volume V_0 is subdivided into droplets having all the same radius r, then the area of the droplet system will be $A = 3V_0/r$. Assuming that the liquid has initially the shape of a sphere with radius r_0 and surface A_0 one can write $A_0 r_0 = Ar$. Transforming 1 liter of liquid into droplets with the radius $r = 1\,\mu$m will lead to the surface $A = 3000\,\mathrm{m}^2$. For water this corresponds to the surface energy $A \cdot \sigma \approx 218.3$ J. The number of droplets is in this case $N = 3V_0/(4\pi r^3) \approx 2.4 \cdot 10^{14}$.

Devices for the generation of droplets or sprays are described as droplet generators, atomizers, or spray nozzles. Nebulizers produce very fine sprays or aerosols, for example for medical applications. For certain technological applications and for scientific investigations it is essential, that the droplet size can be controlled very precisely. The present work deals mainly with physical and chemical properties and behavior of single droplets. The generation of single droplets or relatively simple droplet systems will therefore be described in more detail in the following sections. Emphasis is put on techniques, which allow the production of droplets with controlled size, velocity, temperature, or distance from neighboring droplets. Important devices for the generation of technical sprays, which are usually characterized by a wide spectrum of sizes, will be mentioned briefly in the next section.

2.2 Spray Generation

Pressure Atomizers. Sprays are often produced by discharging a liquid through a small aperture or nozzle into the air or a combustion chamber. The liquid emerges as a thin jet which disintegrates into a spray. In this process energy associated with the pressure is converted into kinetic energy and surface energy of the droplets. Numerous different types and designs of pressure atomizers are discussed in the literature [48, 70, 71].

Rotary Atomizers. Rotary atomizers make use of a disk, cup, or wheel rotating at high speed. The liquid flows from the center to the periphery of the rotating surface, where it is discharged in form of a thin sheet, which disintegrates through the action of aerodynamic forces of the surrounding air or an additional gas blast. Under certain operation conditions droplet formation occurs from liquid threads or directly from the periphery. Correlations for predicting the performance of rotary atomizers are given in the literature [70, 125].

Ultrasonic Atomizer. A thin liquid film on a solid surface vibrating with a high frequency will show a pattern of capillary waves. At large amplitudes the wave crests become unstable and a mist of tiny droplets is observed above the surface of the liquid. The ultrasonic atomizer can deliver very fine sprays, which often have been used for medical applications. The droplets in the produced sprays have a very low velocity.

Another concept of ultrasonic atomization is the standing wave atomizer. This technique has been proposed for the atomization of viscous fluids and liquid metals.

Electrical Droplet Generator. It is well known, that a liquid surface in presence of an intense external strong electrostatic field becomes unstable and is disrupted into a fine spray of charged droplets. This effect has been used to develop electrostatic atomizers mainly for paint spraying, the application of agricultural sprays and printing devices. Tang and Gomez conducted an experimental study on the feasibility of using electro-sprays for the targeted delivery of inhaled drugs [126]. Electrostatic atomizers require low energies and produce sprays with a rather narrow distribution function in a wide range of droplet sizes. Important aspects of electrostatic spraying have been described in the literature [17, 127, 128]. Gomez and Tang studied the charge and size of heptane droplets produced with an electro-spray system [129]. A compilation of papers on electro-sprays can be found in Ref. [130].

2.3 Droplet Stream Generator

2.3.1 Physical Principle and Technical Performance

The droplet generator described in this section is based on the breakup of cylindrical liquid jets, which are formed by discharging a liquid through an orifice as sketched in Fig. 2.1. The disintegration process due to instabilities of the liquid jet has been described in Sect. 1.8. It is assumed, that the radius r_{jet} is determined by the diameter of the orifice with $r_{jet} = d_{ori}/2$. With Eq. 1.36 for inviscid and Eq. 1.39 for viscid liquids the most probable droplet size can be calculated as shown in Sect. 1.8. After disintegration of the jet a droplet stream or droplet chain forms. The velocity v of the droplets is approximately the same as the velocity of the jet v_{jet}. The droplet temperature is determined by the temperature in the liquid reservoir just in front of the orifice plate. The droplets have different sizes, which are near the most probable droplet size.

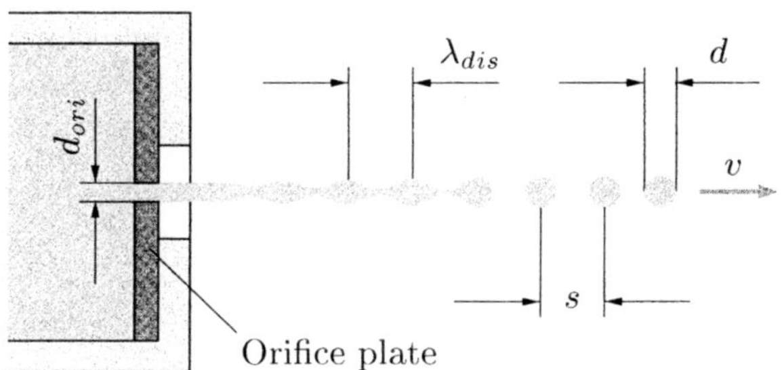

Fig. 2.1. Schematic view of the disintegration of a liquid jet produced by forcing the liquid through an orifice with diameter d_{ori}

Regular disturbances with the wavelength λ_{dis} can be observed along the jet, when the jet is disturbed regularly with the frequency f_{dis}. Then all the droplets have the same radius $r = d/2$ and a monodisperse droplet stream is obtained with the distance $s = \lambda_{dis}$ between neighboring droplets. The wavelength λ_{dis} of the regular disturbances should be near the wavelength λ_{opt}; it may differ only to some extent from λ_{opt} when the regularity of the droplet stream should be sustained. The droplet radius r is determined by the radius r_{jet} of the jet and by the wavelength λ_{dis}. It follows

$$r = \left(\frac{3}{4}\, r_{jet}^2 \lambda_{dis}\right)^{1/3} . \tag{2.1}$$

The flow rate $\dot{V}_{jet}$ through the orifice may be expressed by jet parameters or by the volume of a droplet multiplied with the number f_{dis} of droplets produced per time [131]. Hence

$$\dot{V}_{jet} = \pi r_{jet}^2 v_{jet} = \frac{4}{3}\pi r^3 f_{dis} . \tag{2.2}$$

For the droplet radius it follows

$$r = \left(\frac{3\,\dot{V}_{jet}}{4\,\pi f_{dis}} \right)^{1/3} . \tag{2.3}$$

Such jet disintegration processes can be observed for certain regimes of Ohnesorge number $Oh = \eta/\sqrt{\varrho\sigma d_{ori}}$ and Reynolds number $Re = v_{jet}d_{ori}\varrho_{liq}/\eta_{liq}$, which may be found in the Ohnesorge diagram shown in Refs. [48, 132].

It is desired to choose the parameters of the droplets in the droplet stream precisely and independently. Therefore the droplet generators based on this principle have to fulfill several requirements. To produce droplets of defined size the liquid jet has to be disturbed regularly as mentioned above. Several technical solutions are possible. A hypodermic needle producing the jet may be put into vibrations [133]. The same effect may be achieved with an orifice plate, which is excited mechanically by a piezoelectric ceramic; this generator is therefore often called vibrating orifice generator [131, 134, 135]. Even external sound waves can disturb the jet in the desired manner. It is in many cases sufficient to bring the appropriate disturbances into the bulk liquid within the generator, for instance by means of a piezoelectric ceramic to apply the desired mechanical disturbances to the jet. In Fig. 2.2 an example of a droplet generator is shown schematically.

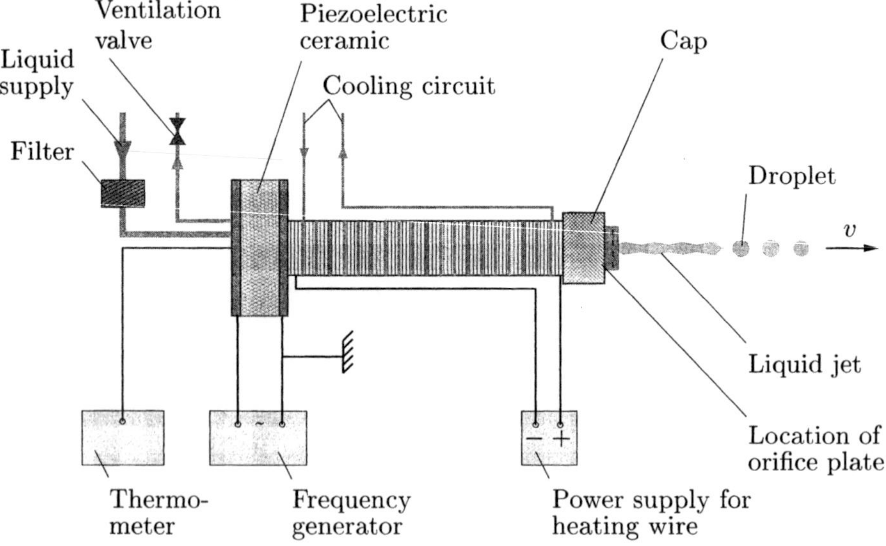

Fig. 2.2. Schematic view of droplet stream generator including peripheral devices

The droplet generator shown on the photograph in Fig. 2.3 has been used in many of the experiments presented in Sects. 6 and 5. The length of the body between the piezoelectric ceramic and the orifice plate was in this case

Fig. 2.3. Photograph of droplet generator used at ITLR

approximately 15 cm, the diameter of the cap on the right hand side 2 cm. The shown layout, however, may be varied according to the needs of the experimental setup. The design with the long slender body allows to inject the droplets into experimental environments with narrow access. The orifice plate at the tip of the generator may be exchanged easily by unscrewing the cap. Depending on the performed experiment, different plates with one or more orifices and with different orifice diameters may be suitable. The pressurized liquid has to pass a filter before entering the generator in order to avoid clogging of the orifice. Vapor bubbles in the generator are removed by using the ventilation valve. The regular disturbances described above are transmitted into the liquid using a piezoelectric ceramic at the back end of the body. The ceramic is excited by a frequency generator; amplitudes from 20 V up to 100 V have been used. In choosing the excitation frequency, the pressure of the liquid supply, and the orifice diameter, droplet size, droplet velocity, and distance between neighboring droplets can be well defined as will be described in Sect. 2.3.2. Orifice diameters of $3\,\mu\text{m} < d_{ori} < 200\,\mu\text{m}$ have been used and droplet velocities of $5\,\text{m/s} < v < 30\,\text{m/s}$ are common. The initial droplet spacing s_0 is related to the initial droplet diameter r_0; ratios of $2 < s_0/r_0 < 6$ are typical.

Orifice plates with more than one orifice can be used in order to obtain several parallel droplet streams. Using this technique planar or three-dimensional droplet systems may be obtained, as will be described in Sect. 3.3.2. Normally all orifices have the same diameter. However, different diameters are possible to some extent, resulting in different droplet diameters in the produced droplet system.

The droplet generator shown in Fig. 2.2 is in addition equipped with a thermocouple to determine the temperature of the produced droplets. To achieve temperatures different from ambient temperature the body is surrounded with a heating coil and by a thin pipe of a cooling circuit. To measure and adjust the temperature of the liquid a thermocouple is positioned in the liquid just beneath the orifice of the droplet generator. It is assumed, that the droplet temperature is close to the measured temperature. Liquid temperatures up to 300°C have been realized.

2.3.2 Operation Characteristics

2.3.3 Operation Characteristics

Natural Jet Disintegration. The behavior of droplets or droplet streams produced with a droplet stream generator, which is based on the disintegration of a liquid jet, will be described. The droplet behavior is determined or at least influenced by the operation characteristics of the generator. In Fig. 2.4 a photograph of a laminar liquid jet emerging from the droplet generator is shown. The jet is not excited with regular disturbances. The jet disintegrates into droplets, as can be seen on the right hand side of the figure. The jet has approximately the same diameter as the orifice. The droplet size and droplet distance in the disintegrated jet are not uniform. According to Eq. 1.38, the mean droplet size is approximately 1.89 times the orifice diameter.

Fig. 2.4. Liquid jet emerging from an orifice of diameter $d_{ori} \approx 37\,\mu\text{m}$ and disintegrating into droplets. The jet is not excited with disturbances

Enlarged views of the disintegration process are shown in Fig. 2.5 with the jet and the droplets propagating from left to right. In this figure disturbances on the jet can be recognized, as well as droplets just after separation from the jet. Shown are three pictures taken at different times. It can be seen, that the wavelength of the disturbances is not uniform, which results in droplets of different size. The disintegration process is instationary. Techniques which

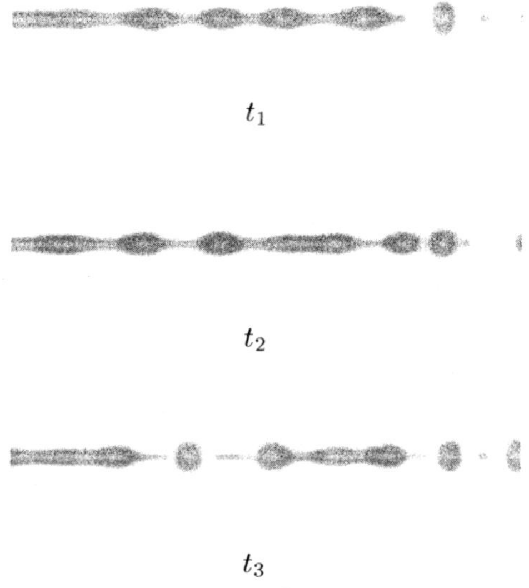

Fig. 2.5. Disintegration pattern of a liquid jet for three different arbitrary times t_i at the same location. In these enlarged views the parameters of the jet are the same as in Fig. 2.4

allow to take such pictures will be described in Sects. 4.2 and 4.4. Regular excitation of the jet is omitted if droplet streams or sprays with finite but narrow size distribution are desired. In this case the average droplet size is determined by the orifice diameter, the droplet velocity, and by the flow rate. To obtain a spray a turbulent gas stream may be used to disturb the the droplet stream just after disintegration. When regular excitations are applied to the jet, a monodisperse spray can be obtained.

During the disintegration of the jet the temperature of the liquid can be influenced by the heat exchange with the surrounding air. The size of the formed droplets is reduced, when liquid evaporates. These processes may cause errors of measurements, when droplet size and temperature must be known very precisely, as for example in phase transition studies. Errors due to these effects can be avoided or at least reduced, when the emerging jet is surrounded with a heatable tube. When the temperature within the tube is equal to the temperature measured in the droplet generator, the vapor pressure will increase until saturation is reached within the tube.

Excited Jet Disintegration. When regular disturbances of a suitable frequency are applied to a disintegrating liquid jet, droplets of uniform size with uniform distance between neighboring droplets can be obtained. A disintegrating excited jet is shown in Fig. 2.6 with the same flow rate, and therefore the same jet velocity as in Fig. 2.4. In comparison with the non-excited jet the disturbances grow faster in the excited jet. This results in an earlier complete disintegration of the jet; that means, that the length l_{jet} from the orifice to the location of the first occurrence of a droplet is shorter.

Fig. 2.6. Liquid jet emerging from an orifice of diameter $d_{ori} \approx 37\,\mu$m and disintegrating into droplets of uniform size. The jet is excited with regular disturbances of the frequency $f_{dis} = 78\,$kHz

Measurements of the jet length l_{jet} have been performed for different excitation frequencies f_{dis}. Results for a jet consisting of a mixture of one part pentane and two parts hexadecane are shown in Fig. 2.7. These measurements have been performed within the large frequency range $0.5\,\text{kHz} < f_{dis} < 116\,\text{kHz}$. It has to be emphasized, that the regular disturbances do not grow for all frequencies and then end in a monodisperse droplet stream. Only for frequencies causing disturbances with a wavelength near the optimum wavelength λ_{opt}, given by Eq. 1.40, a monodisperse droplet stream can be expected. At this optimum frequency the length of the jet l_{jet} should

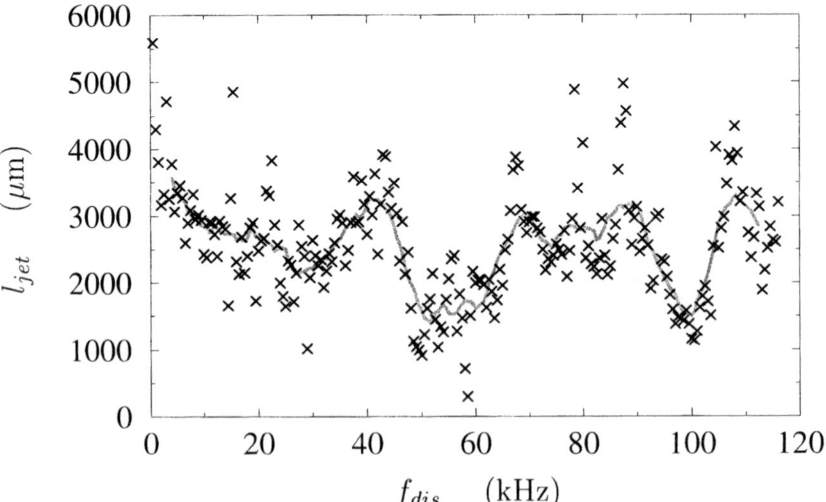

Fig. 2.7. Length l_{jet} as a function of the excitation frequency f_{dis}. The solid line represents the running average between 15 neighboring points. The mean velocity is $v_m = 18.9\,$m/s with the standard deviation $\sigma_v = 0.213\,$m/s, the liquid is a mixture consisting of one part pentane and two parts hexadecane

have a minimum as the disturbances grow fastest. Figure 2.7 reveals a minimum of the jet length between $f_{dis} = 50\,\mathrm{kHz}$ and $f_{dis} = 60\,\mathrm{kHz}$ and for $f_{dis} \approx 100\,\mathrm{kHz}$, which is approximately twice the optimum frequency. At other frequencies a monodisperse droplet stream may be obtained under certain conditions.

In many cases a small ligament of liquid is observed between the separated droplets. These ligaments transform into tiny droplets. Depending on the length and diameter of the ligament one or more so-called satellite droplets form. These satellite droplets collide in many cases with a neighboring droplet. When the satellite droplets have all the same size and interdroplet distance a monodisperse droplet stream is obtained.

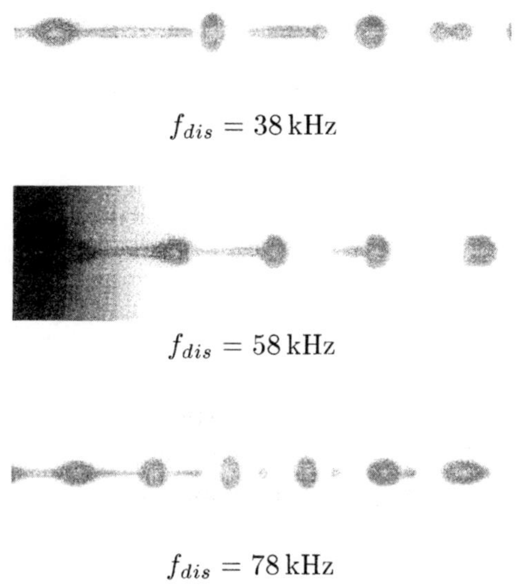

Fig. 2.8. Disintegration of a liquid jet. Shown are three pictures with different excitation frequencies f_{dis}. In these enlarged views all parameters except the excitation frequency are the same as for the jet shown in Fig. 2.6. The frames are taken approximately in the distance l_{jet} from the orifice. This distance varies with the excitation frequency

Figure 2.8 shows photographs of the disintegration process in a jet excited with three different frequencies. Shown are for each frequency a few wavelengths of the disturbances on the liquid jet at the distance l_{jet} from the orifice plate. The location of the frames varies, since l_{jet} varies with the excitation frequency. This has been shown in Fig. 2.7. For all three frequencies a monodisperse equally spaced droplet stream is obtained. On the picture in the center with $f_{dis} = 58\,\mathrm{kHz}$ the cap of the droplet generator can be recognized on the left hand side. This frequency is near the optimum frequency with a short length l_{jet}. No separate ligament forms between the droplets, as it is the case with both other frequencies. The satellite droplets found at the lower and higher frequency are regular and collide with the larger droplet downstream, which results finally in a monodisperse droplet stream.

As can be seen from Fig. 2.8 for higher frequencies a narrower droplet spacing and smaller droplets are obtained. It should be stressed, that it is essential in experimental applications of the droplet stream generator, to check carefully if the droplet stream is really monodisperse.

With some restrictions the droplet stream parameters such as radius r, velocity v, temperature T and spacing s between neighboring droplets can be adjusted individually. The droplet radius is given by Eq. 2.3. The radius can be changed by varying the excitation frequency f_{dis} as shown in Fig. 2.8. When all other parameters are held constant the spacing s changes according to the equation

$$s = \lambda_{dis} = \frac{v_{jet}}{f_{dis}} , \tag{2.4}$$

which is obtained from Eqs. 2.1 and 2.2. It should be emphasized, that the volume flux $\dot{V}_{jet}$ through the orifice remains constant, when the velocity and the radius of the jet are the same. This follows from Eq. 2.2. The droplet radius r as a function of the excitation frequency f_{dis} is shown in Fig. 2.9 for three different values of the volume flux. Within the plotted range of

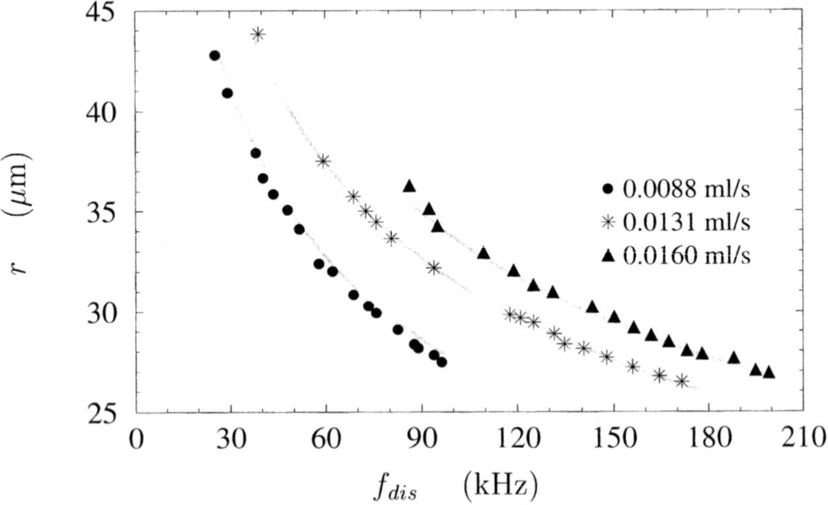

Fig. 2.9. Droplet radius r as a function of the excitation frequency f_{dis} for different measured liquid feed rates $\dot{V}_{jet}$. The symbols represent results of size measurements

the excitation frequency monodisperse droplet streams could be obtained for each volume flux. The results presented by the solid lines have been derived from measurements of $\dot{V}_{jet}$ and using Eq. 2.3. These experimental results have been compared with direct optical measurements of droplet size. The measurements were performed close above the orifice, in order to avoid a significant evaporation of the droplets. Different techniques for size measurements will be described in Sect. 4.6.3.

Changing the excitation frequency is a convenient way to vary the droplet radius. If a variation of the droplet radius at constant droplet distance s is desired the radius r_{jet} of the liquid jet has to be adapted appropriately, as can be seen from Eq. 2.1. The orifice plate has to be exchanged, since the radius of the jet is determined by the orifice diameter in this case. A different orifice diameter results in a different pressure drop across the orifice and in a different volume flux.

To obtain a certain velocity the pressure in the liquid supply has to be adjusted in an appropriate way. An increase of the velocity is obtained by increasing the pressure in the liquid supply. Except the volume flux $\dot{V}_{jet}$ all other parameters can be held constant. Then the excitation frequency has to be increased according to Eq. 2.4.

The distance s between the droplets can be changed by varying the excitation frequency. This results, however, in a change of droplet radius. For constant droplet radius the orifice diameter has to be adjusted.

The droplet temperature T, which can be regulated with the heating wire and the cooling device respectively, is controlled by the thermocouple beneath the orifice plate. As the viscosity of the liquid is a function of temperature, the pressure drop over the orifice changes and therefore the pressure of the liquid supply has to be adjusted accordingly to hold the other droplet parameters constant.

Each droplet parameter can be chosen separately by adjusting the orifice diameter, the pressure of the liquid supply, the heating and cooling device, and the excitation frequency as described. In Table 2.1 five different combinations of the operation parameters of such generators are listed. The data are from Ref. [131].

Table 2.1. Typical values parameters for droplet stream generator

d_{ori} (μm)	f_{dis} (kHz)	λ_{dis} (μm)	r (μm)	s/r	v (m/s)
10	38 – 172	99 – 447	40 – 25	11.0 – 4.0	17.0
23	34 – 88	105 – 271	57 – 43	4.0 – 2.2	9.2
23	51 – 237	74 – 345	64 – 39	5.2 – 1.8	17.6
81	7 – 23	252 – 829	136 – 200	4.1 – 1.9	5.8
120	5 – 10	430 – 860	270 – 212	3.3 – 2.0	4.3

The disintegration process of the jet can be influenced by modulating the excitation frequency [136]. The possibilities for modulation are numerous, therefore only a few examples will be described. The frequency may be wobbled in a small range, in which in any case a regular disintegration is

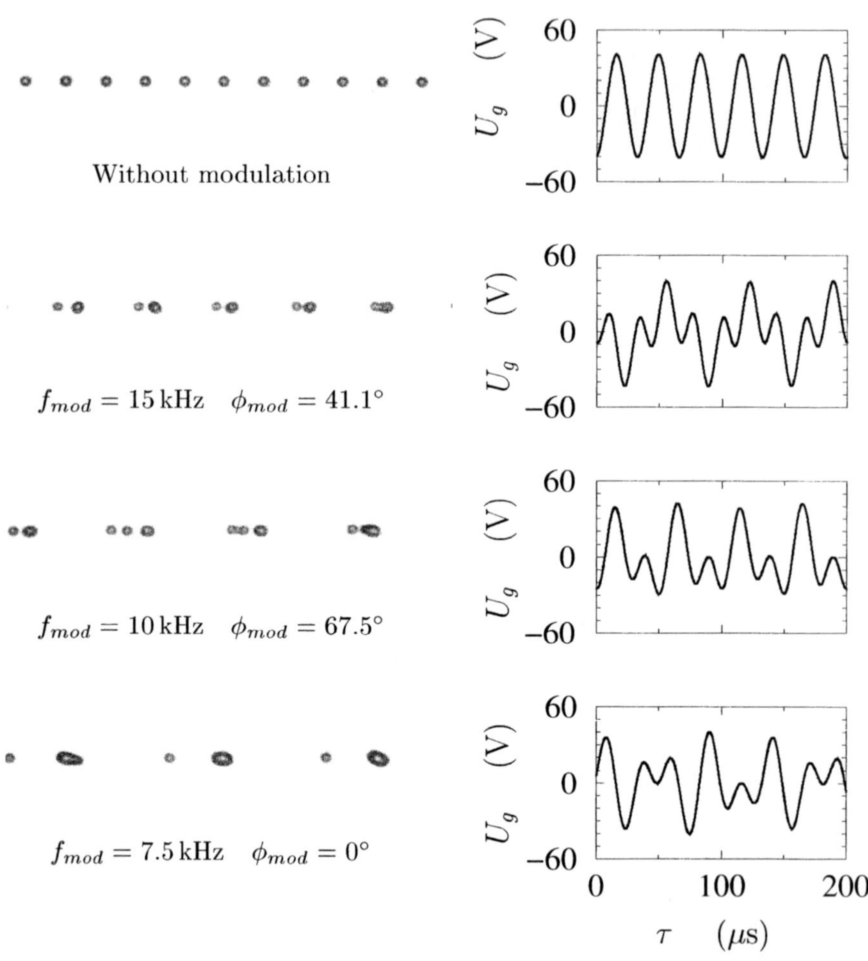

Fig. 2.10. Effects of different amplitude-modulated excitation signals on the behavior of droplet streams. The droplet stream with no modulation is shown at the top. The modulation shown here results in an increase of the droplet spacing

obtained. This results in small variations of the droplet size and distance along the droplet stream. This technique is useful for optical measurement techniques, to obtain the superposition of the scattered light from different droplets with slightly different size. The superposition of light scattered by droplets with different diameters is desirable in some experiments to obtain a smoothed intensity distribution.

For other purposes an amplitude-modulated excitation signal may be applied, as shown in Fig. 2.10. The phase and amplitude of the modulation signal could be varied continuously in the experiments. The frequency f_{mod} of the modulation signal was an integer fraction of the original excitation frequency $f_{dis} = 30\,\text{kHz}$. Examples for different modulation frequencies f_{mod} with different phases ϕ_{mod} are shown in Fig. 2.10. On the left hand side part of droplet streams obtained with different modulation are shown, with the droplets moving from left to right. On the right hand side the corresponding excitation signals are plotted as a function of time. At the top the droplet stream obtained without modulation is shown. In the other cases with modulation regular packages of droplets of different size are obtained. In the cases shown here each package coagulates to a large droplet resulting at last in a monodisperse droplet stream with larger droplets and larger spacings. Here the new spacing s_{mod} is obtained from the original spacing s by the relation

$$s_{mod} = s \frac{f_{dis}}{f_{mod}} , \tag{2.5}$$

as can be seen from Fig. 2.10. As the volume flux $\dot{V}_{jet}$ remains constant the new droplet radius r_{mod} is related to the original radius r by

$$r_{mod} = r \left(\frac{f_{dis}}{f_{mod}} \right)^{1/3} . \tag{2.6}$$

For some experiments the ratio s/r is the most important parameter. Without modulation s/r may be varied in the range of approximately $2 < s/r < 6$. Amplitude modulation allows to increase s/r. From Eqs. 2.5 and 2.6 it follows

$$\frac{s_{mod}}{r_{mod}} = \frac{s}{r} \left(\frac{f_{dis}}{f_{mod}} \right)^{2/3} . \tag{2.7}$$

With other values of the modulation frequency and phase shift it may be achieved, that the packages of droplets do not coagulate and form larger droplets. A large variety of other frequency and phase combinations are possible, which cannot be described here. In any case the effects of modulation on the droplet stream has to be observed and checked by photographic or better video techniques, which will be described in Sects. 4.2 and 4.4. Only the observation allows to control the results of modulation. Different aspects of amplitude-modulated excitation signals are discussed in Ref. [136].

Coherence length of droplet streams. Monodisperse droplet streams loose their regularity and finally their monodispersity at larger distances h from the generator exit. Disturbances of droplet velocity or spacing caused by the excitation and disintegration process are enlarged due to interaction with the ambiance, for example trailing droplets tend to coagulate earlier with their preceding neighbor droplets. The initially constant spacing between the droplets becomes irregular and the coherent structure of the droplet streams is lost. The length of the existence of the coherent structures is larger under rarefied conditions than under normal ambient conditions [137].

With an increase of distance h the irregularity of the spacings increases until neighboring droplets coagulate and form larger droplets. As a consequence monodispersity is lost. The time interval between neighboring droplets $t_s = s/v$ is a measure for the spacing s if the velocity v of the droplet is constant. The time interval t_s was measured using an electronic counter in combination with a microcomputer. The start and stop signals were obtained when the droplets crossed a laser beam. Typical distributions of t_s at different distances h from the orifice of the droplet generator are shown in Fig. 2.11.

The distributions of the time interval t_s of Fig. 2.11(a) are obtained at distances h, for which the droplets are still monodisperse. Very close to the generator exit the distribution is very narrow, which indicates a coherent droplet stream. For larger distances the distributions broaden due to increasing irregularities. In Fig. 2.11(b) a distribution is presented, which is caused by coagulation of neighboring droplets. This distribution is no longer symmetrical. The minimum time interval $t_s = 0$ indicates the collision of the droplets; in the experiment the minimum time interval is, however, determined by the diameter of the laser beam and therefore non-zero. Before the onset of collisions the distributions may be assumed to possess approximately Gaussian shape. For the mean value t_{sm} of the time interval t_s one has the relation

$$t_{sm} = \frac{1}{n} \sum_{i=1}^{j} t_{s,i} n_i \ . \tag{2.8}$$

A measure for the coherence is given by the standard deviation σ_s of the time interval. It holds

$$\sigma_s = \sqrt{\frac{1}{n-1} \sum_{i=1}^{j} (t_{s,i} - t_{sm})^2 n_i} \ . \tag{2.9}$$

In the equations for t_{sm} and σ_s the total number of events is n, the number of events in class k is n_k, whereas the number of classes is j. In the experiments the total number of events was in the range of $2000 < n < 30000$. For two different orifice diameters the mean value t_{sm} and the standard deviation σ_s are shown in Fig. 2.12 as functions of the distance h from the generator exit. The excitation frequency was adapted to the orifice diameter to obtain a fast disintegration of the jet. For $d_{ori} = 31\,\mu\text{m}$ the excitation frequency was

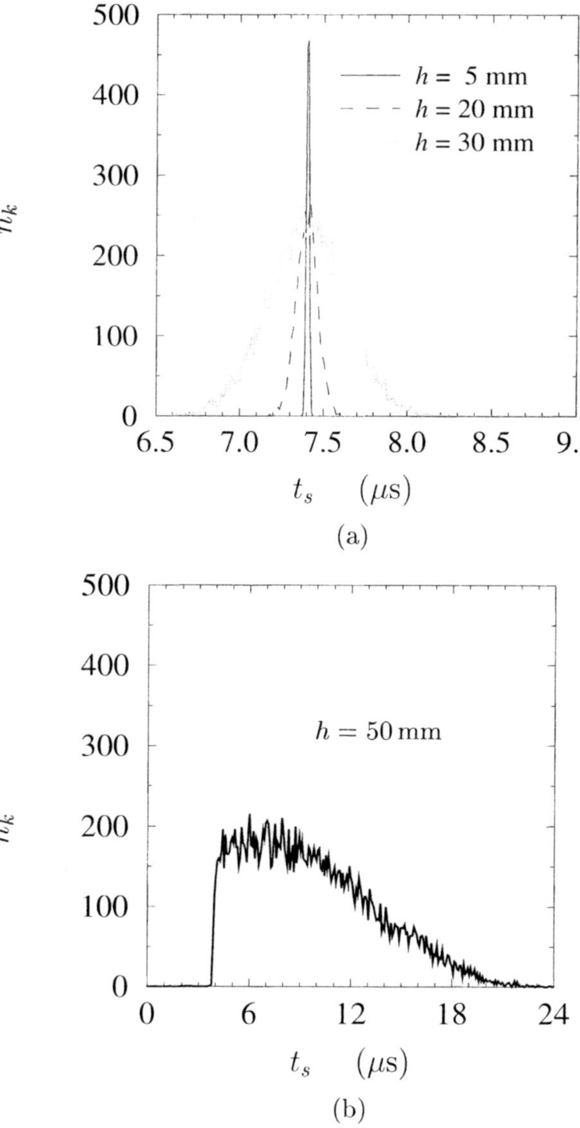

Fig. 2.11. Distribution of time intervals t_s between subsequent droplets at different distances h from the generator exit. The orifice diameter was $d_{ori} = 31\,\mu\text{m}$ and the excitation frequency $f_{dis} = 135.6\,\text{kHz}$. The results have been obtained with a droplet stream of ethanol droplets. In (a) it is shown, that the distributions become wider with increasing h, but remain symmetrical. This results in a constant mean value of t_s and in an increase of the standard deviation σ_s. For larger values of h the rather broad distribution shown in (b) is obtained. This distribution, which is caused by collisions between droplets, is shifted towards larger values of t_s. A consequence is an increase of the mean value t_{sm}

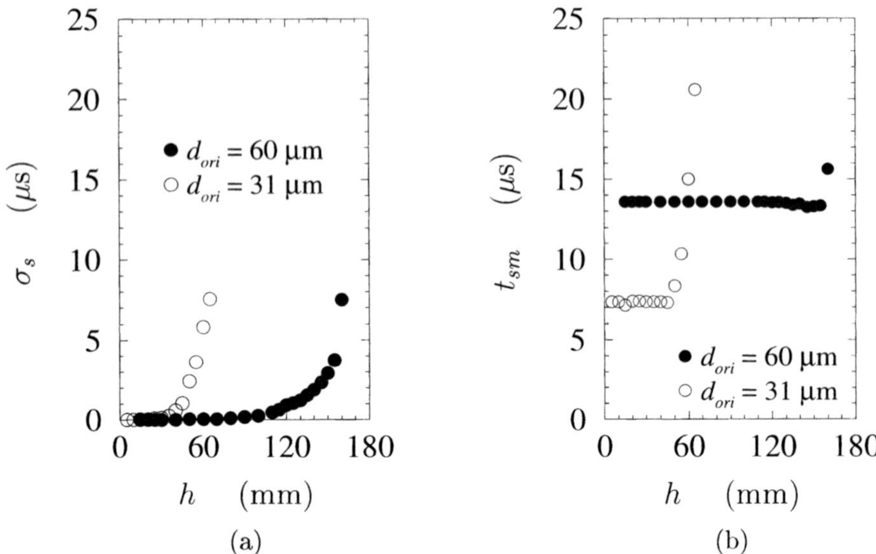

Fig. 2.12. Shown are (a) the standard deviation of the time interval σ_s and (b) the mean value of the time interval t_{sm} as functions of the distance from the generator orifice for two different orifice diameters

$f_{dis} = 135.6\,\text{kHz}$, for $d_{ori} = 60\,\mu\text{m}$ the lower frequency $f_{dis} = 73.5\,\text{kHz}$ was chosen. The mean value, shown in Fig. 2.12(b), remains constant as long as the droplet streams remain monodisperse; collisions of droplets result in an increase of the mean value. In Fig. 2.12(b) the distance h_{sm} at which the mean value of the time interval increases can clearly be recognized. For distances larger than h_{sm} coagulation between droplets is observed. The standard deviation, shown in Fig. 2.12(a), is approximately zero for small distances h. This indicates a regular droplet stream with practically constant spacing s between the droplets. The distance h_{sl} at which the standard deviation exceeds a certain limit σ_{sl} may be called coherence length.

The distances h_{sm} and h_{sl} are influenced by the excitation and disintegration process; this can be seen in Fig. 2.13. Here both distances are plotted as functions of the excitation frequency f_{dis}. Both distances correspond with each other. The higher the values of h_{sm} and h_{sl} the more regular is the disintegration and the better the coherence.

Some investigators have studied the performance of droplet streams at low ambient pressures. In a vacuum variations in size, speed, or direction of the droplet streams can be caused only by irregularities in the breakup process. Muntz and Dixon report angular dispersion of approximately 10^{-6} rad for droplet streams under vacuum conditions [138].

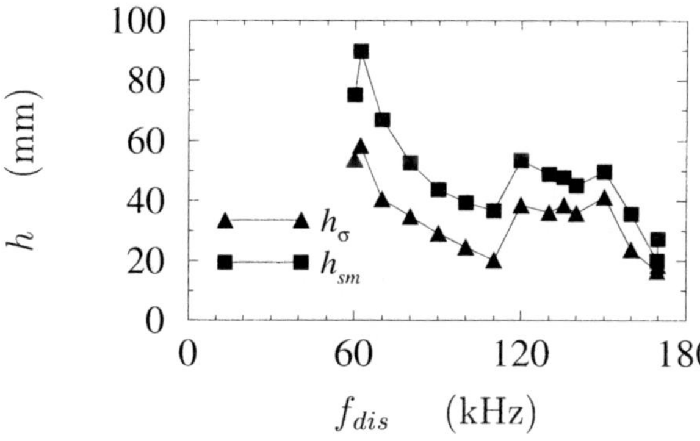

Fig. 2.13. Special values of h as a function of excitation frequency f_{dis}. The height h_{sm} indicates, where the mean value t_{sm} starts to increase, the height h_{sl}, where the standard deviation σ_s exceeds the limit $\sigma_{sl} = 0.5\,\mu\text{s}$

Self Excitation. The droplet stream generator may generate the excitation frequency itself without an external frequency generator. A sketch of a suitable experimental setup is shown in Fig. 2.14. In a certain distance h from the orifice of the droplet generator the passing droplets are detected. This distance must be so large, that the liquid jet disintegrates and droplets form. In order to detect the passing of a droplet a light ray is focused on the droplet path with a lens. A second lens focuses the light on a sensor. Each time, when a droplet passes, the light ray is interrupted and the signal of the sensor decreases. The output signal of the sensor is amplified and then used as excitation signal of the frequency generator.

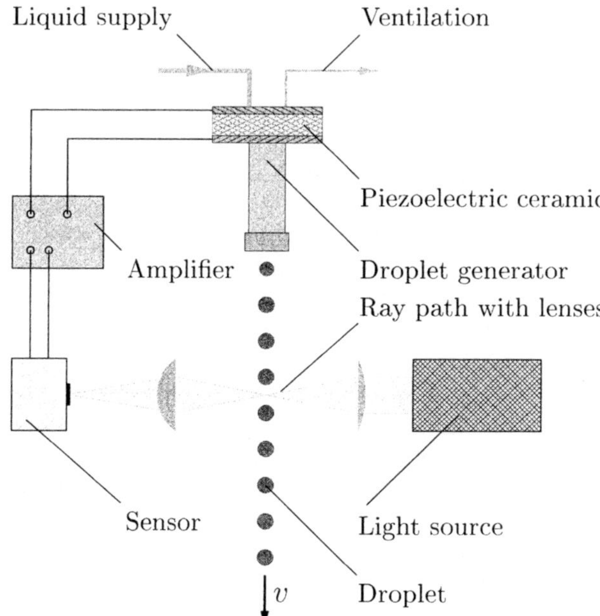

Fig. 2.14. Schematic view of experimental setup for generating the excitation frequency by the droplet stream itself due to a feedback effect

Immediately after starting of the generator a monodisperse droplet stream with a stable excitation frequency is obtained with the configuration shown Fig. 2.14. The frequency obtained is near the optimum frequency, which depends on the properties of the liquid, the droplet velocity, and the orifice diameter. With the droplet velocity the wavelength can be calculated. This result may be compared with the theoretical value obtained from Eq. 1.39. The exact frequency is determined by the distance h from the droplet generator and by the droplet velocity. It should be emphasized, that only frequencies near the optimum are possible. With increasing h the frequency decreases until a critical value is obtained. Then the number of droplets within the distance h increases by one, resulting in a sudden increase of the frequency.

With the described setup monodisperse droplet streams can be produced without frequency generator and the disintegration can be controlled by monitoring the output signal of the sensor. The frequency obtained, the coherence length, and the starting mechanism of the described arrangement should be studied in more detail.

2.4 Droplet on Demand Generator

Single droplets or a series of droplets can be produced with the so-called droplet on demand generator. This device, which is derived from ink-jet printing systems with a piezoelectric ceramic as actuator, can be used for a large variety of different droplet liquids such as water, fuels, inks or even liquid metals. Examples of different designs for various technical applications will be described briefly in Chap. 7.

A common design of a droplet on demand generator is shown schematically in Fig. 2.15. The most important part is the piezoelectric ceramic tube, which is embedded in a larger metal tube with one electric contact of the piezoelectric ceramic grounded. The piezoelectric ceramic is driven by an electric pulse generator. Depending on the polarity of the voltage the inner diameter of the piezoelectric ceramic tube widens or narrows resulting in a droplet issuing through the orifice plate. Pulse height and pulse duration can be varied to change the velocity of the droplets and to adapt the pulses to the geometrical layout. Several pulses in sequence allow to produce a series of droplets. Pulse frequencies up to a few kilohertz may be reached. Detailed theoretical descriptions of this process are given in Refs. [139, 140].

In the design of Fig. 2.15 the liquid reservoir is in direct mechanical contact with the generator. To avoid impurities, which may cause clogging of the orifice, the liquid has to be filtered during or before filling the reservoir. When a single electrical pulse is applied to the piezoelectric ceramic a ligament of liquid emerges through the orifice. Depending on the configuration and the settings of pulse height and pulse width this ligament disintegrates in many cases into two or more droplets. Theoretical descriptions of this process of droplet formation are given in Refs. [141, 142].

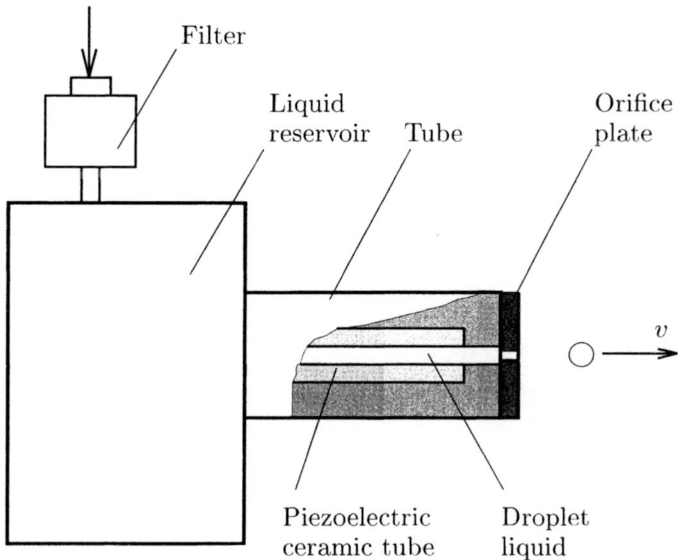

Fig. 2.15. Schematic view of droplet on demand generator

In Figs. 2.16 to 2.18 photos of droplet patterns obtained with different settings of the pulse height and width are shown. In each figure six pictures of the droplet development process are presented with time increasing from top to bottom. The time difference between the frames is $\Delta t \approx 16.7\,\mu$s. The orifice diameter was $d_{ori} = 50\,\mu$m, the pulse rate $f_p = 200\,$Hz. As for each pulse the droplet formation is very regular the video technique described in Sect. 4.4 has been used; therefore each picture shows a different ligament. In Fig. 2.16 the ligament ejected disintegrates into a large droplet and a small one. These two droplets coagulate after a very short time. The velocity of the final droplet, which oscillates at the beginning, is $v \approx 2.8\,$m/s. With the pulse rate of $f_p = 200\,$Hz the initial distance between neighboring droplets is $s = v/f_p \approx 14\,$mm. In comparison with the droplet stream generator this distance is very large; chain effects can therefore be avoided.

The stop distance can be estimated with Eq. 1.47. The droplet size is mainly determined by the orifice diameter. For a given orifice diameter, pulse height and pulse width determine the shape, velocity, and size of the ligament ejected and influence the droplet parameters. In Fig. 2.17 the development of a droplet with higher velocity resulting in a higher droplet distance of $s \approx 36\,$mm is shown. The small secondary droplet will later coagulate with the large droplet, which shows oscillations with a rather large amplitude in a higher oscillation mode. In the sequence of Fig. 2.18 the velocity is approximately $8\,$m/s. The ligament disintegrates first into one large and two small satellite droplets. The small droplets coagulate and form one satellite droplet.

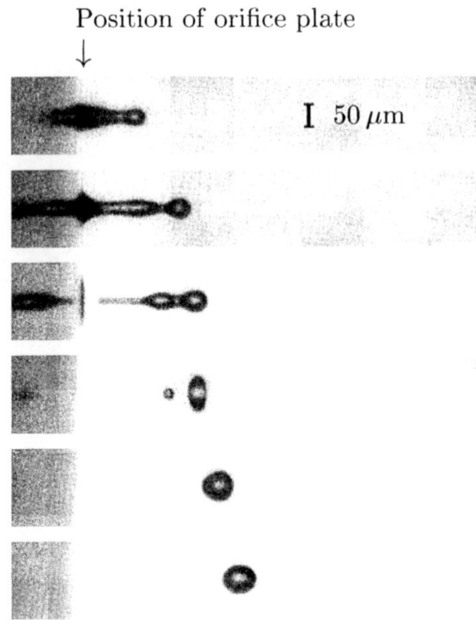

Fig. 2.16. Photo of liquid jet issuing from droplet on demand generator. The position of the orifice is the same for all frames. The droplet velocity is approximately $v \approx 2.8\,\mathrm{m/s}$

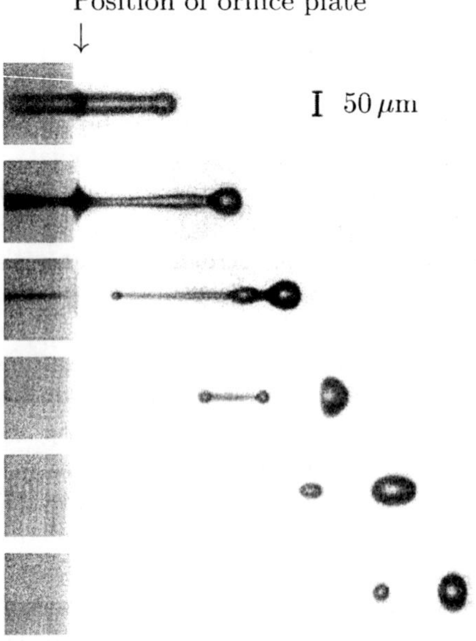

Fig. 2.17. Photo of liquid jet issuing from droplet on demand generator. The position of the orifice is the same for all frames. The droplet velocity is approximately $v \approx 7.2\,\mathrm{m/s}$

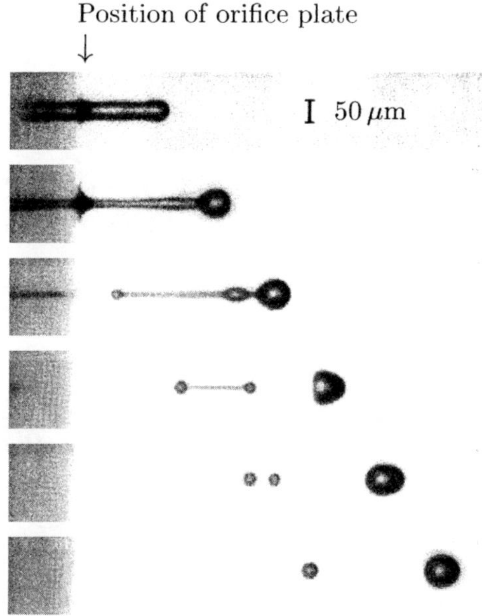

Fig. 2.18. Photo of liquid jet issuing from droplet on demand generator. The position of the orifice is the same for all frames. The droplet velocity is approximately $v \approx 7.6\,\mathrm{m/s}$

However, there is no coagulation observed between the satellite droplet and the large droplet. The two droplets have of course different stop distances.

To conclude it can be said, that the droplet size is mainly determined by the orifice diameter, but depending on the pulse settings. The exact value of the droplet size has to be measured. It depends on the pulse settings if one or more droplets are produced. If only one droplet per pulse is desired the pulse height and the pulse width have to be set carefully. Experimental setups with such a droplet on demand generator will be presented in Sect. 6.2.2.

2.5 Dropper

The most common and elementary way to produce single droplets can be observed, when a liquid is slowly discharged from a faucet, a burette, or a similar device. This process is depicted in Fig. 2.19. A free-falling droplet forms, when the weight of the liquid exceeds the force resulting from surface tension. Mechanical equilibrium between these two forces can be expressed in the form

$$mg = \pi d_{min} \sigma \;, \tag{2.10}$$

where d_{min} is approximately the diameter of the dropper exit. From this relation is follows for the droplet diameter

$$d = \sqrt[3]{\frac{6\, d_{min} \sigma}{\varrho_{liq} g}} \;, \tag{2.11}$$

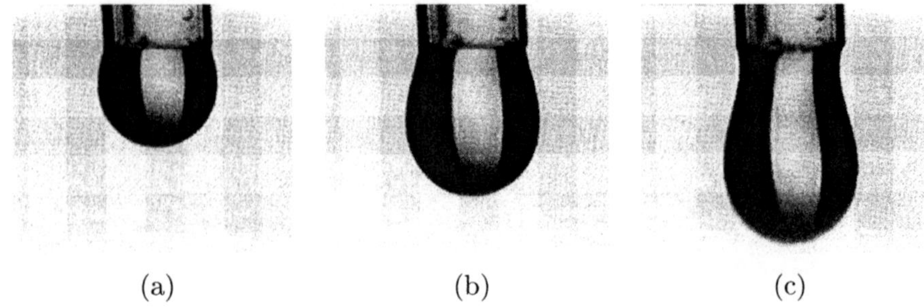

(a) (b) (c)

Fig. 2.19. Generation of single droplets by a dropper

when it is assumed that the released droplet is a sphere of diameter d. This equation can be written in a dimensionless form by introducing the Bond number defined by

$$Bo = \frac{d_{min}^2 \varrho g}{\sigma} \; . \tag{2.12}$$

It can easily be seen that the Bond number is proportional to the ratio of the gravitational force and forces resulting from surface tension. The resulting equation for the droplet diameter can be written in the dimensionless form

$$\frac{d}{d_{min}} = \sqrt[3]{\frac{6}{Bo}} \; . \tag{2.13}$$

More details can be found for example in Ref. [55].

3. Droplet Systems

3.1 Introduction

Droplets in the form of sprays occur in nature for example as fogs, clouds, and rain as well as in numerous technical applications, such as fuel combustion or spray painting. For the simulation of these complex droplet systems it is important to understand the behavior of simpler systems, such as regular three-dimensional or two-dimensional arrays. Regular droplet configurations can be generated with droplet stream generators or with droplet on demand generators. Single droplets can be studied either as moving droplets or as suspended droplets. Freely falling water droplets have been used to study evaporation rates. Motionless droplets have been investigated with different suspension techniques. In the classical suspension technique larger droplets are suspended by thin fibers, threads or filaments. Physicists have accomplished stable levitation with different contact-free forces. A stable position is obtained by subjecting the levitated material to restoring forces vertically and horizontally.

Various techniques for stationary positioning of liquid or solid materials in a gas or vacuum have been described in the literature. These techniques compensate the droplet weight with electrostatic or electrodynamic forces, light pressure or acoustical pressure. Small spherical metallic particles have been suspended by dielectrophoretic forces. Some investigators have combined different suspension and levitation techniques, for example electrostatic and acoustic levitation techniques. Physical aspects of different levitation mechanisms have been discussed in Ref. [143].

3.2 Sprays

Sprays are droplet systems, which consist of a large number of droplets dispersed in a carrier gas, often air or a mixture of air and the vapor of the droplet liquid. Different mathematical means are available for describing the configuration as well as, mechanical and thermodynamic properties of a spray system. As a direct method the spray may be considered as a collection of material particles. The motion of the individual particles follows from the formalism of classical mechanics. Using modern computer capacities this computational task may be solved for a number of droplets, which is limited but may be representative for the whole system or a subsystem.

Often it is easier to describe a spray in terms of a distribution function, which contains statistical information on the microscopic state of the droplets, for example droplet size, shape, position, velocity, temperature, composition and so on. This distribution function, in general, will vary with time. For simplicity it is assumed in the following that the macroscopic properties of the spray can be characterized by the droplet radius $r = d/2$ the droplet position $\boldsymbol{x} = (x_1, x_2, x_3)$ and velocity $\boldsymbol{v} = (v_1, v_2, v_3)$. The properties of a single droplet can be represented by a point in the space $\boldsymbol{x}$, $\boldsymbol{v}$, r. For a statistical description of the spray system a distribution function $F(\boldsymbol{x}, \boldsymbol{v}, r, t)$ with the property

$$\Delta N = \int\limits_{\Delta \boldsymbol{x}} \int\limits_{\Delta \boldsymbol{v}} \int\limits_{\Delta r} F(\boldsymbol{x}, \boldsymbol{v}, r, t) d\boldsymbol{x} d\boldsymbol{v} dr \tag{3.1}$$

is introduced, where ΔN is the probable number of droplets with positions, velocities and radii within the intervals $\Delta \boldsymbol{x} = (\Delta x_1, \Delta x_2, \Delta x_3)$ about $\boldsymbol{x}$, $\Delta \boldsymbol{v} = (\Delta v_1, \Delta v_2, \Delta v_3)$ about $\boldsymbol{v}$, and Δr about r. The abbreviations $\int \ldots d\boldsymbol{x}$ and $\int \ldots d\boldsymbol{v}$ stand for 3-fold integrations. It can easily be seen that F can be interpreted as a number density in the higher-dimensional space $(\boldsymbol{x}, \boldsymbol{v}, r)$. When the intervals indicated by Δ become so small that F can be assumed to be constant it follows

$$F(\boldsymbol{x}, \boldsymbol{v}, r, t) = \frac{\Delta N}{\Delta x_1 \Delta x_2 \Delta x_3 \Delta v_1 \Delta v_2 \Delta v_3 \Delta r} . \tag{3.2}$$

Macroscopic properties of a spray can be expressed in terms of moments of the distribution function F. The mass flux in x_1-direction $\dot{m}_1$ for droplets with radii $r < r_2$ for example is given by the relation

$$\dot{m}_1 = \int\limits_{r=0}^{r=r_2} \int\limits_{v_1} \int\limits_{v_2} \int\limits_{v_3} \frac{4}{3} \pi r^3 \varrho_{liq} v_1 F(\boldsymbol{x}, \boldsymbol{v}, r, t) dr dv_1 dv_2 dv_3 . \tag{3.3}$$

Changes of the distribution function are caused for example by droplet-droplet and droplet-wall collisions, by evaporation and condensation, or by convective transport. Following the methods of the kinetic theory of gases Boltzmann-like equations have been proposed for describing the evolution of the distribution function of aerosols [144] and sprays [72]. It needs no explanation that solutions of practical problems with this method are mathematically sufficiently formidable to make it necessary to look for simplifications.

When the position and the velocity are not considered of primary interest the variables (x_1, x_2, x_3) and (v_1, v_2, v_3) can be eliminated by integration and one obtains a simpler reduced distribution function. When the distribution F is known, it is possible to calculate for example the reduced distribution function

$$f(r,t) = \int\int F d\boldsymbol{x} d\boldsymbol{v} = \frac{dN}{dr} \ . \tag{3.4}$$

The product $f(r,t) \cdot dr$ represents the number of droplets which at time t are in the interval of radius from r to $r + dr$. The number of droplets in the range of radius from $r = r_1$ to $r = r_2$ is given by the integral

$$N_{12} = \int_{r_1}^{r_2} f(r,t) dr \ . \tag{3.5}$$

The distribution function $f(r,t)$ gives no information about the droplet position and velocity. Setting $r_1 = 0$ one obtains from Eq. 3.5 the number of all droplets whose radii are smaller than r_2. The total surface area A of all droplets with radii between $r = 0$ and $r = r_{max}$ can be expressed as

$$A = \int_{r=0}^{r_{max}} 4\pi r^2 f(r,t) dr \ . \tag{3.6}$$

For technical applications one has often to determine the volume of all droplets in a certain interval of radius. From Eq. 3.4 it follows

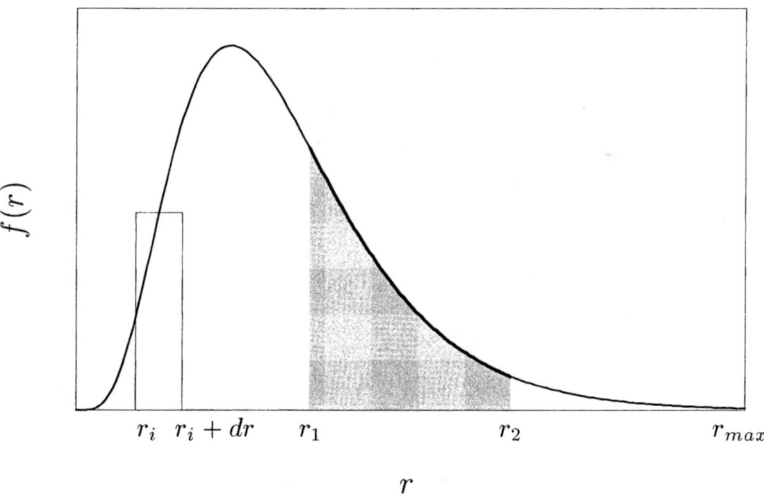

Fig. 3.1. Schematic representation of typical distribution function $f(r)$. The rectangle with the infinitesimal area $f(r_i) \cdot dr$ represents the number of droplets in the interval of radius from r_i to $r_i + dr$. The shaded area is proportional to N_{12} defined by Eq 3.5. The distribution function $f(r)$ approaches zero for $r = 0$ and $r = r_{max}$

$$V_{12} = \int_{r_1}^{r_2} \frac{4}{3}\pi r^3 f(r,t) dr \ . \tag{3.7}$$

A consequence of the factor r^3 in Eq. 3.7 is that the smaller droplets contribute very little to the volume and mass of the liquid phase. Distribution functions for the surface, volume or other functions of radius r can easily derived from $f(r,t)$.

It is well known that different mean values for the droplet size can characterize the droplet size in the spray. These mean values are defined with different moments of the distribution function f. A very important mean diameter is the so-called Sauter mean diameter, which is defined by

$$d_{Sauter} = 2\frac{\int r^3 f(r,t) dr}{\int r^2 f(r,t) dr} \ . \tag{3.8}$$

Different analytical expressions have been proposed in the literature as approximations for the distribution function f [71]. Experimental results for distributions are always obtained in histogram form due to the finite resolution of the measurements

Multi-droplet systems have been studied both experimentally and theoretically by numerous researchers from different viewpoints. Theoretical approaches for describing the evolution of size and chemical composition in aerosols due to coagulation, chemical reaction, condensation, and particle sources have been presented by different authors [145, 146]. In combustion research a great number of papers have been devoted to the investigation of spray combustion [48, 71, 72]. The distribution function f becomes a delta-function, when all droplets have exactly the same size. Such monodisperse droplets are often used for studying elementary properties of droplets.

3.3 Droplet Streams

3.3.1 Monodisperse Streams

Monodisperse droplet streams offer the possibility to study droplets with exactly the same size, velocity and spacing, which have undergone the same history, for example heating and evaporation. A detailed description of the generation and behavior of such monodisperse streams has been given in Sect. 2.3.

The droplet properties, that characterize a monodisperse equally spaced droplet stream are given in Fig. 3.2. All droplets have the same initial values of velocity v, diameter d, temperature T and spacing s between neighboring droplets. In most cases these parameters have changed for instance due to deceleration, evaporation, or heating in a flame, when the droplets pass the plane of observation, which is perpendicular to the droplet stream and located in the distance h from the orifice of the droplet generator.

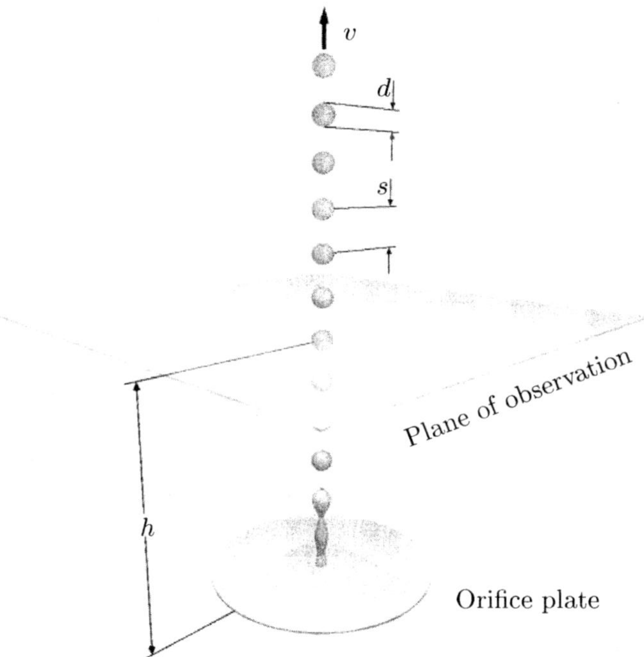

Fig. 3.2. Monodisperse droplet stream originating from an aperture in the orifice plate of the droplet generator. All droplets have initially the same velocity v, the same diameter d and the same distance s between each other. The plane of observation is located at the level h above the orifice of the droplet generator

All droplets traversing the plane of observation have the same history. Therefore parameters like velocity, size, temperature, or composition are the same for each droplet at the location of the plane. Measurements of the droplet parameters in this plane with various techniques will be described in Chap. 4. To study the transient behavior of the droplets, measurements are performed in different planes with different distances h from the orifice. The plane of observation must have a minimum distance h_0 from the orifice, since the droplets do not form immediately, when the liquid jet leaves the orifice. At this distance h_0 the liquid jet has disintegrated completely and droplet oscillations can be neglected.

When planning experiments, one should take into account, that the droplet parameters may change before the droplets have reached the plane at the distance h_0. Changes in size and temperature may be avoided by a small heatable tube, which surrounds the droplet stream from the orifice to the plane at distance h_0. Inside the tube the air will become saturated with the vapor of the droplet liquid. This helps to avoid evaporation, which may change the droplet composition for instance of multicomponent fuel droplets.

Temperature changes of the droplets can be avoided to a large extent, when the temperature of the tube is approximately equal to the temperature, which is measured with the thermocouple inside the droplet generator. The initial droplet temperature is then practically equal to the temperature inside the droplet generator.

From measurements along the droplet stream at different distances from the orifice the temporal evolution of the investigated parameters can be calculated, if in addition the droplet velocity $v(h)$ is measured along the stream. The time a droplet needs to travel from h_0 to h is given by the relation

$$t = \int_{h_0}^{h} \frac{1}{v(h)}\, dh \; . \tag{3.9}$$

For performing this transformation between distance and time it may be convenient to approximate the measured values of the velocity v by a regression curve.

Here it becomes evident, that the temporal evolution of the droplet parameters can be studied by performing measurements at different distances from the orifice of the droplet generator. A droplet with the velocity $v = 10\,\mathrm{m/s}$ will travel a distance of 1 cm during the time of 1 ms. A variation of h in steps of 1 mm will result in a time resolution of 0.1 ms.

When optical measurement techniques are employed, which evaluate for example scattered light, a superposition of the light scattered by many droplets can be used for evaluation, since all droplets passing the plane of observation at h have the same history. This effect allows to use less sensitive sensors or cameras or a laser with less power.

The periodicity of the droplets is an essential advantage for various measurement techniques. The periodicity of the droplets allows for example a special illumination techniques. This technique, which makes use of a stroboscopic effect, will be described in detail in Sects. 4.3 and 4.4.

In a droplet stream a droplet is influenced aerodynamically by its neighbors. The Reynolds number for example is reduced approximately by a factor of ten, resulting in an increase of the stop distance of the droplets. Internal circulation may be influenced and the break-up of droplets can be reduced. In varying the droplet distance the influence of neighboring droplets on the evaporation or combustion rate can be studied for instance. For the study of droplets which are not influenced by its neighbors single droplets are used, as described in Sect. 3.4. For the investigation of stronger influence of neighboring droplets, which is important in sprays, droplet arrays can be employed [147]. These will be described in the following Sect. 3.3.2.

Connon and Dunn-Rankin [148] studied droplet streams in the range of ambient pressures from 1–70 bar. This research may be important for the investigation of droplet behavior near the critical point. This topic will be discussed briefly in Sect. 6.6. Muntz and Dixon have designed an apparatus for the investigation of droplet streams under vacuum conditions. They obtained regular droplet streams containing a very large number of droplets (see also Sect. 2.3). Other applications of monodisperse droplet streams may be found in Sect. 7.

3.3.2 Droplet Arrays

More complicated droplet systems can be studied, when more than one droplet stream is used. Binary droplet collisions can be studied with two or eventually more streams as described in Sect. 5.4. In this case a separate droplet generator is used for each droplet stream. The minimum spacing s_a between neighboring streams is then limited by the geometrical dimensions of the droplet generators. With separate droplet generators the spacing is too large to observe mutual interaction of the neighboring streams for instance in the evaporation or burning rate.

Parallel monodisperse droplet streams with smaller spacings s_a can be obtained by employing a droplet stream generator of the type described in Sect. 2.3, but with more than one orifice in the orifice plate. The geometrical arrangement of the orifices in the plate determines if the resulting droplet array is two- or three-dimensional. If the orifices are arranged in a line two-dimensional or planar droplet arrays are obtained as shown schematically in Fig. 3.3(a). Other arrangements of the orifices result in three-dimensional droplet arrays, as shown schematically in Fig. 3.3(b) for three droplet streams. Numerous variations of the arrangements are possible. Such two- or three-dimensional droplet systems may be considered as simple models for real sprays.

In Sect. 6.4.1 the influence of neighboring droplet streams on the burning rate of the droplets is shown [149]. For such measurements the initial droplet parameters have to be adjusted and controlled in the same way as for a single droplet stream. Therefore the variation of the orifice diameters has to be very

(a) (b)

Fig. 3.3. Schematic representation (a) of planar droplet array consisting of five parallel droplet streams and (b) of three-dimensional droplet array. The plane of observation is located at the level h above the exit of the droplet generator

small. It has to be emphasized, that difficulties in operating such systems increase with the number of droplet streams. In Sect. 6.4.1 results for up to five streams are shown. Orifice plates with different orifice diameters in one plate are possible. In this case the difference between the diameters is limited if monodisperse droplets are desired in each stream. For the chosen excitation frequency a monodisperse disintegration has to be possible for each stream. Orifice plates with several hundreds of orifices have been realized. In such cases it is very difficult to control all droplet parameters, especially coagulation between neighboring streams may occur. However, a nearly monodisperse spray with a very narrow size distribution may be obtained.

3.3.3 Deflection with Electrical Field

For certain experiments with droplet streams a larger spacing s between the droplets is required. For example the drag coefficient or the evaporation and burning rate may be influenced by neighboring droplets. Some researchers have used electrical forces to pulse out single droplets for their studies. Schneider et al. use vibrating capillary tubes to produce streams of uniform sized and equally spaced droplets [150]. The droplets could be charged by applying an electrical potential difference between a charging electrode and the capillary tube. The charged droplets enter the electric field between two electric plates and are deflected toward a collector. For producing a single droplet a voltage pulse is applied to the charging electrode during the formation of the droplet. This droplet, which will have a different charge and follow a trajectory which differs from the main stream, can be considered as single droplet which is not influenced by neighboring droplets. By operating two such systems simultaneously Adam et al. studied binary droplet collisions [19]. More details may be found in Sect. 5.4.2.

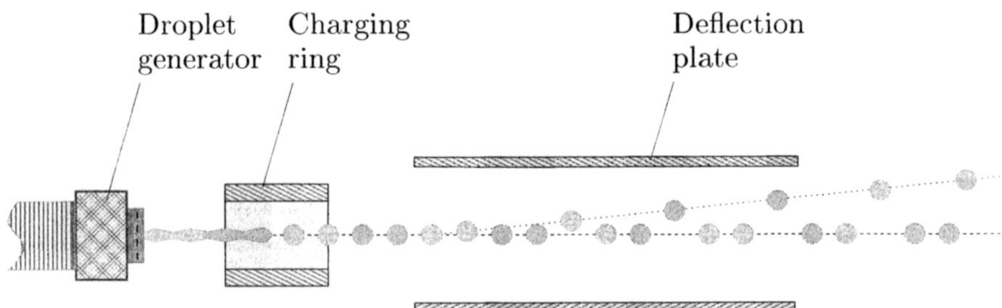

Fig. 3.4. Sketch of setup to charge and deflect droplets of a monodisperse droplet stream

3.4 Single Droplets

3.4.1 Single Moving Droplet

The deflection technique of the previous section allows to produce single moving droplets by applying a single electric pulse. Other techniques are the droplet on demand generator or the dropper as described in Sects. 2.4 and 2.5 respectively. The dropper technique has been used for example by Kinzer and Gunn, who studied freely falling water droplets. The evaporation rate of such droplets was determined by measuring the terminal velocity. The droplets passed through a charge inducing ring electrode. The droplets became slower as their size was reduced by evaporation [151].

3.4.2 Aerodynamically Suspended Droplets

Kinzer and Gunn describe an apparatus for supporting large water droplets in the millimeter range by aerodynamic forces in the air flow through a tapered glass tube mounted vertically. Water droplets with maximum diameters of approximately 5 mm could be suspended in this manner [151]. Nordine and Atkins developed an aerodynamic levitation technique, which they used for suspending solid spheres and in an exploratory study liquid droplets [152].

3.4.3 Droplet Suspension with Thin Fibers

In the classical suspension techniques single droplets are suspended at the end of filaments such as thin wires or quartz fibers. Since the suspended droplet is stationary the evaporation process can be investigated in great detail. Because of the thickness of the wires it is however difficult to suspend droplets with diameters below 1 mm. The filament also distorts the droplet shape from spherical and conducts heat from the droplet. Droplets with diameters of $1-2$ mm have been suspended on thin threads with diameters in the range of a fraction of a millimeter for studying evaporation and combustion phenomena. For the ignition of fuel droplets electrical sparks have been used in many investigations [153]. At higher pressures higher voltages have to be applied, which may lead to fragmentation of the droplet, when the Rayleigh limit is reached (see Eq. 1.15 in Sect. 1.4). Resistance coil ignitors have therefore been used at higher pressures [154]. Niioka and Sato report on combustion of fuel droplets at pressures up to 20 MPa using a suspension technique with filaments with a diameter of 400 μm and a sphere at their tip [155]. The droplet diameters were between 1 and 1.9 mm. Bradley et al. measured evaporation rates, vapor pressures and diffusion coefficients by suspending liquid and solid particles by means of a microbalance with a high sensitivity [156, 157].

3.4.4 Electrostatic Levitation

The levitation with electrostatic fields offers the possibility of combined positioning and positioning sensing. The classical Millikan oil drop apparatus consists of two parallel flat electrodes having the distance d. A typical Millikan chamber is shown in Fig. 3.5. The electrodes have rounded edges in

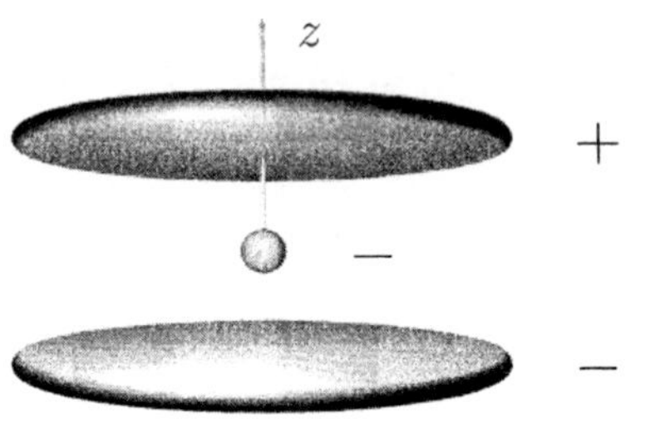

Fig. 3.5. Schematic view of capacitor plates for classical Millikan experiment. A small electrically charged particle is injected into the space between the plates. The voltage across the capacitor plates can be adjusted for suspending the particle

order to minimize high field effects and distortions of the electric field. By applying the dc potential U the uniform electric field $E = U/d$ is set up in the region between the electrodes. A small electrically charged droplet is injected into the electric field between the two electrodes. The equation of motion for a droplet of mass m with the electric charge q is

$$m\ddot{z} = -6\pi\eta r\dot{z} - mg + m_G g + qE \ , \tag{3.10}$$

where $(m - m_G)g$ represents the difference between the gravitational force and the buoyant force. The aerodynamic drag has been described by Stokes law. The solution of this equation has the form

$$\dot{z} = A\exp(-t/\tau_v) - \tau_v(\frac{m - m_G}{m}g - \frac{qE}{m}) \ , \tag{3.11}$$

where A is a constant, and $\tau_v = m/6\pi\eta r$ is the relaxation time of the velocity, as defined by Eq. 1.44. Inspection of Fig. 1.12 in Sect. 1.9.1 reveals, that for droplets with diameters between 1 and 100 μm the relaxation time τ_v ranges from a few microseconds to a second. This means that for most observations the term with $\exp(-t/\tau_v)$ can be neglected. In this case one has

$$\dot{z} = \tau_v(\frac{m_G - m}{m}g + \frac{qE}{m}) \ . \tag{3.12}$$

For $\dot{z} = 0$ one obtains the condition of equilibrium $qE = (m - m_G)g$ or $qdE + Edq = 0$. A change of dq of the electrical charge on the droplet can be compensated by change of the electrical field

$$dE = \frac{(m - m_G)g}{q^2}dq \ . \tag{3.13}$$

In the classical Millikan configuration the charged droplets are not stabilized laterally, the lateral drift has to be monitored and corrected for. Newer techniques provide methods for laterally focusing the particles and for suspending the particles in a well defined small volume. Horizontal drift of the particles may be eliminated by applying an additional horizontal focusing field. Automatic positioning systems have been described in the literature [158-161]. The high electric fields required can lead to breakdown and discharge in the background gas [162]. The problem of breakdown can be reduced by working at gas pressures in excess of 1 bar. Static electric fields have been used to levitate uncharged particles by polarization of the sample.

Some investigators have used an electrified pin to capture the droplet and to center is along the vertical axis with the electric field produced between the electrified pin and the ground plate. The applied voltage is adjusted manually until a droplet is caught. At this point in time an automatic feedback system may be activated. Newer versions of the Millikan apparatus can be pumped out to remove residual vapors or to backfill the chamber with a desired gas [163]. Electrostatic suspension has been used for example to study evaporation rates of tiny droplets [11, 12].

3.4.5 Electrodynamic Levitation

Electrodynamic levitation is a method of containing small electrically charged particles in dynamic equilibrium by alternating electric fields. This technique has been employed for various physical and physico-chemical investigations [164]. The electrodynamic levitator consists of three hyperbolic electrodes with azimuthal symmetry about the z-axis, as shown schematically in Fig. 3.6. This electrode configuration gives sinusoidally time varying forces, whose strength is proportional to the distance from the central origin. The levitated specimen is balanced at the geometric center of the levitator by applying a dc voltage to the endcap electrodes, whereas focusing is achieved by connecting the ring electrode to an ac source. The motion of the droplet of mass m with charge q is described by the differential equations

$$m\frac{d^2z}{dt^2} = -6\pi\eta r\frac{dz}{dt} + q(E_z + E_0) - mg \tag{3.14}$$

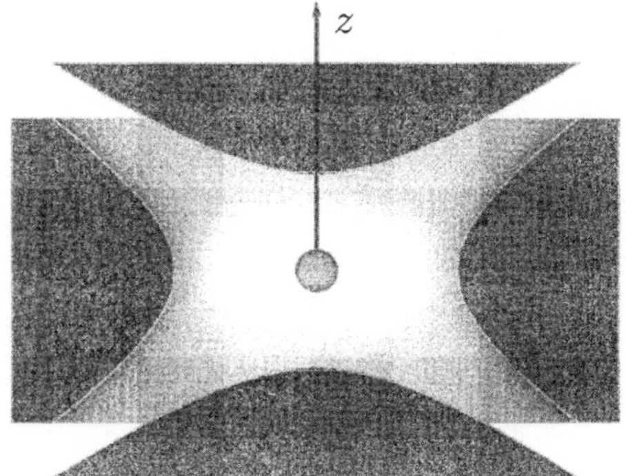

Fig. 3.6. Cross-sectional view of electrode system for electrodynamic levitation

and

$$m\frac{d^2r}{dt^2} = -6\pi\eta r\frac{dr}{dt} + qE_r \ , \qquad (3.15)$$

where z and r are coordinates in vertical and radial direction, E_z and E_r represent components of the electric ac field and E_0 the dc field. It has been assumed, that the aerodynamic drag force can be described by Stokes law. The two equations of motion of a particle in the two directions z and r are mutually independent. It can be shown that each of the two different equations is transposable into a modified Mathieu equation. The stability regions of these equations have been described by Wuerker et al. [165]. Different designs of electrodynamic levitators may be found in the literature [166-168].

3.4.6 Electromagnetic Levitation

Electromagnetic levitation has been proposed for material processing in a containerless fashion. The use of this method, which works by eddy current induction, is limited to highly conducting materials. The applied electromagnetic fields are produced with current carrying coils. The levitated material is constrained in a potential well of the applied electromagnetic field. The sample suspension is usually accompanied by a stirring action [143, 162]. Suspension of molten as well as solid metals has been accomplished with electromagnetic fields. Okress et al. report on the realization of stable levitation by two coaxial coils connected in series and carrying alternating currents. The necessary supporting and lateral restoring forces are due to eddy currents induced by the alternating currents in the coils. Heating and melting of the metal charge depends upon different parameters such as the available power, frequency and and electromagnetic coupling between the levitated specimen and the primary loops. Experimental investigations were conducted with solid and molten aluminum, solid molybdenum and copper, silver and titanium [169].

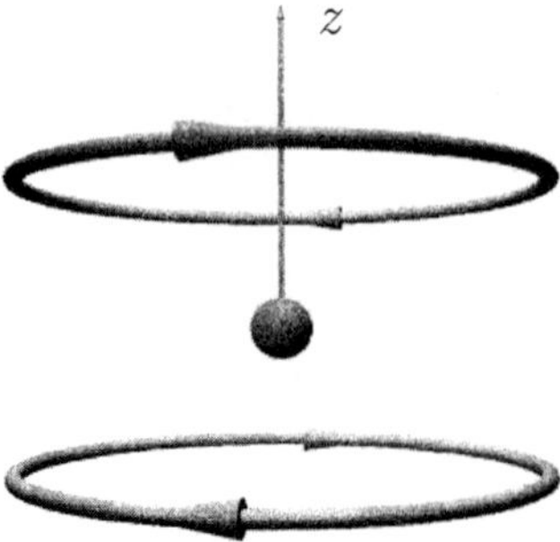

Fig. 3.7. Electrical currents for electromagnetic levitation of conductive sphere

3.4.7 Dielectrophoretic Levitation

Static electric fields have been used to levitate uncharged particles by induced polarization of the sample. Dielectrophoretic levitation requires the use of dielectric fluids as host medium. Jones and Kraybill demonstrated the levitation of single uncharged metallic spheres, multisphere chains and particles of material with high dielectric constant [170].

3.4.8 Ferrofluid Levitation

Ferrofluid consist of ferromagnetic particles suspended in a liquid. Rosenzweig reports on stable levitation of a glass sphere in a ferrofluid as the host medium [171].

3.4.9 Optical Levitation

Droplets can be levitated and stabilized using radiation pressure forces described in Sect. 1.13.6. The droplets can be stabilized by employing one or two divergent laser beams whose waist diameter in most cases is larger than the droplet diameter. Forces act in direction of the beam and perpendicular to it. For transparent droplets lasers operating in the TEM_{00} mode are used. In this case the forces perpendicular to the beam act to stabilize the droplet on the laser axis.

Two laser beams are needed, when the beam direction is horizontal [115, 172]. Then forces perpendicular to the beam axes are equal to the gravity forces on the droplet. The second beam is used to stabilize the droplet in horizontal direction.

Only one laser beam is needed to stabilize the droplet, when the beam direction is vertical [173, 174]. Then the droplet is stabilized horizontally by the forces perpendicular to the beam axes and the forces in beam direction compensate the gravity forces. This situation in shown schematically in Fig. 3.8. Ashkin and Dziedzic studied levitation under rarefied conditions [113, 175]. Based on the generalized Lorenz-Mie theory (GLMT) Ren et al. studied theoretically the influence of a highly focused laser beam with waist diameters smaller than the droplet diameter [176]. For highly reflecting particles lasers operating in the TEM_{01}^* mode have to be used to stabilize the particle or for instance a droplet of a molten metal with the forces perpendicular to the beam. Roosen and Slansky calculated these forces under such conditions using geometric optics [177]. More than one vertical laser beam can be used to stabilize and manipulate a levitated particle [178, 179]. Even bubbles can be levitated and stabilized by radiation pressure forces [180]. Different applications of laser radiation pressure were presented by Ashkin [181]. Experimental setups and examples of measurements on optically levitated droplets will be presented in Sect. 6.

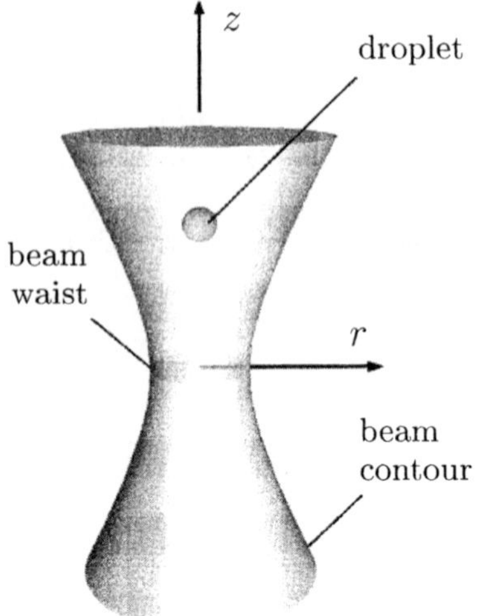

Fig. 3.8. Schematic view of droplet levitated optically in a vertical laser beam. The laser beam is directed in z-direction

3.4.10 Acoustic Levitation

A further possibility to study single droplets is to levitate them in an acoustic field generated in the gap between an oscillating surface and a reflector. The reflected wave interferes with the primary wave. For acoustic levitation the distance between transmitter and reflector is adjusted to create a system of standing waves with nodes and antinodes. An example for the mean density distribution in such a wave system is shown in Fig. 3.9. This interferogram of the wave system has been obtained by bringing the levitator into the ray path of a Mach-Zehnder interferometer. The mirrors of the interferometer have a diameter of approximately twenty centimeters. Before applying the acoustic field they were adjusted exactly parallel to each other to obtain a uniform intensity distribution without fringes. With this illumination the transmitter and reflector appear as shadows on a screen. The negative in Fig. 3.9 shows the density distribution in a system of standing acoustic waves with four density or pressure nodes, which are represented by the black fringes The distance between transmitter and reflector is $5 \cdot (\lambda/2)$.

A particle in the field of a sound wave is subjected to hydrodynamic forces. The magnitude of these forces, which is equal to the average flux of momentum through any closed surface containing the particle, is given by the equation $F_i = \oint \Pi_{ik} dA_k$, where $\Pi_{ik} = p\delta_{ik} + \varrho_{gas} v_i v_k$. In this equation p is the static pressure, δ_{ik} the Kronecker symbol and v_i, v_k are components of the gas velocity. This method allows stable levitation of liquid droplets as well as of solid bodies. The acoustic radiation force on bodies in a standing wave field has been discussed by different authors [182, 183]. For the acoustic radiation force on a small solid sphere King [184] has derived the equation

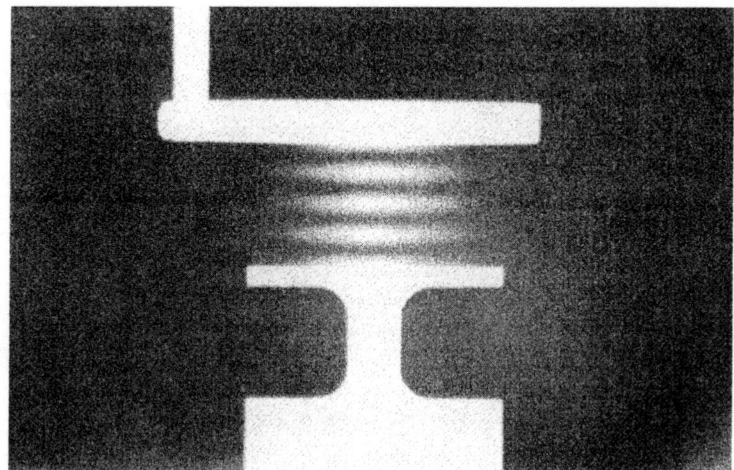

Fig. 3.9. Interferogram of standing wave system between transmitter and reflector of acoustic levitator system. The white fringes correspond to velocity nodes or pressure and density loops. The diameter of the lower plate is 38 mm. This plate is excited with the frequency 36800 Hz. At 20°C the wave length is approximately 9.2 mm, the sound level was larger than 160 dB. With the shown configuration water droplets of approximately 2 mm in diameter can be levitated (from G. Funcke, ITLR)

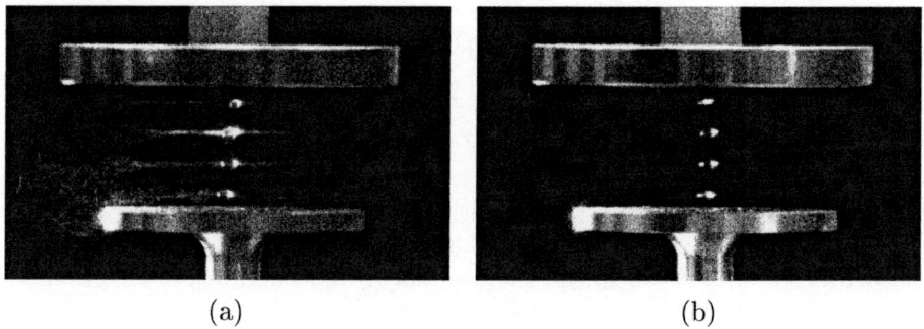

(a) (b)

Fig. 3.10. A spray produced with an ultrasonic aerosol generator is introduced into the gap between vibrating plate and reflector of an acoustic levitator. The injected droplets are levitated in the pressure nodes and start to coalesce (a). After a short time all spray droplets have coagulated (b) (from G. Funcke, ITLR)

$$F = \frac{\pi p_1^2}{\varrho_{gas} c_s^2} k r^3 \frac{\varrho + 2(\varrho - \varrho_{gas})/3}{2\varrho + \varrho_{gas}} \sin(2kz) \ . \tag{3.16}$$

For $\varrho \gg \varrho_{gas}$ one obtains

$$F = \frac{5\pi}{6} \frac{p_1^2}{\varrho_{gas} c_s^2} k r^3 \sin(2kz) \ . \tag{3.17}$$

In this relation p_1 represents the pressure amplitude, $k = 2\pi/\lambda$ the wave number, ϱ_{air} the density of the ambient air, c_s the velocity of sound, and r the radius of the spherical particle. In Eq. 3.17 it has been assumed, that the density of the particle is much higher than the density of the gas. This force can be related to the potential of the acoustic radiation forces [185]. A droplet placed in an acoustic resonance chamber of length L will experience a maximum of the acoustic levitation force at the pressure nodes with $L = \lambda/8,\ 3\lambda/8,\ 5\lambda/8, \ldots$ and no force at the corresponding antinodes. Deviations from King's theory for higher sound intensities have been studied in Ref. [182]. In mechanical equilibrium the gravitational force acting on the levitated particle is balanced by the acoustical radiation force

$$\frac{5\pi}{6} \frac{p_1^2}{\varrho_{gas} c_s^2} k r^3 \sin(2kz) \geq \frac{4}{3} \pi r^3 \varrho g \tag{3.18}$$

or

$$\frac{5}{6} \frac{p_1^2}{\varrho_{gas} c_s^2} k \sin(2kz) \geq \frac{4}{3} \varrho g \ . \tag{3.19}$$

In typical levitation systems the ultrasonic field is generated by a piezoelectric transducer with rotational symmetry. The gap dimensions are usually a few centimeters, frequencies range from 15 kHz to 40 kHz. In order to avoid overlapping of successive nodes of the acoustic field the droplet diameter should fulfill the condition $d < \lambda/2$.

Efforts have been made to develop both, systems for earth-based environment and systems for the microgravity environment of earth orbit. Acoustic levitation allows for instance to study sophisticated processes as the evaporation of droplets containing solutions of water and a salt as shown in Fig. 3.11. It can be seen, that the shape of the levitated droplet is not spherical, since the droplet is deformed by the acoustic field. This deformation is larger for the larger droplet on the right hand side (b). With increasing time, from top to bottom, salt crystals grow in the droplet. A more detailed description will be given in Sect. 6.3.1.

An enhancement of the levitation force can be achieved by curving the surfaces of the driver and reflector [185]. Details of the design of ultrasonic transducers and acoustic levitation systems may be found in the literature [185-188]. The determination of the shape of liquid droplets levitated in standing waves has been described by Trinh et al. [189]. An acoustic levitator with electronic and optical instrumentation for studying a variety of fluid dynamic and physical properties of fluids has been presented by Trinh [190].

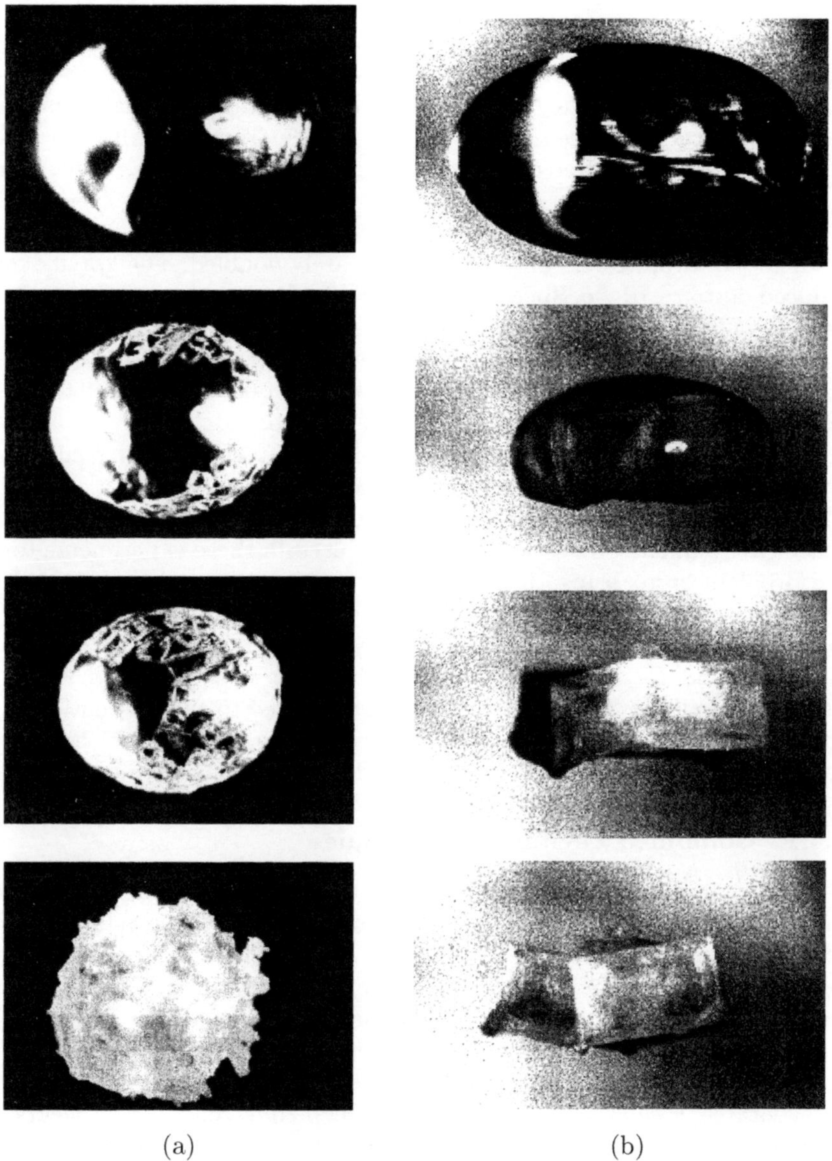

(a) (b)

Fig. 3.11. Evaporation of acoustically levitated droplets, consisting of a saturated solution of water and salt (NaCl). In series (a) initially tiny crystals contaminated the solution serving as germs for crystallization, whereas in (b) it was tried to avoid these impurities. The initial shape of the droplets is approximately an ellipsoid. The maximum diameters are approximately 2 mm for the droplet of series (a) and 2.5 mm for series (b) (from R. Stoll, ITLR)

3.4.11 Microgravity experiments

Most theoretical descriptions of physical and chemical processes in droplets are based on spherical symmetry. This assumption is of course not valid at normal gravity, when buoyancy becomes important. Accordingly substantial research efforts have been made to create an environment with low gravity levels and practically free of natural convection. Drop towers, which provide test times ranging from a few seconds up to ten seconds, have been used for example for various combustion studies [191-193]. Using catapult systems the test times of drop towers may be approximately doubled. The gravity can be reduced down to $10^{-6}g$. Special experimental techniques and equipment have been developed for performing these investigations. Other authors achieved a reduction of gravitational effects by accelerating the droplets in a special tube [194]. Aircraft flying parabolic trajectories offer longer test times, but gravity levels below $10^{-3}g$ are difficult to achieve.

Experiments performed in space offer advantages, which facilitate the comparison between measurements and existing theories. Shape oscillations, rotation, collision and coalescence of liquid droplet have been investigated in space environment. Acoustic forces can be used in these experiments to hold the droplets or to bring them together for studying coalescence. Different dyes have been added to distinguish the droplet liquids. The effect of surfactants dissolved in the droplet have been studied. Dietrich et al. designed an experiment for testing droplet combustion in spacelab [195]. Other experiments under low-gravity conditions are devoted to surface-tension-driven convection, the so-called Marangoni effect, to the spreading of liquids over solid and liquid surfaces and to containerless processing technology.

3.4.12 Combined Levitation Techniques

Clancy et al. used electrostatic forces in the design of a positioning system for the investigation of metallurgical processes, such as crystal growth or glass preparation in space [162]. Tian and Apfel applied an electric dc voltage between transducer and reflector of an acoustic levitator [196]. For producing the droplets a thin layer of liquid is applied to the vibrating surface. When the vibration amplitude increases fine droplets are generated as described in Sect. 2.2. Due to the electric field the droplets are charged. These droplets are levitated similarly as shown in Fig. 3.10, but coalescence occurs only initially. The droplets formed by coalescence are larger and carry a larger charge. This leads to electrostatic repulsion between the droplets resulting in the formation of a two-dimensional droplet array. The mechanical equilibrium along the gravity direction can be written as

$$\frac{4}{3}\pi r^3 \varrho_{liq} g = F_{acoustic} + F_{electric} \; . \tag{3.20}$$

The factor r^3 in this equation cancels since the forces $F_{acoustic}$ and $F_{electric}$ are proportional to the droplet volume. The described apparatus has been used for the levitation of ethanol droplets. Perturbances of the evaporation process by the acoustic field are shown to be negligible in this experiment, since the levitation force comes mainly from the electrostatic field and the sound intensity can be low.

A combination of acoustic and aerodynamic levitation has been developed by Weber et al. [197]. A gas jet provides a large force, whereas exact position, rotation, and shape of the levitated specimen are controlled by small acoustic forces. The levitated material could be heated and melted at temperatures up to 2700 K using a cw CO_2 laser beam. These authors used a cylindrical resonator to form a three-dimensional trap. They were able to levitate samples ranging in diameter from approximately 3 to 4 mm and of density up to 9 g/cm^3. An optical pyrometer provided temperature data. The technique was used to study melting and undercooling of different materials. Seaver et al. combined acoustic levitation with a horizontal wind tunnel [198]. They were able to hold millimeter diameter droplets in a moving air stream. Laser-induced fluorescence allowed remote measurements of droplet temperature within $\pm 1.5°$C precision.

4. Experimental and Measurement Techniques

4.1 Introduction

The generation of droplets and various droplet systems have been described in previous sections. For experimental investigations of droplet behavior measurement techniques are necessary, which allow to determine shape, size, velocity, surface tension, temperature, refractive index or other droplet properties. In this section some selected techniques for measuring these droplet parameters will be described briefly. Techniques have been selected, which are very popular or which have been used in experiments presented in Chaps. 5 and 6. However, numerous other techniques exist and are used. Depending on the experiment to be performed different variations of the techniques described here are possible. Therefore the descriptions are focused on the physical principles of the measurement techniques.

Photographic and video observation techniques allow to determine the shape and to some extent the size of droplets. Video techniques based on periodical phenomena allow to study the temporal evolution of shape and size, and in addition the droplet velocity may be determined.

Laser Doppler velocimetry (LDV) and phase Doppler velocimetry (PDV) have become standard methods to determine the velocity and size respectively of solid and liquid particles. With these instruments the velocity or both velocity and size of droplets are measured in single points of the flow field. If the particles are sufficiently small they may be used as tracers for the determination of the fluid velocity. A statistical analysis is suitable to determine these droplet parameters in the case of instationary flows. The same holds for the laser-two-focus velocimetry (L2F velocimetry). Particle image velocimetry (PIV), however, allows to determine the velocity of the fluid or the droplets at a point of time in a plane of the flow. Special modifications of this technique allow to determine in addition the droplet size.

The size is probably the most elementary parameter to characterize a droplet. Due to the importance many different techniques have been developed. A simple method is sampling and weighing a defined number of droplets to determine mean values for mass and size of the collected droplets. However, for many applications nonintrusive methods are preferable. Therefore optical size measurement techniques, which are practically nonintrusive, become more and more popular. These methods are normally limited to spherical droplets, which means that the droplet size is determined by the radius. Deviations from the spherical shape can be detected directly, when photographic or video techniques are employed. Some techniques allow to handle deviations from the spherical shape.

The optical properties of the droplet liquid depend on the complex refractive index $N = n + ik$, whereas the geometrical properties are usually expressed by the Mie-Parameter $\alpha = 2\pi r/\lambda$, which has been introduced in Sect. 1.13.4. The refractive index is a function of the wavelength of the light. The imaginary part k is a measure for the absorption, the real part n determines the refraction of light. In the following the measurement of the real part is considered. An extended phase Doppler system allows for instance to determine the refractive index. Another method based on the evaluation of light scattered in the region of the first rainbow will be described in Sect. 4.7.4. With measurements of the refractive index different liquids can be distinguished, whereas in binary solutions or mixtures concentrations can be determined.

The refractive index is a measure of the droplet temperature, when the composition of the droplet is known. Difficulties may occur for water near the temperature 0°C due to the anomalous behavior of water. Infrared imaging of the droplets allows to determine the surface temperature. More intrusive techniques to determine droplet temperatures are the suspension of droplets on thermocouples or the schlieren method described in Sect. 4.8.

The surface tension influences the droplet behavior essentially during mechanical interactions with gas streams, other droplets, or walls. Two methods to determine the surface tension will be described briefly in this chapter. Both methods are based on the mechanical behavior of droplets. One method uses oscillations of droplets, the other evaluates the shape of pending droplets.

The use of partial wave resonances allows to determine the evaporation or condensation rates immediately with very high accuracy, whereas the phase transition from liquid to solid, i.e. freezing, can be detected by measuring the polarization ratio of the light scattered by the liquid and the solidified droplet.

4.2 Photographic Observation Techniques

Devices for optical observations contain usually different forms of lenses. The well-known basic rules of imaging with a thin lens may be deduced from Fig. 4.1. Let $f_l = f_l'$ represent the focal length of the lens. The distances from the lens plane to object and image plane s_l and s_l' satisfy the well-known lens equation

$$\frac{1}{f_l} = \frac{1}{s_l} + \frac{1}{s_l'} \; . \tag{4.1}$$

The magnification M of an object of size O is given by

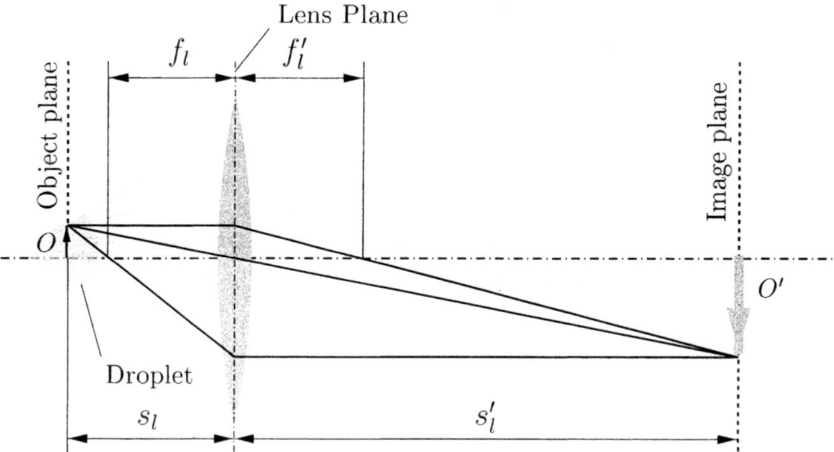

Fig. 4.1. Formation of an image by a thin lens. Rays scattered from a point in the object plane are refracted by a thin lens to pass through one point in the image plane.

$$M = \frac{O'}{O} = \frac{s_l'}{s_l} , \tag{4.2}$$

where O' is the size of the image. Points of the droplet in the object plane result in sharp image points. However, points close to the object plane appear as small spots of finite size on a photographic film or on a video chip positioned in the image plane. The size of these spots increases, when the distance of these points from the object plane increases. The image appears sharp, when the spots are smaller than the grain of the film emulsion or the pixels of the video chip.

Single lenses or lens systems used in experimental arrangements show deviations from the ideal behavior depicted in Fig. 4.1. For the observation of droplets the minimum length of the distance s_l is often determined by certain geometrical dimensions of the experimental setup. Good optical results are usually obtained with high quality photographic lens systems optimized for the application in the setup. Detailed descriptions of imaging may be found in textbooks on optics [96, 97].

Another important aspect is the illumination of the droplet. With diffuse or divergent light the image may give a three-dimensional impression of the object. For the determination of the size and the characterization of the form shadows of the droplets are appropriate. Shadows are obtained, when light parallel to the optical axis illuminates the droplet from the background. The light is diffracted by the edge of the droplet. To obtain sharp shadows this edge has to be imaged on the film or video chip. Using image processing techniques the droplet size and form can be determined automatically.

Short exposure times are required, when the droplets are in motion relative to the imaging system. Droplets moving with the velocity 10 m/s travel a distance of 10 μm during 1 μs. This means that exposure times well below 1 μs are required in order to obtain sharp images. As such short shutter times are not practicable with mechanical shutters, flash lamps with very short duration of the flash are often employed (Nanolite). The flash is triggered while the shutter of the camera is open. Depending on the type of flash lamp used, flash durations of 20 to 200 ns are obtained. The combination of flash lights with video techniques will be described in Sect. 4.4.

4.3 Droplet stroboscope

With monodisperse droplet streams periodic short duration light pulses can easily be obtained from the light scattered by the droplets. For this purpose a laser beam is passed through the droplet stream, as shown in Fig. 4.2. This stroboscopic technique has been described in Ref. [199]. A laser beam

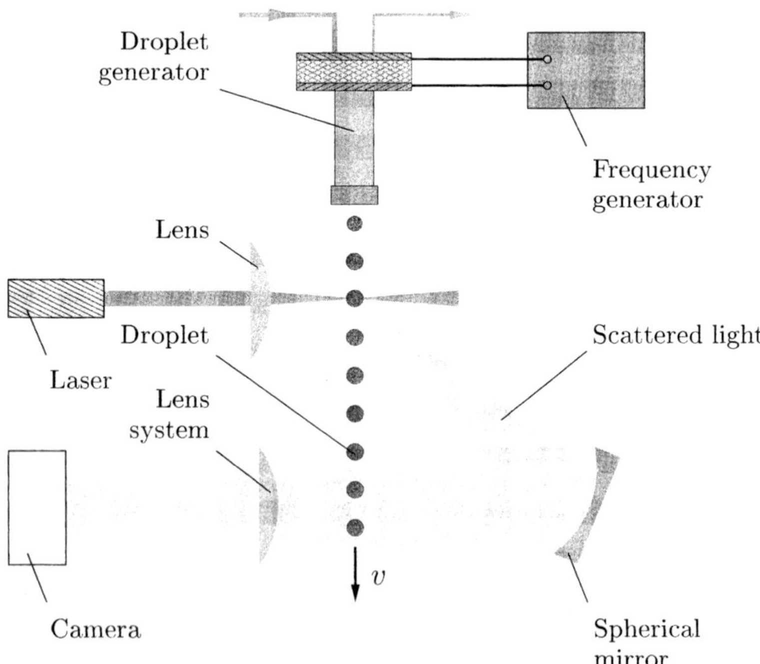

Fig. 4.2. Schematic view of setup for droplet stroboscope. Part of the laser light scattered by the droplets is deflected with the spherical mirror to illuminate droplets. Even with the naked eye standing pictures of the droplets can be observed

is focused by a lens onto the axes of a droplet stream. The focus should be smaller than the droplet diameter. When a droplet passes the laser focus, light is scattered in all directions. Part of the scattered light is collected by a spherical mirror. This mirror is adjusted to illuminate a section of the droplet stream, as can be seen from Fig. 4.2. The light reflected by the mirror may be parallel as sketched in Fig. 4.2, or convergent in order to obtain a higher intensity. With a lens system the illuminated part of the droplet stream is imaged on a camera, which may be a normal camera with a photographic film or a video camera.

The duration of the light pulses generated by the droplets is very short, since the focus of the laser beam is very small. The scattered light is strictly synchronous with the frequency of the droplets. As a result sharp images of the droplets are obtained. The camera may be replaced by a simple screen to observe the droplets even with the naked eye.

Rather complicated phenomena of droplet behavior like oscillations, wall interactions, or coalescence can be observed with this technique on-line as long as each droplet of the stream shows exactly the same behavior. When several droplets of the stream are imaged, one can observe simultaneously different states of regular processes, which each droplet undergoes. The time resolution Δt_{res} is given by the time difference between neighboring droplets, which is the reciprocal value of the excitation frequency $\Delta t_{res} = 1/f_{dis}$.

Fig. 4.3. Image of droplet stream showing coalescence of large droplet with smaller satellite droplet just behind the droplet generator. This image has been obtained with the described droplet stroboscope technique

A result obtained with this technique is shown in Fig. 4.3, where a small satellite droplet coalesces with a larger droplet. The oscillations of the resulting larger droplet can be observed clearly. It should be emphasized, however, that due to possible irregularities of the droplet stream the droplet should pass the laser focus rather close to the orifice in a point where the disintegration process is complete and the droplets have stopped to oscillate.

4.4 Video Observation Techniques

The use of video techniques offers new possibilities. Video cameras together with video recorders or other recording units like frame grabbers in combination with memory cards allow to inspect the images immediately after the exposure. The resolution of a film is in most cases higher than the resolution of a video image. Photographic techniques, however, depend on the development of the film. For VHS video systems the resolution is less than 1000 lines per image, but in many technical or scientific applications the trend is to higher resolutions by employing video cameras with CCD-chips, which have more pixels. Special frame grabbers and memory cards are necessary, when these high resolution cameras are applied. Due to the limited memory only single shots or a sequence of a few shots can be obtained. This limit may be extended in the future.

Single shots in video techniques are performed similarly as with photographic techniques. In normal operation, motion pictures are taken, which can be viewed on-line on a monitor and recorded for later inspection and evaluation. The time resolution is 50 half-frames per second, which is not very high for the observation of transient phenomena of droplet behavior. These difficulties may be overcome by using high speed video cameras, which, however, do not allow on-line observation.

The combination of monodisperse droplet streams and video techniques allows a higher time resolution and on-line observations at a lower price. A sketch of a typical experimental setup is shown in Fig. 4.4. A monodisperse equally spaced droplet stream is illuminated with a flash lamp (Nanolite) to obtain shadowgraphs of the droplets, as described in Sect. 4.2. With a lens system the droplets are imaged on the CCD-chip of a video camera. The output of the video camera can be recorded with the video recorder and observed on-line on the monitor. On the video tape the time code inserter inserts a label into each video frame for individual identification. The essential feature of the setup is the appropriate triggering of the flash lamp. This is performed with a digital frequency generator. First, the lamp has to be triggered when the shutter of the video camera is open. This is achieved with the synchronization signal of the video camera, which opens a gate of the frequency generator. Second, the flash is triggered by the first signal of the excitation frequency of the droplet generator; third, the gate is closed to obtain only one flash per video frame.

In the case of regular processes a quasistationary image can be observed on the monitor. When several droplets are imaged, different states of regular processes can be seen, where the time resolution is given by the reciprocal value of the excitation frequency, like with the droplet stroboscope technique.

It has to be emphasized, that each video frame is not a superposition of many droplet pictures, as it is with the droplet stroboscope technique; here each frame shows an individual situation. On the following frame other

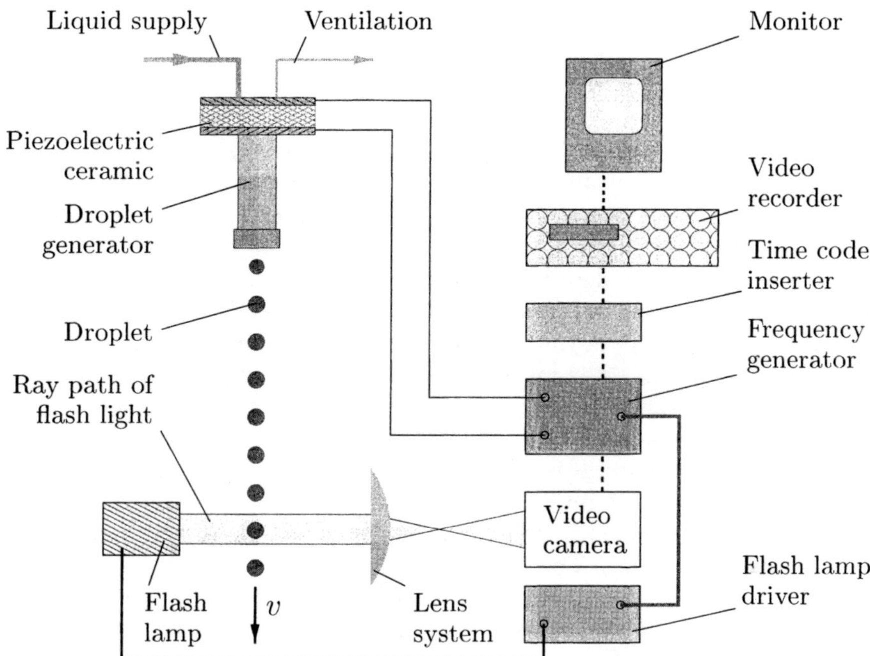

Fig. 4.4. Schematic view of experimental setup for video observation of droplet phenomena with high time resolution

droplets are imaged, where in the case of regular processes these droplets have the same size and position.

Regular processes allow to obtain a motion picture using a stroboscopic effect. In this case each flash is triggered with an additional phase shift $\Delta\varphi$ with respect to the preceding trigger signal. The time resolution between two video frames is then given by

$$\Delta t_{res} = \frac{\Delta\varphi}{360^\circ} \cdot \frac{1}{f_{dis}} , \tag{4.3}$$

where f_{dis} is the frequency for disintegration, as described in Sect. 2.3. Motion pictures and therefore high time resolution can be obtained for many complex mechanical droplet interactions. If the processes become irregular, for instance when a regular droplet stream splashes against a wall, a momentary situation is frozen on each video frame. In this case details of the process can only be obtained, when each frame is evaluated separately.

4.5 Velocity

4.5.1 Laser-Doppler Velocimetry

Laser-Doppler velocimetry (LDV) has become one of the most popular tools for determining flow velocities in quite different flow situations. Detailed descriptions of this technique have been presented in different textbooks [200, 201]. Therefore only a short description will be given here. This technique is based on the determination of the velocity of fine tracer particles, which follow the fluid flow practically without delay. The velocity is determined in single points of the flow field. For the evaluation of the whole flow field the velocity has to be measured at many different points. For instationary flow fields a statistical evaluation for each point may be suitable for the description of the whole flow field.

The spatial resolution is given by the size of the measurement volume, which is defined by the intersection of two laser beams. The beams are produced by splitting one laser beam into two parallel beams, which are focused symmetrically by a lens. The resulting beams cross each other with the angle 2α, as shown in Fig. 4.5. The focus of each beam is in the center of the measurement volume. When a tracer particle passes the measurement volume part of the scattered light is collected by a lens and detected by a sensor.

For the detection of the low intensity of the light scattered by the small tracer particles photomultipliers or avalanche photodiodes are commonly used. Less sensitive sensors can be used for determining the velocity of larger particles or droplets.

The output of the sensor shows the temporal evolution of the scattered light while a particle passes the measurement volume. According to the Doppler effect an observer at the sensor location sees, due to the angle between the two laser beams, a very small difference in the frequencies of the light scattered by the two beams. The oscillating output signal of the sensor is the superposition of the scattered light from both beams.

A more vivid explanation can be given with help of Fig. 4.5. In the cross section of the the two laser beams an interference pattern can be observed on a screen after magnification with an additional lens. The interference fringes are planes of equal intensity, which are perpendicular to the plane of drawing of Fig. 4.5. Particles or droplets passing the measurement volume are illuminated and scatter light, when they pass a bright fringe of the pattern. The result is an oscillating signal, the so-called Doppler signal, which is detected by the sensor. When a particle traverses the measurement volume the amplitude of the oscillations increases first and decreases then according to the Gaussian light intensity distribution of the laser beams.

Knowing the fringe spacing Δs, the velocity component of the particle in the plane of drawing, i.e. perpendicular to the fringes, can be calculated easily from the frequency f_D of the Doppler signal with the relation $v = \Delta s f_D$. The

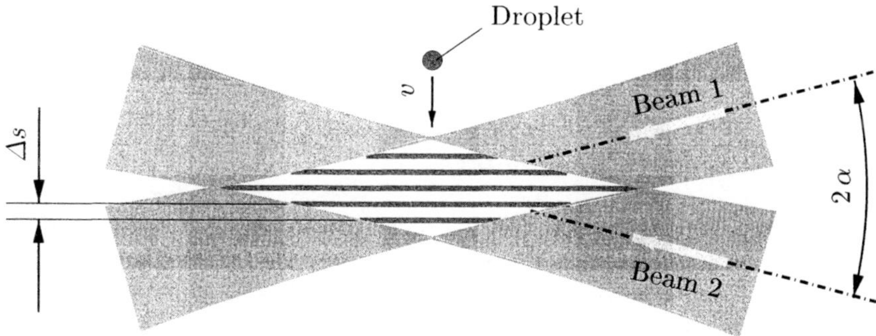

Fig. 4.5. Crossing of two laser beams with angle 2α. Shown is the interference pattern in the cross section schematically with the spacing Δs between the interference fringes

spacing is given by the crossing angle 2α of the laser beams and the wavelength λ of the incident light as $\Delta s = \lambda/(2\sin\alpha)$. Special instrumentation allows to measure droplet velocities in two-phase flows and simultaneously flow velocities with tracers. With modern instruments high data rates with on-line evaluation are possible. In addition instruments are available, which allow to measure other velocity components. Fiber optics allow a miniaturization of the instruments.

4.5.2 Laser-Two-Focus Velocimetry

Laser-two-focus velocimetry (L2F velocimetry) evaluates light scattered by tracer particles or droplets, which are illuminated by laser beams. In this technique two parallel laser beams with a small distance Δs_{2F} are used to generate two pulses of scattered light, when a particle passes both beams. The two laser beams are produced with a beam splitter and a lens. Here the distance between the beam splitter and the lens is exactly the focal length of the lens f_{2F}. This results in two parallel beams, each with a focus in a plane perpendicular to both beams. The location of the foci and the waist of the beams can be calculated with equations given in Ref. [108]. For the spacing between the beams holds $\Delta s_{2F} = 2f_{2F}/\tan(\alpha_{2F}/2)$, where α_{2F} is the angle between the beams after the beam splitter. This instrument determines the velocity component, which is in the plane of the two laser beams and perpendicular to these beams.

The sensors applied are comparable to the sensors used with the LDV technique. If a particle passes one of the laser beams the signal of the sensor starts a clock, which is stopped, when the particle passes the other beam. With the time interval Δt between start and stop signal the velocity component is given as $v_{2F} = s_{2F}/\Delta t$. Similarly as with LDV, velocities at one point of the flow field can be determined, when a laser-two-focus instrument is used. For instationary flows statistical evaluation is appropriate. A detailed description of this instrument has been presented by Schodl [202]. This method is suitable for close-to-wall measurements. Extensions of this technique allow to determine all components of the velocity [203].

4.5.3 Particle-Image Velocimetry

Particle-image velocimetry, or PIV, is based on imaging of solid or liquid particles, which may serve as tracers, when they are small enough. These particles are illuminated in a light sheet obtained with a laser beam and various arrangements of spherical or cylindrical lenses [204]. The components of the flow velocity in the plane illuminated by the light sheet is obtained at one moment. However, the repetition rate is lower than with the LDV or the L2F technique. At least double or even multiple exposures of the particles or droplets are made by using a pulse laser.

The intensity of the scattered light is very low, when small tracer particles are used. In this case a laser of high power will be needed. The frames can be recorded photographically on a film or electronically with a CCD camera. The whole flow field imaged is divided into so-called interrogation areas. For each area a velocity vector is obtained. The interrogation area must be large enough to obtain a significant amount of particle images, but so small, that velocity gradients within the area can be neglected.

The evaluation method depends on the density of particle images on the recordings. For low densities, where individual pairs or traces of particle images can be detected by eye, tracking methods are used for the evaluation. In some cases for each particle imaged a velocity vector can be obtained from the spatial distance between two images of the same particle or droplet and the time interval between the exposures. For medium and high particle image densities optical or digital evaluation techniques are applied to the interrogation area to obtain the distance between the particle images. The optical method was initially based on the evaluation of Young's fringes, which are observed, when the interrogation area on a film is illuminated by a laser beam. In modern instruments the distance between the droplets is obtained digitally by applying correlation techniques to the interrogation area. A practical guide for PIV with a large variety of experimental setups, evaluation methods, and a detailed description of the state of the art may be found for instance in Ref. [204].

Under special conditions the size of droplets may be determined in addition. One method is based on glare points, as described in Sect. 4.6.5. Another method is based on the observation of the far-field of the scattered light, where fringes of the scattering pattern can be seen at each location of an imaged droplet. According to the method described in Sect. 4.6.3, the droplet radius can be determined from the angular spacing of the fringes [205].

Related to PIV is the laser-speckle photography, which allows to obtain for instance density fields in flows. Fomin describes this technique in Ref. [206] in detail.

4.5.4 Young's fringes

The velocity of droplets in equally spaced droplet streams can be determined by evaluating the light diffracted by several droplets. When an ensemble of neighboring droplets inside the stream is illuminated by a laser beam the droplets act as a grid, which produces Young's fringes. From the distance of the fringes the droplet distance can be calculated similarly as with optical evaluation techniques in PIV measurements. The droplet velocity results from the spacing between the droplets and the excitation frequency applied to the droplet stream generator [207].

4.6 Size

4.6.1 Images

An obvious method for determining the droplet size is to take a picture of the droplet using photographic or video techniques, as described in Sects. 4.2 and 4.4. When the magnification M of the imaging system is known the droplet size can be determined from the picture. In order to minimize systematic errors the magnification has to be known very exactly. The magnification can be determined by imaging a microscopic scale or an object of known size. The size of the droplet on the picture has to be measured either manually with a scale or automatically using an image processing system. Semi-automatic image processing systems are also in use. In any case, however, sharpness and contrast of the image enhance the accuracy of the measurement.

With increasing amount of measurements to be performed, the degree of automatization should increase. With image processing systems the contour of the droplet image can be detected automatically. The droplet radius can be determined from the length of the contour line detected or from the area surrounded by the contour line. If the contour line is exactly a circle both methods result in the same droplet radius. Practical images, however, show a little bit wrinkled contour lines, with the result, that a larger droplet radius is derived from the length of the contour line. The ratio of the droplet radius derived with both methods is approximately constant, when the parameters of the image processing system remain the same. This ratio will change, when the shape of the droplets is not spherical anymore. This can be seen from frozen droplets with ice crystals growing on the droplet surface, as shown in Sect. 6.5.

With the computer power of today it is almost impossible or at least very difficult to measure the droplet size on-line. For on-line measurements other techniques like phase Doppler systems are more suitable. From images, however, the shape of the droplets can be determined to some extent. A theoretical description of particle imaging based on the generalized Lorenz-Mie theory and the diffraction theory have been given by Ren et al. [208] together with some experimental results.

4.6.2 Phase Doppler Technique

Phase Doppler velocimetry (PDV) is closely related to laser Doppler velocimetry (LDV)). In addition to the droplet velocity the size of the droplet is determined. In standard phase Doppler systems the illumination is performed in the same way as with LDV. The Doppler signals, however, are detected by at least two sensors. From each sensor a Doppler signal is obtained. The sensors observe the scattered light at the same scattering angle θ, but at least one of the sensors is not located in the observation plane, it is elevated at an angle ϕ with respect to the observation plane. From the Doppler signals the droplet velocity can be derived as described in Sect. 4.5.1. The Doppler signals detected at different elevation angles ϕ show a phase shift with respect to each other, which depends on the droplet velocity and on the droplet radius if the droplet is spherical. This phase shift increases with increasing difference between the elevation angles of the sensors. When this difference is larger than the period length of the frequency of the Doppler signals, the phase shift is ambiguous. In this case a third sensor is needed with a smaller difference of the elevation angle with respect to one sensor. It should be mentioned, that the phase shift may depend on the refractive index and may to some extent show the resonance structure caused by partial wave resonances.

Various configurations of experimental setups are in use. The influence of various droplet parameters like non-sphericity, inhomogeneities or the influence of the droplet path through the measurement volume have been studied. A comprehensive description of phase Doppler systems is given by Albrecht et al. [209]. Descriptions of PDA systems with different properties for various purposes may be found in many relevant scientific journals.

Another instrument with a single detector, which is based on a laser Doppler system has been described by Massoli et al. [210]. In this case the Doppler signal is produced with two crossing laser beams, one with clockwise, and the other one with counter clockwise circular polarization of the light. The size information is obtained from the polarization ratio of the scattered light at scattering angles of $\theta \approx 90°$. This ratio is obtained from the modulation of the Doppler signal. When the size is known, information about the absorption in the droplet can be obtained [211].

4.6.3 Interference Method

In the forward hemisphere the intensity distribution of the light scattered by a droplet shows regular fringes for scattering angles large enough, that diffraction phenomena can be neglected. These fringes can be explained by the interference of rays of order zero with rays of order one, as described in Sect. 1.13.1. The angular distance $\Delta\theta_{sf}$ between neighboring fringes is a measure of the droplet radius r, as described by König et al. [131]. An intensity distribution calculated with the Mie theory, which shows these fringes, is depicted in Fig. 4.6 for the droplet radius $r = 50\,\mu\text{m}$. The angular fringe spacing $\Delta\theta_{sf}$ is indicated. The calculated intensity distribution is shown for scattering angles $40° \leq \theta \leq 43°$. This theoretical intensity distribution can be obtained in experimental investigations with a linear CCD camera, which detects the scattered light in an angular range of 3° around the mean scattering angle θ_m, which is in this case 41.5°.

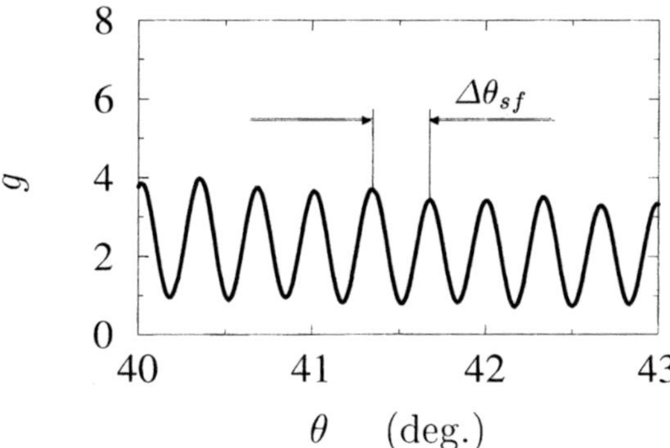

Fig. 4.6. Intensity distribution in the forward hemisphere of a droplet with radius $r = 50\,\mu\text{m}$ and refractive index $n = 1.4$. The angular distance between neighboring regular fringes is denoted by $\Delta\theta_{sf}$

This method for determining the droplet size has the advantage to be independent of the intensity of the incident light. Based on geometrical optics the fringe spacing $\Delta\theta_{sf}$ is obtained as a function of droplet radius r as shown by Anders [212]. The result can be written as

$$\Delta\theta_{sf} = \frac{\lambda}{r\left[cos(\theta_m/2) + \dfrac{n\,sin(\theta_m/2)}{\sqrt{1+n^2-2\,n\,cos(\theta_m/2)}}\right]}\,, \tag{4.4}$$

where n is the refractive index, λ the wavelength of the incident light and θ_m the already mentioned mean scattering angle, which is also called observation angle. The fringe spacing $\Delta\theta_{sf}$ calculated with this formula is shown as a function of droplet radius r in Fig. 4.7. It can be seen, that the method is more sensitive for smaller radii.

The following diagrams, which show the influence of the refractive index n and the observation angle θ_m on the fringe spacing, have been obtained

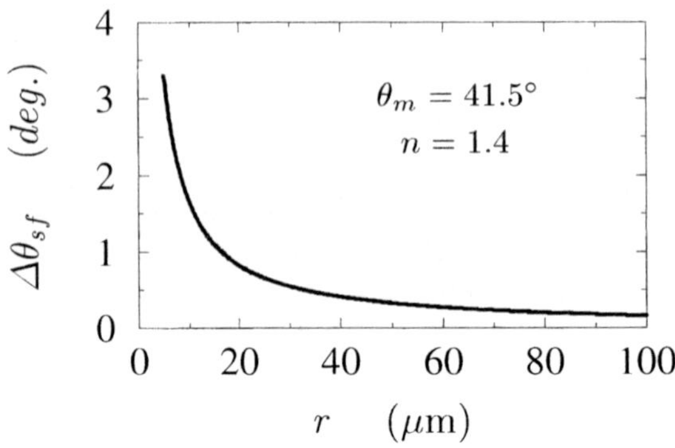

Fig. 4.7. Fringe spacing $\Delta\theta_{sf}$ as a function of droplet radius r

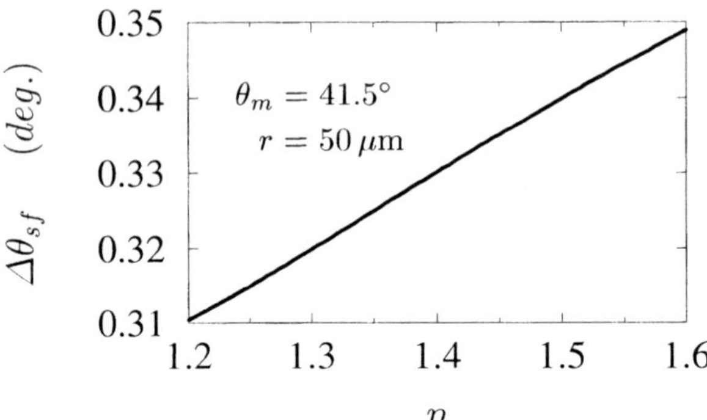

Fig. 4.8. Fringe spacing $\Delta\theta_{sf}$ as a function of refractive index n

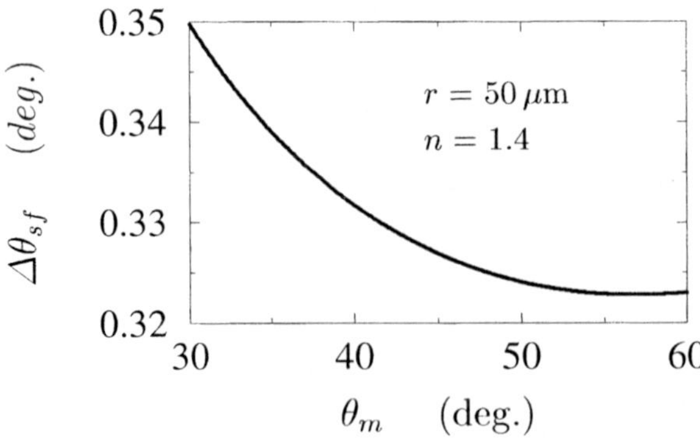

Fig. 4.9. Fringe spacing $\Delta\theta_{sf}$ as a function of observation angle θ_m

with Eq. 4.4. The dependence on the refractive index n, shown in Fig. 4.8, is approximately linear for the plotted range. The influence of the observation angle θ_m shown in Fig. 4.9 decreases with increasing observation angle. The influence of the wavelength is linear as can be seen directly from Eq. 4.4.

In experimental configurations the droplets are illuminated with a monochromatic laser beam polarized perpendicular to the observation plane. The scattered light is in many cases detected with linear or planar CCD cameras. The signals obtained can be evaluated on-line in the case of linear CCD cameras. When planar CCD cameras are used, video tapes are obtained, which can be evaluated after the experiment using image processing utilities.

This size measurement technique is a very suitable tool for investigations on monodisperse droplet streams, because a quasistationary intensity distribution of the scattered light is obtained on the camera. Therefore the detection of the light is easier: longer shutter times of the camera can be used to obtain sufficient intensity. To enhance the contrast the droplet streams can be adjusted conveniently with respect to the measurement volume until optimal results are obtained. More details about configurations or about the influence of Gaussian intensity distribution in the laser beam can be found in the literature [213, 214].

4.6.4 White-Light Method

In this technique nearly white light, for instance of a Xenon lamp, is used for illumination. The scattered light is observed in most cases at a mean scattering angle of $\theta_m = 90°$. The observation is performed with two photomultipliers from opposite sides. With masks in the illumination and in the receiving optics a measurement volume is optically defined. The masks of the two receiving optics have different size, which results in a different size of the measurement volume. A comparison of the signals detected by both receiving optics allows to decide if the droplet is fully illuminated; errors due to border effects can be eliminated. The height of the signal peaks is a measure for the droplet size.

The instrument has to be calibrated. For calibration purposes droplet on demand or droplet stream generators can be applied, which allow to produce droplets of known size. A smooth calibration curve is obtained as the white light consists of a spectrum of wavelengths, which cause an averaging of the ripples and resonances obtained with monochromatic light, when the droplet radius is varied. A detailed description of this method and different experimental setups have been presented by Umhauer [215-217].

4.6.5 Glare Point Method

Glare points, already described in Sect. 1.13.2, allow to determine the droplet size. In the previous section it has been described how the droplet radius can be determined from the interference fringes in the far-field of the scattered light. These fringes are caused by the interference of rays of order zero with rays of order one. The glare points originate also from rays of order zero and order one, but they can only be observed in the near-field of the scattered light. The near-field is obtained by imaging the droplet as described in Sect. 1.13.2. Then the glare points appear as bright spots in the image plane.

From geometrical optics it follows, that the distance between the glare points in the image plane is proportional to the droplet radius. These bright spots can be detected for instance with a CCD camera, the distance between them may be determined with autocorrelation techniques.

The droplet size and velocity of single moving droplets can be measured simultaneously, when a position sensing diode (PSD) is used as sensor. With a PSD sensor the first moment of the light intensity distribution along one extension of the sensor surface is obtained directly as analog output signal. How the sensor has to be orientated and how the resulting output signal has to be evaluated has been described in Ref. [218]. A detailed description of the method including results of measurements has been given by Herrmann [104]. To some extent related to the glare point technique is the so-called pulse displacement technique. This method has been described by Hess and Wood [219, 220].

4.6.6 Rainbow

The optical explanation of the rainbow phenomena has been presented in Sect. 1.13.3. From the intensity distribution of the light scattered in the region of the first rainbow the droplet size can be determined. Details of this method have been described in Ref. [98]. Refinements of the technique and measurements have been presented by van Beeck and Riethmuller [221, 222]. In Fig. 4.10 an example of a calculated intensity distribution in the region of the first rainbow is shown. The calculations are based on Mie theory with plane waves of the incident light. The main maximum and the first super-

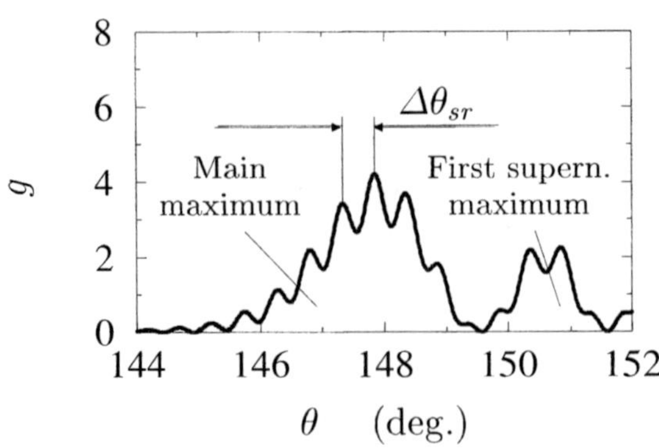

Fig. 4.10. Intensity distribution in the rainbow region of droplet with radius $r = 50\,\mu$m and refractive index $n = 1.4$. The wavelength is $\lambda = 514.5$ nm. The angular spacing between maxima in the ripple structure is denoted by $\Delta\theta_{sr}$

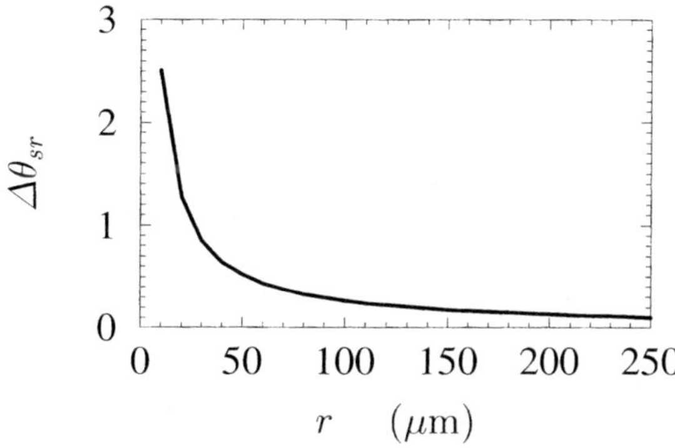

Fig. 4.11. Angular spacing $\Delta\theta_{sr}$ between neighboring ripples as a function of droplet radius r

numerary maximum can be seen both superimposed by a ripple structure. The droplet radius can be determined from the angular distance between neighboring rainbow maxima according to Airy's theory of the rainbow. To determine the angular positions of the rainbow maxima the intensity distribution has to be filtered in order to eliminate the ripple structure. The angular spacing $\Delta\theta_{sr}$ between the ripples, however, is a function of droplet radius [98]. This can be seen from Fig. 4.11. Here the fringe spacing $\Delta\theta_{sr}$ between the ripples is represented as a function of droplet radius r for droplets with the refractive index $n = 1.4$. A strong increase of $\Delta\theta_{sr}$ can be observed for small radii. This results in a higher sensitivity for smaller droplets. The angular spacing between supernumerary rainbow maxima is also a function of droplet size.

A comparison of the droplet size derived from $\Delta\theta_{sm}$ with the value obtained from $\Delta\theta_{sr}$ allows the detection of nonsphericity of droplets. The far-field intensity distribution in the rainbow region can be determined for instance with a linear CCD camera. Another technique uses a pinhole in front of a photomultiplier to obtain the intensity distribution. With this method the droplet size and velocity are obtained.

A comprehensive description of all these methods has been given by van Beeck [100]. From the angular position of the first rainbow the refractive index of the droplet can be obtained, as described by Roth [105]. This technique will be described in Sect. 4.7.4.

4.6.7 Other Techniques

Sampling methods. For determining droplet size or other droplet properties samples are often drawn from the droplet system, which are studied in a separate measurement device or instrument. Such sampling methods are therefore more or less intrusive. Both, the droplet system and the probe may be influenced, when such techniques are employed [223, 224]. Doyle et al. for instance determined the size of droplets by adding a small amount of dye to the test liquid and collecting droplets on a white filter paper target [225]. The size of the obtained spot was taken as measure of droplet size. A sample method to determine the mean droplet diameter has been described already in the introduction of this chapter. Other instruments are for instance cascade impactors or condensation nuclei counters [226].

Laser diffraction spectroscopy. In previous sections different techniques have been described, which allow to determine the size of single droplets. To investigate more complex systems like sprays the whole spray has to be scanned with these methods, in order to obtain for instance the size distribution of the droplets. With such a scanning procedure very detailed information of the spray is obtained. However, this procedure is very time consuming and therefore expensive.

The size distribution of a spray can be obtained from the scattered light in the close forward direction, where the intensity distribution of the scattered light is mainly determined by diffraction phenomena [227]. For this purpose an ensemble of droplets, for instance a spray or a droplet cloud, is illuminated with an expanded laser beam. The scattered light in the close forward direction is detected by a sensor, which is located in the focal plane of a lens. In this case the lens acts as Fourier lens and the far-field of the scattered light is obtained [104]. A monolithic photodiode array with annular ring detector elements of different width has been used to obtain a mean radial intensity distribution of the diffracted light [228]. With the described optical setup the light intensity distribution is obtained for polydisperse droplet systems. The measurement problem is to reconstruct the droplet size distribution from the measured light intensity distribution. This problem has been studied theoretically for instance by Hirleman [228, 229]. The possibility to obtain the size distribution immediately with one shot allows to apply this method for instance in industrial quality tests of spray nozzles in production processes. The performance of ten different laser diffraction instruments has been tested and compared in Ref. [230]. The influence of particle shape on the results has been studied in Ref. [231].

Dispersion Quotient Method. The dispersion quotient method is principally based on the extinction of light of different wavelengths by a particle or droplet cloud. In the experiments two or more monochromatic light beams of different wavelength are superimposed before they pass through exactly the same droplet collective. The light extinction for each wavelength is determined by using interference filters in front of the detectors. With this method the mean particle size is obtained.

If more detailed information is of interest, for instance about the size distribution or the real part of the refractive index, more than two wavelengths have to be used. This method is well-suited for very small particles forming a collective of high particle density. Time resolved measurements of particle size can be performed and a calibration is not needed. More details may be found in Refs. [232, 233].

4.7 Refractive Index

4.7.1 Introducing Remarks

The complex refractive index $N = n + ik$ is one of the most important parameters, which characterize the optical properties of a droplet. Several techniques will be described to determine the real part n of the refractive index in the following, just called refractive index. The refractive index n depends on the droplet liquid as can be seen from Fig. 4.12(a). For this reason a determination of the refractive index allows to distinguish in many cases between different liquids. Measurements of the refractive index allow to determine concentrations in binary systems, since the refractive index depends on the concentration in mixtures of liquids, as can be seen in the example of Fig. 4.12(b).

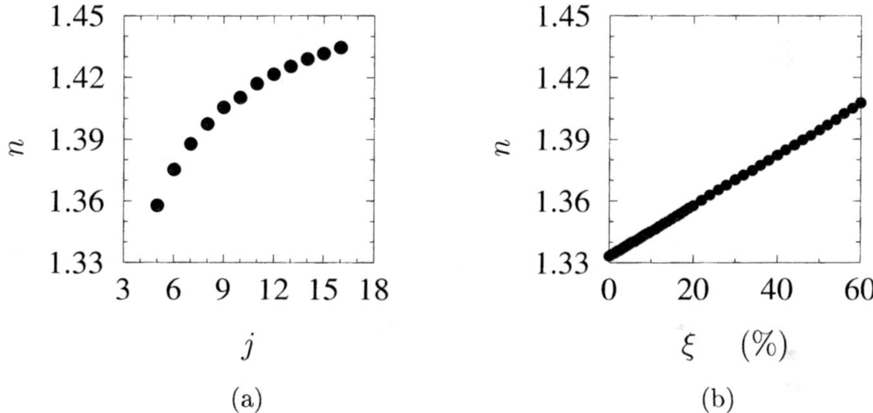

Fig. 4.12. Refractive index n for hydrocarbons with different numbers j of carbon atoms (a) and (b) refractive index of a sulfuric acid-water mixture as a function of the mass fraction ξ of sulfuric acid

An increase of temperature results normally in a decrease of density. There are some exceptions, like water with a maximum density at 4°C. The refractive index is a function of density. The knowledge of the refractive index allows therefore to determine the temperature, as can be seen from Fig. 4.13. Figure 4.13(a) shows that the refractive index of water approaches a maximum for a temperature close above the freezing point, where the density of water has its maximum. A function which describes the the temperature dependence of the density of water may be found in Ref. [20]. The data of Figs. 4.12 and 4.13 are from Ref. [3]. It should be mentioned, that for temperature measurements with an accuracy of a few degrees the refractive index has to be measured with a very high accuracy. This follows from Fig. 4.13.

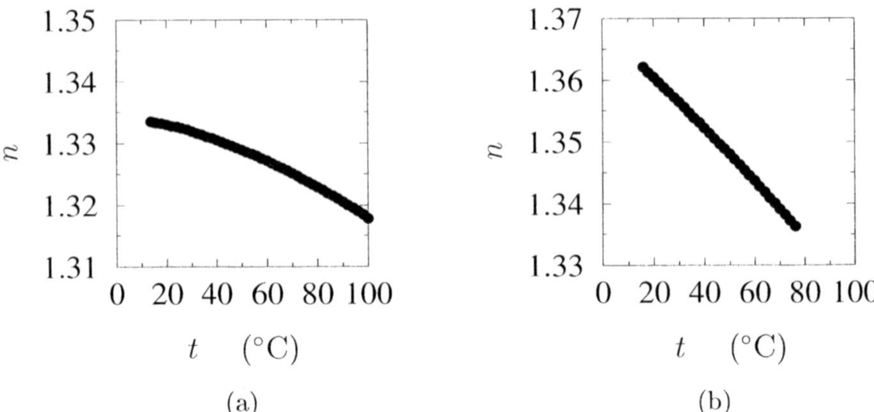

Fig. 4.13. Refractive index n as a function of temperature (a) for water and (b) for ethanol

The techniques described here assume a homogeneous droplet with constant refractive index. The influence of gradients within the droplet on the scattered light and on these techniques in general, have been described by different authors [234-241]. These investigators have also attempted to determine gradients within droplets.

4.7.2 Phase Doppler Based Instruments

This method is mainly based on a phase Doppler system, which allows to determine simultaneously droplet size and velocity as described above. Modifications of the method allow to determine in addition the refractive index n. One system, the so-called extended phase Doppler system, uses two receiving devices, which allow to observe the scattered light for two different scattering angles θ. Observed is in this case light dominated by rays of order one, which is influenced by refraction and hence contains information about the refractive index [242]. From each receiving device two Doppler signals are obtained which allow to calculate a phase difference for each device. The ratio between these two phase differences depends only on the refractive index, but is independent of droplet size. Such an extended phase Doppler instrument has been described in detail by Naqwi et al. [243].

The interaction between polydisperse sprays has been investigated by Brenn et al. using an extended instrument [244]. In these investigations the mixing of two sprays has been studied. The sprays, which consisted of pure water and of a solution of sugar in water, were produced with two separate nozzles. The droplet size, velocity, and refractive index were measured in the mixing region of the spray system. The accuracy and resolution has been studied theoretically and experimentally [245].

Another method to determine the refractive index has been proposed by Onofri et al., the so-called dual burst technique [246]. In this technique only one pair of detectors is used, as for a normal phase Doppler system, but the Doppler signals derived from light dominated by reflection and from light dominated by refraction are evaluated both. Only the refracted light is influenced by the properties of the droplet material, whereas the Doppler signals of the reflected light allow to determine the droplet velocity and size. The signals of the refracted light allow to determine the refractive index. In addition the absorption or the imaginary part k of the refractive index can be measured. More details of the technique and of the configuration of the system are found in Ref. [246].

A method based on an experimental setup mentioned at the end of Sect. 4.6.2 has been described by Massoli et al. [247]. Evaluated is in this method the intensity of the component of the scattered light with horizontal polarization. The circularly polarized laser light used for illumination shows a top-hat intensity distribution. The intensity of the scattered light at a scattering angle of $\theta = 33°$ is a function of droplet size and nearly independent of the refractive index. Oscillations in the angular pattern of the scattered light were smoothed out by using a relatively large aperture of the receiving optics. For determination of the refractive index the light scattered at $\theta = 60°$ is detected. The ratio of the light intensities detected at $\theta = 33°$ and at $\theta = 60°$ is a function of droplet size and refractive index. Hence the size and the refractive index can be determined both.

4.7.3 Glare points

It has been explained in Sects. 1.13.2 and 4.6.5, that glare points can be used to determine the droplet size. The ratio of the brightness of the glare points is a function of the angle of observation and of the refractive index. Schaller et al. presented a technique, which allows to determine the refractive index of spherical droplets by observing the glare points [248]. These authors presented experiments with water droplets and with cylindrical liquid jets having different initial liquid temperatures and hence different refractive indices. Experiments on liquid jets are described in Ref. [249], where the observation angle at which the brightness ratio of the glare points is unity is considered as a function of the refractive index.

4.7.4 Rainbow

Rainbows are observed in the far-field of light scattered by spherical droplets. The scattering pattern can be observed on a screen as described in Sect. 1.13.4 and illustrated by Fig. 1.27. Patterns of the light scattered in the region of the first rainbow are shown in Fig. 4.14 for water and iso-propanol. The main maximum of the first rainbow, the supernumerary maxima, and the ripple

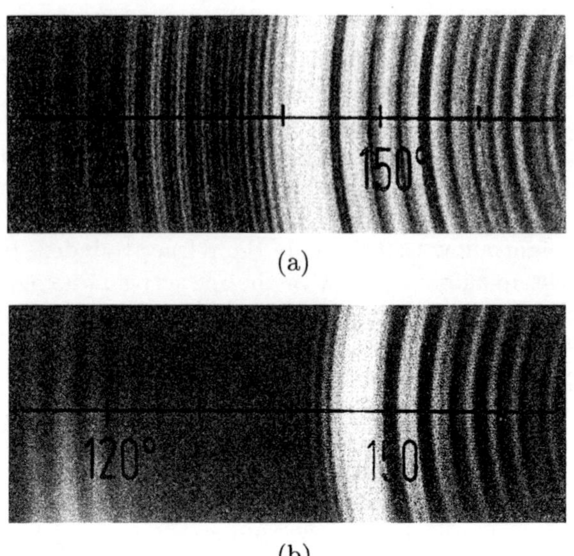

Fig. 4.14. Scattering patterns in the rainbow region (a) for water and (b) for iso-propanol. It can clearly be seen, that for iso-propanol the main maximum of the first rainbow is at a larger scattering angle

structure can easily be identified. The main maximum appears in the case of water at a smaller scattering angle than in the case of iso-propanol. This effect is due to the lower refractive index of water. A theoretical interpretation of rainbow phenomena on the basis of geometrical optics has been given in Sect. 1.13.3. The dependence of the angular position of the first rainbow on the refractive index is given by Eqs. 1.124 and 1.125 according to the theories of Descartes and Airy respectively. Rainbows in nature can be seen clearly at a defined position, though the raindrops are polydisperse. The influence of droplet size on the position of the first rainbow in nature is small as raindrops are rather large. In this case the theory of Descartes is a good approximation.

For smaller droplets the size dependence becomes more and more important, and the theory of Airy has to be used to describe the position of the rainbow. In Fig. 4.15 the positions of the rainbow according to Descartes θ_{rg} and according to Airy θ_{ra} are shown as functions of droplet radius. For increasing droplet radius the rainbow position θ_{ra} approaches θ_{rg} and the influence of droplet size becomes negligible.

In experiments the complex angular distribution of the scattered light in the region of the rainbow is obtained with a linear CCD camera. Such intensity distributions show a ripple structure and the position of the rainbow is in addition influenced by partial wave resonances.

There is no unique way for the definition of the rainbow position. One possibility is to take the position of the absolute maximum of the intensity distribution as rainbow position. Calculations on the basis of Mie theory show, that the absolute maximum is not coincident with the positions of the rainbow according to Descartes θ_{rg} and Airy θ_{ra}.

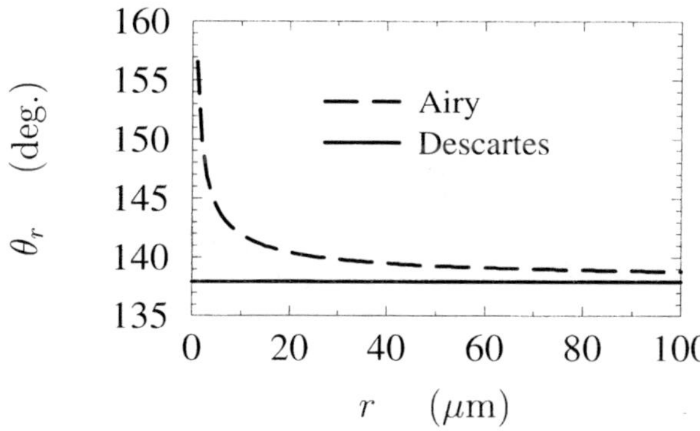

Fig. 4.15. Calculated angular positions of the first rainbow according to the theories of Descartes and Airy

Different techniques, which allow to define a position of the rainbow from calculated or measured intensity distributions have been described by Roth [105]. One possibility is to calculate the correlation of the intensity distribution with a suitable comparison function. The angular position of the maximum of the correlation function defines the rainbow position θ_{rc}. Another technique defines the position of the rainbow on the basis of the first moment of the intensity distribution. The latter technique is important, when the intensity distribution in the rainbow region is determined with a PSD sensor, as described in Sect. 4.6.5. When a digital filter is applied to the intensity distribution in the region of the rainbow, the ripple structure can be eliminated. The result for a droplet with radius $r = 50\,\mu$m and refractive index $n = 1.4$ is shown in Fig. 4.16. Both the position of the maximum θ_{rfm} and the position of the point of inflection θ_{rfw} of the filtered distribution, which are indicated by vertical lines in Fig. 4.16, are two further possibilities to define an angular position of the rainbow.

The dependence of rainbow positions on the refractive index is shown in Fig. 4.17. Results of the theories of Descartes and Airy are compared with

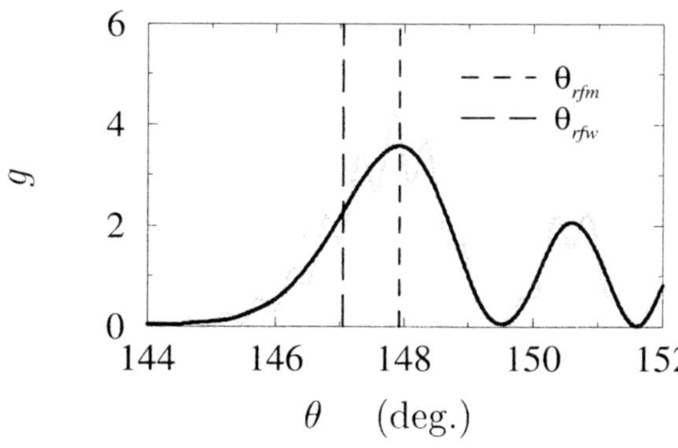

Fig. 4.16. Calculated intensity distribution with corresponding filtered distribution. The filtered distribution shows no ripple structure. The vertical lines indicate the angular positions θ_{rfm} and θ_{rfw}

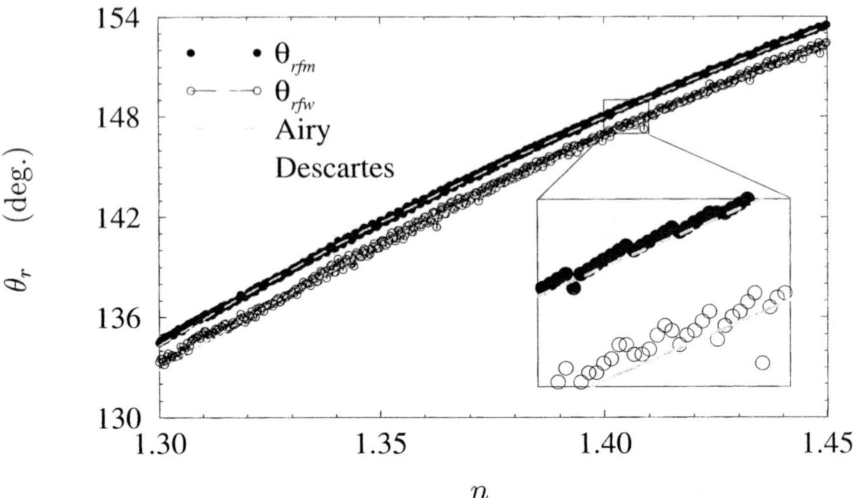

Fig. 4.17. Rainbow positions θ_{rfm} and θ_{rfw} as a function of refractive index. In the enlarged view it can be seen that the values for θ_{rfm} are close to the results for the rainbow position according to the theory of Airy and the values for θ_{rfw} are close to the results for the rainbow position according to the theory of Descartes. The droplet radius was $r = 50\,\mu$m

calculations on the basis of the Mie theory. The results shown for the angular positions θ_{rfm} and θ_{rfw} have been derived from filtered intensity distributions. In the enlarged view it can clearly be seen, that the rainbow position θ_{rfm} is close to the results of Airy's theory and the rainbow position θ_{rfw} is close to the theory of Descartes. The rainbow position θ_{rfm} shows smaller deviations from the corresponding theoretical curve than the rainbow position θ_{rfw}. It should be mentioned, that the rainbow position θ_{rfw} is nearly independent of droplet size in a wide range of radius. Therefore size measurements are not necessary to determine the refractive index. More precise measurements are obtained, however, when θ_{rfm} is used as position of the rainbow. Then the droplet size has to be determined in addition. If the size measurements depend on the refractive index an iterative process leads to correct values for size and refractive index. For an iteration the result obtained with θ_{rfw} may be used as start value for the refractive index.

It depends on the application which position of the rainbow is suitable. Linear CCD cameras can be used as sensors. These cameras have been applied for instance for measurements on droplet streams, as described in Ref. [250], and in combination with a phase Doppler system, as described in Ref. [251]. A method using a photomultiplier is described in Ref. [221]. More details about measurements using light scattered in the rainbow region may be found in Refs. [100, 105].

4.8 Temperature

The methods described above, which are used to determine the refractive index, allow to determine the droplet temperature. The refractive index is a function of density and therefore of droplet temperature. An example of this dependence is depicted in Fig. 4.13 of Sect. 4.7.1. Here in this section additional methods for the determination of droplet temperature are described, namely a technique using thermocouples, a schlieren method, a method based on infrared radiation and the application of thermochromic liquid crystals.

4.8.1 Thermocouples

Johnson studied the surface temperature of water droplets suspended from a small thermocouple in a test chamber [252]. The humidity and the temperature of the surrounding air could be controlled and measured. A relation for the droplet temperature was obtained by equating the expressions for the radial conduction of heat by the air to the droplet and the enthalpy change of the droplet. Hall determined temperature gradients within burning droplets with a diameter of 1 mm using small thermocouples [253].

4.8.2 Schlieren Method

In a study of thermodynamic properties of freely falling water droplets Kinzer and Gunn describe an apparatus especially devised for measuring the droplet temperature [151]. The water droplet, whose temperature is to be measured, falls into a glass cell containing water at a known adjustable temperature. Should the temperature of the droplet be equal to the temperature of the water bath the index of refraction is equal everywhere. When the droplet temperature is different, local gradients of the index of refraction will occur, which are observed using an optical schlieren system. Light from a slit source is transmitted through the transparent glass cell and focused on a wire in front of the objective of a telescope used to view the cell. The cell appears dark when the index of refraction of the water is the same everywhere in the cell. A droplet with a different temperature appears as luminous globule. The temperature of individual droplets is determined by allowing successive droplets to fall into the water bath and adjusting the bath temperature until luminosity effects are no longer visible. Kinzer and Gunn, who studied droplets with diameters ranging from 0.1 to 3.5 mm, report an uncertainty of 0.5 K among different observers. Many measurements of the equilibrium temperature of drops falling in different environments revealed that the equilibrium temperature is equal to the ventilated wet bulb temperature to within ± 0.3 K.

The described method of measurement has been used to measure transient temperatures before thermal equilibrium was established. A water droplet was levitated aerodynamically (see section 3.4.2). At a certain instant after formation the droplet was allowed to fall into the water bath below by shutting off the air stream. By repeating this procedure the temperature of the droplet could be measured as a function of time.

4.8.3 Infrared Method

A method based on the infrared radiation emitted by the droplet has been presented by Naudin for measuring droplet temperatures [254]. The temperature of the droplet surface is obtained by comparing the infrared radiation of the droplet with the radiation from a black body of known temperature located behind the droplets. With a CCD camera, sensitive in the infrared, images the droplets moving in front of the black body are taken. The moving droplets are traced by the system. The evolution of surface temperature of evaporating droplets has been studied. Ganzevles and van der Geld studied thermocapillary convection within droplets resting on a solid surface. In these experiments the temperature at the interface between gas and liquid was determined with an infrared camera [40].

4.8.4 Thermochromic Liquid Crystals

Thermochromic liquid crystals (TLC) change their color when their temperature changes. Dependent on the type of crystals this color-play can be observed in individual temperature ranges. Two forms of TLCs have been used. These are the pure form as neat liquid crystals and the microencapsulated form, where the crystals are encapsulated in thin-walled plastic beads, with diameters ranging from 5 to 3000 μm. In special applications the beads are used as tracer particles. For this purpose beads with neutral buoyancy can be provided. Transient temperature measurements on suspended droplets have been performed with this technique by Richards and Richards [255].

4.8.5 Other Methods

Temperature measurements based on non-elastic light scattering are described in the literature. Vehring and Schweiger studied the cooling of water droplets in the initial unsteady phase of evaporation using a Raman scattering technique [256]. Exciplex fluorescence methods where applied by Seaver and Peele to measure temperatures of acoustically levitated water droplets [257]. Wells and Melton applied exciplex fluorescence thermometry to investigate the heating of decane droplets falling through a hot quiescent oxygen-free environment in a chamber [258].

4.9 Surface Tension

Many different methods for determining the surface tension of liquids are described in the literature [7-9]. It can be distinguished between static and dynamic methods. In both types a solid can be involved in the system or not.

One static method, the pendant drop method, will be described briefly here. A droplet pending at a vertical capillary of a dropper as described in Sect. 2.5 is imaged as shown in Fig. 4.18. The geometrical parameters needed for determining the surface tension are indicated. To measure the surface tension the maximum diameter d_{max} is determined from the image of a pendant droplet as shown in Fig. 4.18. In addition a second diameter d_s is measured at a distance of d_{max} from the tip of the droplet. The gravity forces and forces due to surface tension are balanced. From mechanical equilibrium one obtains for the surface tension

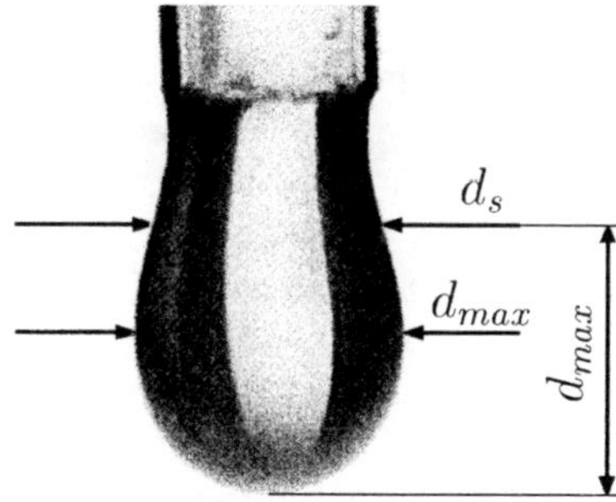

Fig. 4.18. Image of a pendant droplet. The maximum diameter d_{max} and the diameter d_s at the distance d_{max} from the tip of the droplet are indicated

$$\sigma = \frac{\varrho \, g \, d_{max}^2}{H} \, , \tag{4.5}$$

where ϱ is the density of the liquid, g the gravitational acceleration and H a shape factor. The shape factor H is a function of the ratio d_s/d_{max}. This function has been determined experimentally and theoretically. A tabulation of this function may be found for instance in Ref. [7].

A method without interaction of a solid has been described by Tian [259]. In this work the surface tension is determined from the oscillation frequency of acoustically levitated droplets.

4.10 Size Change Rates

The rate dr/dt at which a droplet changes its radius determines the amount of vapor evaporating from or condensing on the droplet per time and surface area. This follows from the relation

$$\frac{dV}{dt} = 4\pi r^2 \frac{dr}{dt} \ . \tag{4.6}$$

The rate dr/dt can be determined by differentiating the temporal evolution of the droplet radius. For determining the change of volume, however, the actual droplet radius has to be known in addition. A large number of size measurements have to be performed when evaporation or condensation processes are studied. Sankar et al. presented direct methods to determine the rate dr/dt instantaneously [260]. These methods are based on the phase shift of the rainbow ripple structure, which is observed, when a droplet changes its size.

Resonances. The method described here uses the change of rainbow position due to partial wave resonances for the measurement of the derivative dr/dt. Partial wave resonances, which have been described briefly in Sect. 1.13.7, depend on the Mie parameter $\alpha = 2\pi r/\lambda$ or for constant wavelength λ on the droplet radius r and the real part of the refractive index n. Patterns of partial wave resonances calculated for different refractive indices and radii are compared with measured resonance structures for instance of evaporating droplets. This rather difficult and time consuming comparison allows to determine size and refractive index of the droplet very precisely [95, 120].

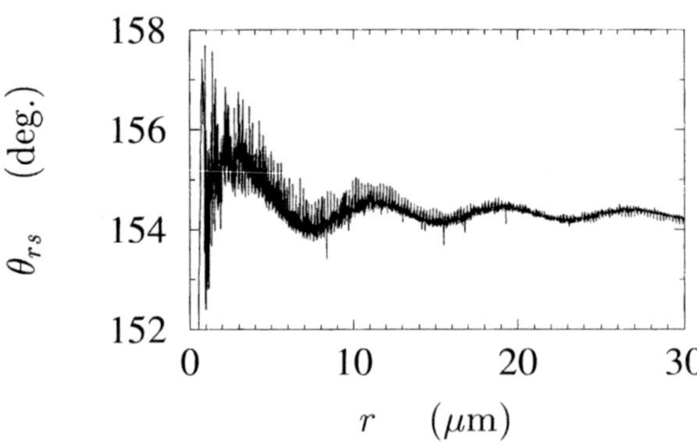

Fig. 4.19. Rainbow position θ_{rs} as a function of droplet radius. The corresponding intensity distributions have been calculated with the Mie theory in a window of $\Delta\theta_w = 17.06°$ for incident light with the wavelength λ = 514.5 nm

In Fig. 1.41 it has been shown, that the angular position of the first rainbow is influenced by partial wave resonances. It has been explained in Sect. 4.6.5, that the analog output signal of this linear sensor is proportional to the position of the first moment of the light intensity distribution along the sensor. A PSD sensor is therefore a very suitable device to detect the rainbow position θ_{rs}. Values of θ_{rs}, which have been calculated with the Mie theory, are shown in Fig. 4.19 as a function of droplet radius for the refractive index $n = 1.43584$. The position θ_{rs} shows long period and very short period oscillations both with decreasing amplitude for increasing droplet radius. In

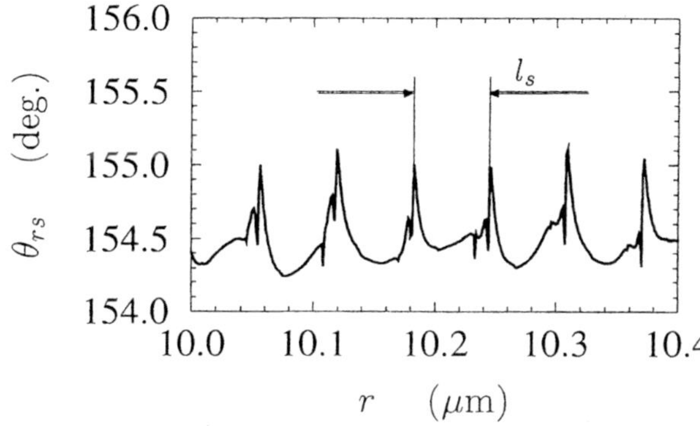

Fig. 4.20. Rainbow position θ_{rs} as a function of droplet radius in an enlarged view. The period length is l_s

the following the oscillations with the very short period, which are caused by partial wave resonances, are of interest. In the enlarged representation of Fig. 4.20 these oscillations can be seen in detail. These oscillations are dominated by one period length l_s. This period length is the distance between two resonances of apparently the same type, which may be called corresponding resonances.

In the oscillations of the rainbow position not all types or orders of resonances can be found; this facilitates the search for corresponding resonances. The period length l_s, which is nearly independent of droplet radius, can be obtained from results shown for instance in Fig. 4.19. For this purpose an algorithm based on a fast-Fourier-transform (FFT) is used.

The dependence of l_s on the droplet radius is obtained, when only the first 2^q data points are used, where $q = 1, 2, 3, \ldots$. The mean period length within the range of 2^q points is obtained from the position of the appropriate maximum of the result of the FFT. This period length is associated with the mean radius in this range. To get the next value of the period length the window of 2^q points is shifted by one or even more points and again a FFT is performed. This procedure is repeated until the whole data is scanned and the evolution of l_s with droplet radius r has been determined. With the data of Fig. 4.19 the variation of l_s with r shown in Fig. 4.21 is obtained. As can be seen, the period length is very small and only minor variations are found for different radii. The distance between resonances has been described theoretically in Ref. [261]. There it has been shown, that the period length or distance between corresponding resonances is a function of the refractive index and the droplet size and is connected with the resonance type.

In practical application, however, only a few types or orders of resonances are found. The influence of the droplet radius can therefore be neglected in most cases. The theoretical period length $l_{s,th}$ is then given approximately by the relation

$$l_{s,th} = \frac{\lambda}{2\pi} \cdot \frac{\tan^{-1} \sqrt{n^2 - 1}}{\sqrt{n^2 - 1}} \, . \tag{4.7}$$

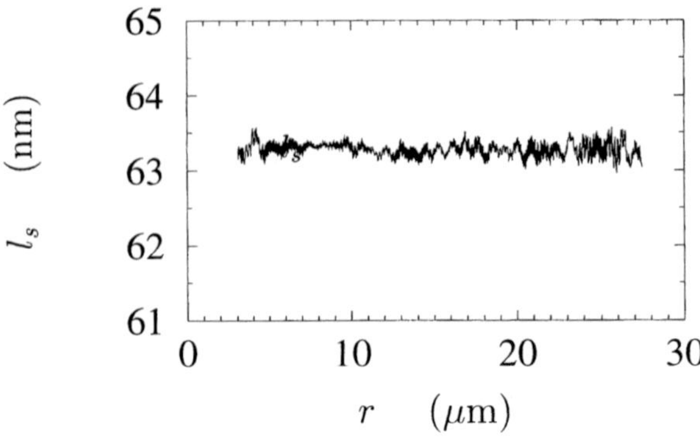

Fig. 4.21. Period length l_s as a function of droplet radius

In Fig. 4.22 the theoretical period length $l_{s,th}$ is shown as a function of the refractive index n [261]. The mean period length obtained from the results of Fig. 4.21, which is indicated by the symbol, is very close to the theoretical result .

In experiments the position θ_{rs} is detected by means of a PSD sensor. An optically levitated droplet may for instance condense or evaporate. Due to the size change of the droplet, oscillations of θ_{rs} are obtained as a function of time. The results of the measurements are evaluated with a FFT using the same procedure as described above. Then one obtains the frequency f_{rs} of these oscillations as a function of time. With the theoretical period length $l_{s,th}$ obtained from Eq. 4.7, which can be interpreted as radius change per oscillation, the size change rate $dr/dt = f_{rs} l_{s,th}$ is obtained as a function of time. Results of such experiments will be presented in Sect. 6.3.1.

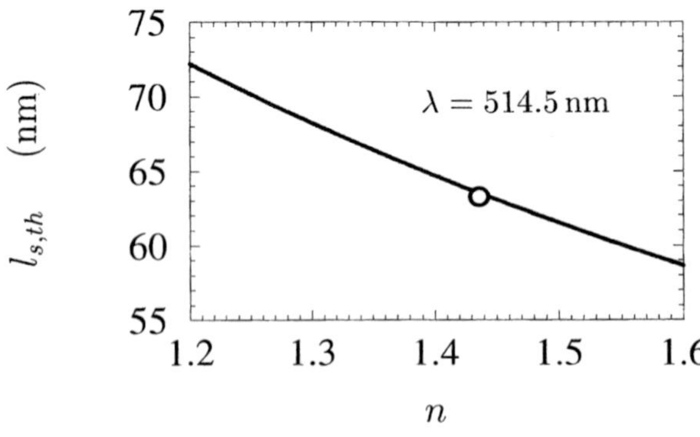

Fig. 4.22. Theoretical period length $l_{s,th}$ as a function of refractive index n. The open circle indicates the mean value of the results shown in Fig. 4.21

5. Experiments to Study Mechanical Interactions

5.1 Introduction

Liquid droplets may experience mechanical interactions with gases, liquids, or solids. Droplet-gas interactions occur in gas flows, sprays, clouds, and in rocket nozzles. Interactions between droplets and liquids are observed as collisions between droplets in sprays, in emulsions, or between droplets and liquid films. Interactions with solid material are observed, when droplets impact on solid walls. The behavior of droplets subjected to an external flow has been studied in numerous theoretical and experimental investigations. Experiments have been performed for different droplet liquids in pipe or duct flows, in jets, in wind tunnels, and shock tubes.

5.2 Droplet-Gas Interaction with surrounding fluids

Flows of a medium consisting of a suspension of liquid droplets in a gas are important for many engineering applications. Droplet-gas interactions through viscous drag and heat transfer produce changes of the velocity and the temperature of the droplets. The mass of the droplets may be changed by mass transfer due to condensation, evaporation or chemical reactions. For the calculation of the motion of the droplets one has to know the forces with which the surrounding fluid acts on the droplets. These forces are usually expressed in terms of the drag coefficient C_D.

The computational simulation of droplet motion in two-phase systems depends essentially on the drag coefficient of liquid drops. For the steady motion of single rigid spherical particles C_D depends only on the Reynolds number. Results of measurements for this case are known as 'standard drag coefficient'.

In many real situations the motion of the particle is not steady. This case has been studied for example by Temkin and Kim and by Temkin and Metha [61, 62]. An example for the discussion of the role of instationary effects, which are described by the so-called Basset force term, is given in Ref. [262]. A few correlations which include the influence of ambient turbulence or mass transfer have been proposed by other authors. Such correlations for the drag of isolated droplets can be used for the simulation of dilute sprays [263, 264].

In dense sprays groups of droplets have to be considered as a whole. A critical discussion of drag coefficients in interacting droplet systems has been given by Poo and Ashgriz [265]. These investigators come to the conclusion, that a single correlation cannot describe the drag coefficient in the whole spray region, and they distinguish three regions with completely different correlations for the drag coefficient. There exists an extensive literature on different aspects of droplet behavior in dense sprays [266] and on bubbly liquids [267, 268]. Other important topics are thermophoretic forces [269, 270], drag in rarefied gases [271], drag of droplets in their own vapor [272, 273], effects of turbulence [274], forces acting on flow tracer particles [275] the drag in droplet streams [147, 276, 277] and the interaction between droplets and shock waves [278, 279].

5.2.1 Wind tunnels for the investigation of droplet systems

Experimental information on the behavior of two-phase systems is very useful for testing the validity of theoretical assumptions or models needed for the numerical simulation. Wind tunnels are important tools, which offer an accurate and economical means for such research. A few typical devices will be mentioned in the following. The test section of a small open circuit wind tunnel used at ITLR for studying the interaction between a fully developed turbulent air flow and droplet streams is shown schematically in Fig. 5.1. A typical velocity profile is depicted in Fig. 5.2. The cross section has the dimension 80 mm × 80 mm, the mean velocity in the test section ranges from approximately 3 to 25 m/s, the Reynolds number based on the width of the cross section is in the range from 15000 to 128000. The walls of the test section were made of acrylic glass, in order to allow easy access for optical observations [280]. Qualitative observations of the droplet streams can be

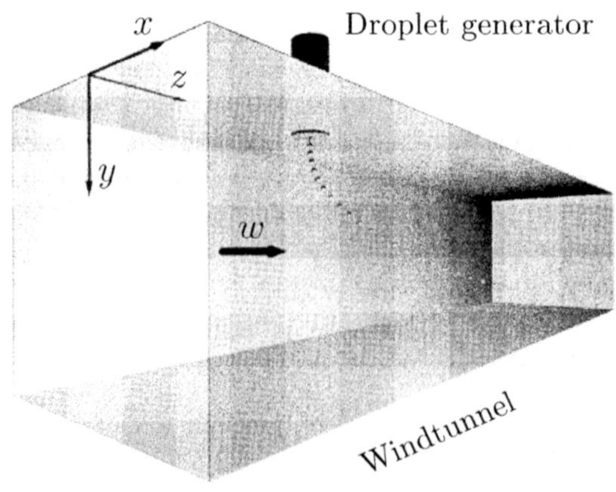

Fig. 5.1. Schematic of wind tunnel test section for the investigation of relaxation effects. Droplets are injected with a droplet stream generator or a droplet on demand generator. The direction of the initial velocity of the droplets is vertical to the axis of the wind tunnel. The velocity profile in the cross section of the droplet generator is shown in Fig. 5.2

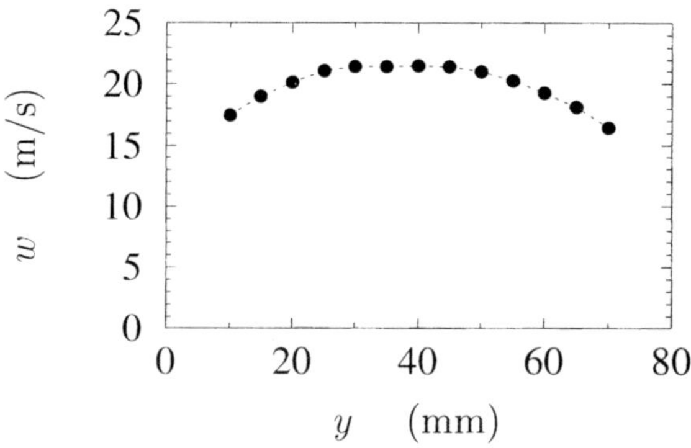

Fig. 5.2. Profile of velocity component in direction of z −axis in cross section of the droplet generator. The cross section of the wind tunnel has the dimensions 80 mm × 80 mm

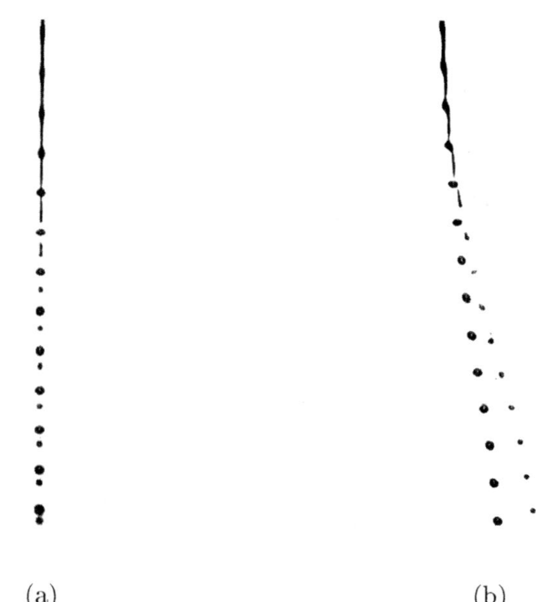

Fig. 5.3. Shadow photograph of droplets injected (a) into still air and (b) into air with velocity distribution according to Fig. 5.2. The stronger deflection of the satellite droplets is clearly visible

made with the naked eye, photographic recording can be achieved by employing a flash light. In Fig. 5.3 a shadow photograph of the injection of a droplet stream with two droplet sizes (a) into still air and (b) into air with a velocity distribution according to Fig. 5.2 is shown. It can clearly be seen, that the smaller satellite droplets are stronger deflected than the main droplets. A similar droplet stream is shown with a smaller magnification in Fig. 5.4. It can be recognized, that both, the smaller and the larger droplets are influenced by the effect of turbulence.

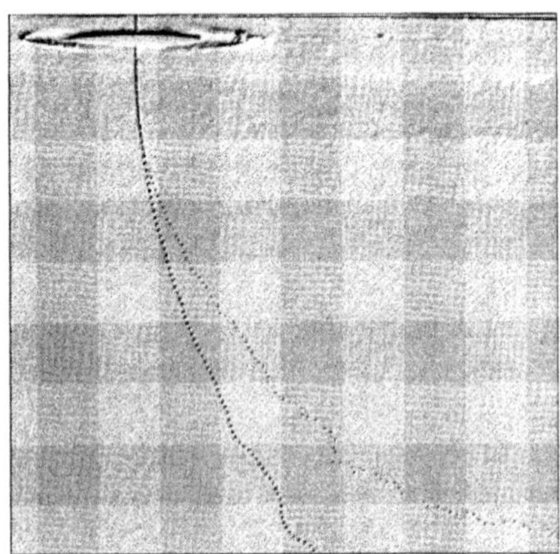

Fig. 5.4. Shadow photograph of injected droplet stream under similar conditions as in Fig. 5.3, but with a smaller magnification. Effects of turbulence on main and satellite droplets are visible

A wind tunnel for aerosol inhalation studies had been designed by Chung et al. [281]. This facility which allows measurements of aerosol inspirability at different wind speeds and directions includes a tunnel, an aerosol generator and a particle injector. The compact open-circuit wind tunnel system has a $1\,\mathrm{m}^2$ working section. The aerosol generator provides particles ranging from 0.2 to $5\,\mu$m. Uniform distributions of the particle concentration were obtained with a four-hole aerosol injector.

The exposure of working persons to aerosols in occupational environments is usually evaluated with personal aerosol samplers which are expected to collect the same particle concentration as the working person inhales. Data obtained with samplers of different design may differ from each other appreciably. The performance of the personal aerosol sampler is of course influenced by the presence of the human body on which it is mounted. Thus it has been assumed that in wind tunnel tests of personal samplers the human body must be simulated with samplers mounted on a full-size manikin. This requirement leads to wind tunnels having large cross-sections. Tests in large wind tunnels are expensive, on the other hand it is difficult to obtain sufficient uniformity of the flow field and aerosol concentration.

Witscher et al. presented a study in which the full-size manikin was replaced by a simplified smaller torso allowing the use of of smaller, less expensive wind tunnels [282]. The new equipment was developed and tested in a closed-loop wind tunnel with a working section 1.22 m high and 1.83 m wide, air velocities between 0.003 and 0.3 m/s, and monodisperse solid particles having a diameter of 70 μm were used to generate the test aerosol. The flow pattern could be visualized with smoke stream lines. The air flow in the test section was seeded with oil droplets having a diameter of 0.7 μm in order to determine the velocities of the air with a laser Doppler anemometer. The preliminary tests reveal that the data obtained with the simplified body were similar to those taken with the full-size manikin.

Ramachandran et al. used scaling relationships and known empirical equations for the design of small-scale samplers which can be tested in small wind tunnels [283]. For examining their theoretical results they used an open-loop wind tunnel with a 0.3 m $\times$ 0.3 m test section. The air velocity ranged from 0.15 to 20 m/s with a background free stream turbulence of 0.25 %. Higher levels of turbulence could be generated by using a lattice-type screen at the entrance of the test section. Velocity profiles in the test section were determined with an anemometer probe. Test aerosols with diameters in the range between 1 and 37 μm were used. The authors come to the conclusion, that aerosol samplers can be tested rapidly and reproducibly in small wind tunnels.

Under certain meteorological conditions clouds may contain liquid water in form of supercooled droplets [284]. In the absence of freezing nuclei the water may be supercooled by tens of centigrades. Supercooled droplets impinging on solid structures such as aircraft will tend to freeze. Ice accumulation on the surface of aircraft can affect the aerodynamic performance of all types of lifting components such as the wings, the horizontal tail and the vertical tail, rotary wings including propellers. Experiments designed to investigate ice accretion on the surface of aircraft are important for the development of ice protection technologies. Different types of ground testing facilities have been developed in order to simulate controlled aircraft icing conditions to investigate different types of ice accretion [285]. This simulation of icing simulation is important for aircraft safety and certification. Flight testing solely under suitable natural icing conditions would be extremely expensive and time consuming.

Different aspects of the complex ice accretion process are simulated in ground facilities. The objective of one type of experiments is to study the flow field and aerodynamic coefficients for iced aircraft components. These experiments which use detailed replicas of ice shapes can be performed with dry air. A second type of experiment uses mixtures of air and water droplets at temperatures above the freezing point in order to study the trajectories of the droplets and their interaction with the aircraft surface. For complete simulation of all phenomena the temperature in the wind tunnel must be below the freezing point. Icing wind tunnels are similar to conventional aerodynamic wind tunnels but include a refrigeration system and a system for injecting water droplets into the air stream.

Special experimental arrangements have been developed for meteorological investigations. Gunn and Kinzer determined the terminal velocities of falling droplets in stagnant air [286]. Vertically mounted hypodermic needles were used to produce droplets of mass from 0.2 to 100000 μg. The velocity was measured by permitting the electrically charged droplets to fall through ring-shaped inducing electrodes. The same authors presented an apparatus for supporting droplets in a stream of upward flowing air confined in a tapered glass tube [151]. Cotton and Gokhale designed a vertical wind tunnel in which water droplets up to 9 mm in diameter could be supported by aerodynamic forces [287]. They used this arrangement to study mechanical interactions between droplets. Various processes occurring in atmospheric clouds have been studied by Mitra et al. in a vertical wind tunnel having a 250 mm × 250 mm test section [288].

Wind tunnels with vertical observation sections allow the levitation of droplets with diameters from $50\mu m$ up to a few millimeters [47, 288]. In this configuration the fall velocity of the suspended droplet can be compensated by the velocity of the air stream. Stability of the levitation is achieved by a slight convergence of the air flow. The particles can be removed and collected by inserting a small obstacle into the test section. The resulting local decrease of the air speed causes the particle to fall.

5.2.2 Interaction with Acoustic Field

In a variety of natural and technical processes droplets are under influence of acoustic fields. Examples are droplets in the Earth's atmosphere, containerless processing, atomization and combustion, as well as acoustic levitation. In all these situations droplets are excited by acoustic forces and will experience forced oscillations, as shown in Fig. 5.5. The droplet response depends on the ratio of liquid and gas density, the ratio of the natural droplet frequency for the second vibration mode, and the frequency of the acoustic field.

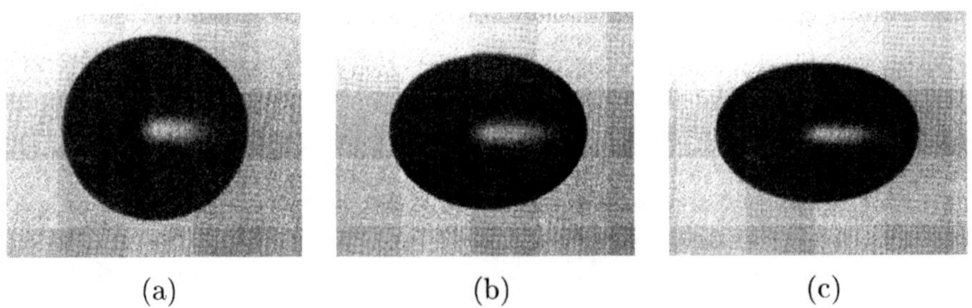

Fig. 5.5. Droplet oscillation forced with frequency $f = 100$ Hz. The neutral shape shown in (b) deviates from spherical due to the acoustic field, the maximum deformations are shown in (a) and (c). The droplet liquid is water, the droplet size approximately 1.5 mm

An approximate theory for the equilibrium shape of levitated droplets has been compared by Trinh and Hsu with experimental results [189]. More recent investigations include larger droplet deformations and comparisons of numerical results with different experimental measurements [289-291]. A brief summary of the literature on the deformation of levitated droplets may be found in Ref. [290].

Resonant shape oscillations have been used to measure the surface tension of the droplet liquid [292]. The oscillations were excited by applying different modulation techniques to the drive voltage of the levitator. Results were obtained for water, hexadecane, silicone oil, and aqueous solutions of glycerine. Apfel used an acoustic standing-wave field in a column of one liquid in order to trap an immiscible droplet of another liquid [293]. Experiments were performed with superheated droplets.

Magill et al. describe an apparatus for laser heating of acoustically levitated spherical samples under high pressure for melting and resolidification. Cronin and Brill used acoustic levitation for obtaining IR-spectra [294]. A survey article on acoustic levitation has been published by Lierke [188].

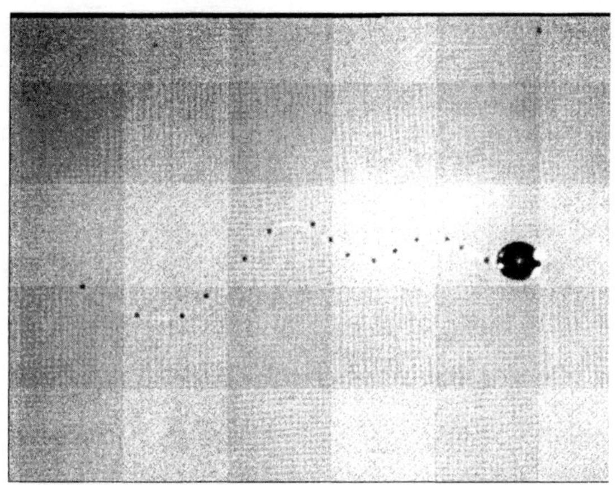

Fig. 5.6. The large droplet, which is levitated acoustically, is fed by a chain of small droplets produced by a droplet on demand generator. The trace of the droplets can be seen due to illumination with a continuous light source and a longer exposure time. The shadows of the droplets of the chain have been obtained by additional illumination with a series of light flashes

An example of coalescing droplets is shown in Fig. 5.6. The large droplet is levitated acoustically. The smaller droplets, which are produced periodically with a droplet on demand generator, are injected horizontally into the space between the two plates of the levitator, where they come into contact with the large droplet and coalesce.

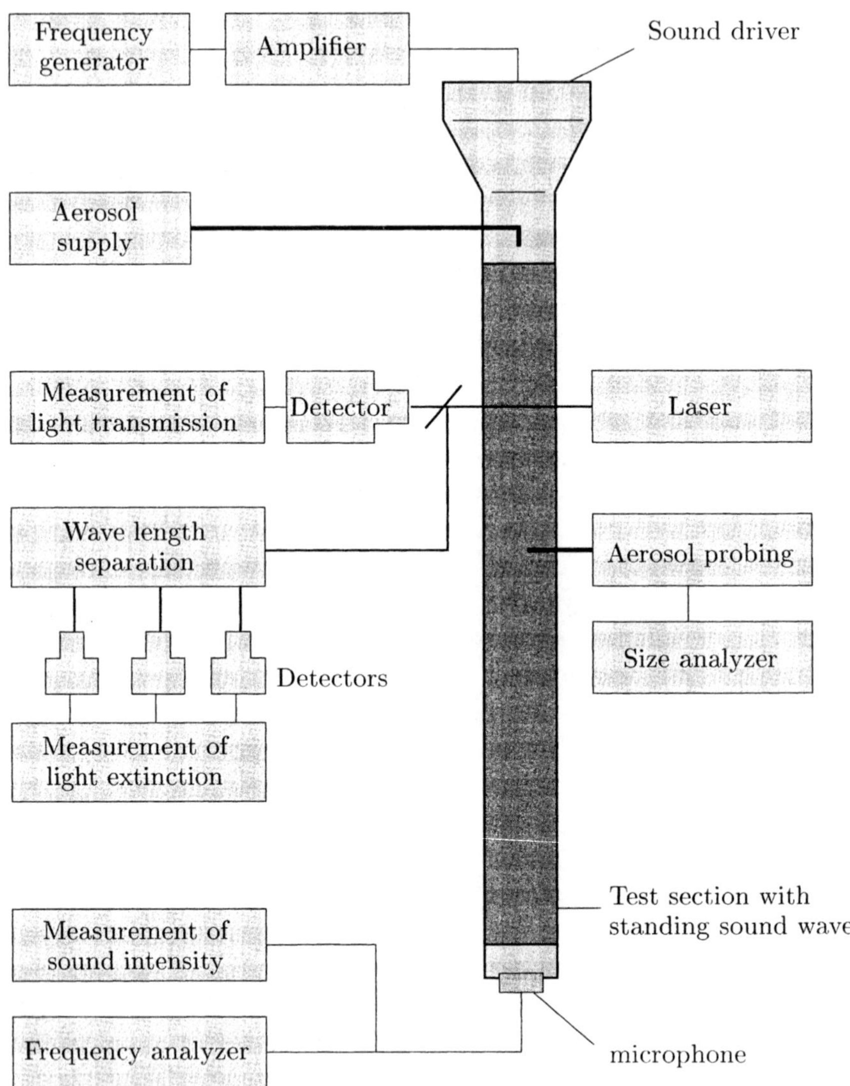

Fig. 5.7. Apparatus used at ITLR for studying the interaction between sprays or aerosols and an acoustic field generated with a loudspeaker at the top of the test section, which is made of acrylic glass, in order to provide optical access. Three measurement systems are available for determining different properties of the dispersed phase

It has been shown in Sect. 3.4.10 that a droplet placed in an acoustic field will experience a force towards the nodes, which according to Eq. 3.17 depends on droplet size. This effect results in relative velocities and encounters between the droplets, which will enhance droplet coalescence [295]. For droplet sizes from 0.1 to 100 μm and sound frequencies below 10 kHz the acting forces result from viscous and pressure effects.

An apparatus for studying the behavior of sprays or aerosols in the presence of an acoustic field is shown schematically in Fig. 5.7 [296]. An acrylic glass tube serves as test section. Sound waves are generated with a loudspeaker at the top of the test section. The intensity of the acoustic field can be determined with a calibrated microphone. Three different optical measurement systems are installed. An extinction technique is used to determine changes of the droplet concentration. Absolute droplet concentrations are obtained with an extinction system based on three different wavelengths. The size distribution is determined with an API-Aerosizer. In this system the droplets are accelerated in a supersonic nozzle. The size of the droplet follows from time-of-flight measurements in the accelerated flow.

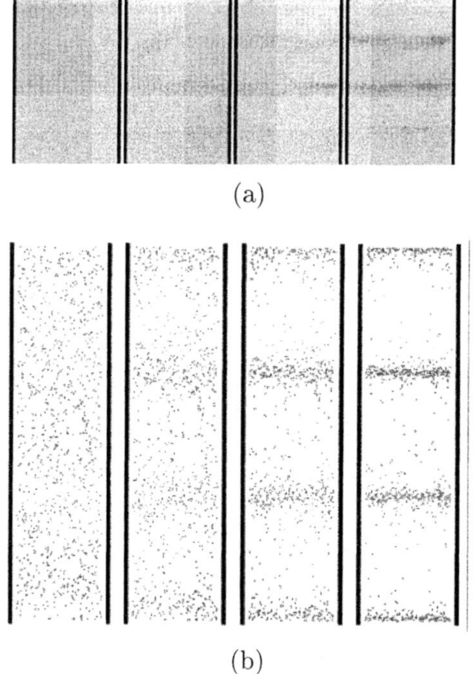

Fig. 5.8. In (a) photographs of the acrylic glass tube presented in Fig. 5.7 are shown. The tube has been filled homogeneously with a fine aerosol consisting of solid particles. The four frames show the process of particle accumulation in the nodes of the applied acoustic field during a time of approximately one second. In (b) a numerical simulation of this experiment is presented (from G. Funcke, ITLR)

The described apparatus has been used to study the evolution of the concentration distribution of an aerosol in the presence of an acoustic field.

The photographs in Fig. 5.8(a) show that the solid particles accumulate in the nodes of the acoustic field. The results of the numerical simulation in Fig. 5.8(b) have been obtained with a Lagrange type description of the particles. The most important forces result from viscous effects and pressure gradients [296].

It is well known, that strong acoustic fields have an effect on the coagulation process in aerosols or fine sprays. The application of this effect for enhancing the separation efficiency of industrial aerosol collection filters has been extensively discussed in the literature [295, 297]. The agglomeration coefficient β of a monodisperse system is defined by the well known equation

$$\frac{dn}{dt} = -\beta n^2 \, , \tag{5.1}$$

where n represents the number of droplets per volume [298]. Numerical and experimental results for the agglomeration coefficient, as represented in Ref. [296], are shown in Fig. 5.9. It can be seen that the agglomeration coefficient has a sharp increase starting at approximately 120 dB. It should be mentioned that the numerical calculation of agglomeration coefficients can easily be extended to the interaction of droplets with different size [296]. For very high sound intensities the sound waves steepen to form shock waves. Drift velocities under such condition have been investigated by Rudinger et al. [299].

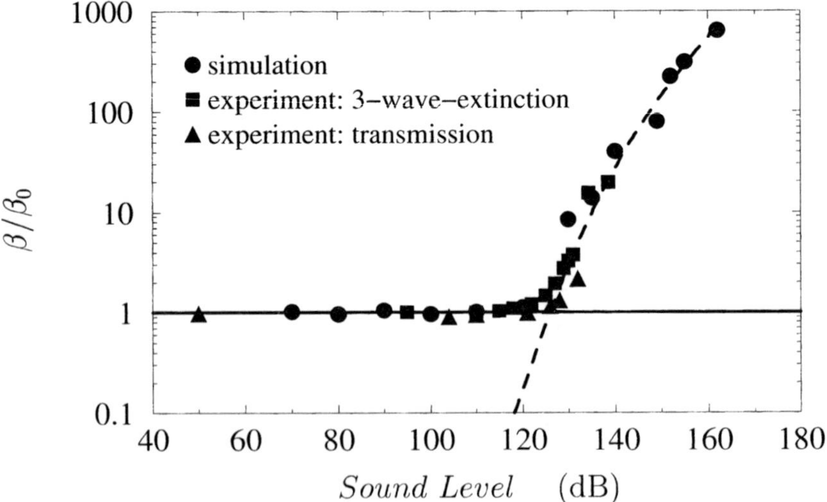

Fig. 5.9. Dimensionless form of the agglomeration coefficient as a function of the sound level (from G. Funcke, ITLR)

5.2.3 Shock Tube Experiments

Shock tubes are used to produce a short duration gas flow of high velocity and temperature. The so-called gasdynamic shock tube consists of a tube, which is separated into two sections by a diaphragm. The initial pressure distribution along the tube is shown in Fig. 5.10(a). By increasing the pressure p_4 in the high pressure section the diaphragm is caused to rupture. The subsequent motion of the gas in the tube is explained in Fig. 5.10(b). After rupture of the diaphragm a shock wave moves with supersonic velocity v_s from left to right. The test time Δt is the time between arrival of the shock wave and arrival of the interface, which separates driver gas and driven gas. The velocity of the interface v_2 is equal to the velocity of the test gas between shock wave and the interface. The flow Mach number of the test gas is defined as $Ma_2 = v_2/c_2$, where c_2 is the sound velocity of the shocked gas.

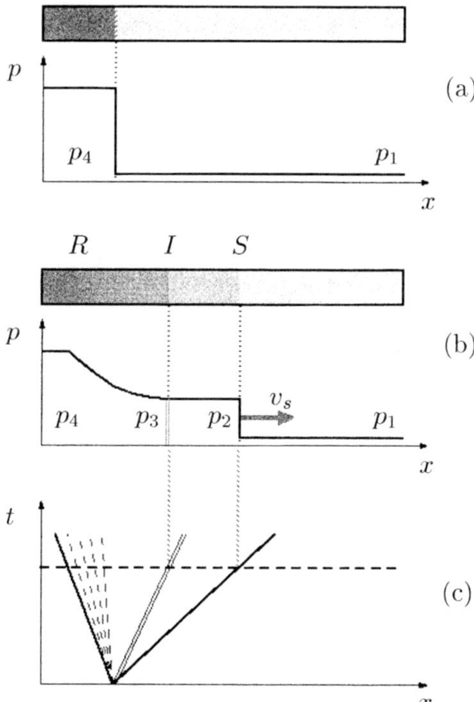

Fig. 5.10. Pressure distribution in shock tube (a) before and (b) after rupture of the diaphragm. Typical values of the test gas pressure p_1 in the driven section range from a few millibar to a few bar. The ratio of the driver pressure and the test gas pressure p_4/p_1 is usually large. The wave pattern after rupture of the diaphragm (c) consists essentially of a shock S traveling with velocity v_S into the low pressure section and a rarefaction fan R moving into the high pressure section. The boundary conditions at the interface I between driver and driven gas are $p_3 = p_2$ and $v_3 = v_2$

The shock tube is a very useful instrument for producing sudden changes of the velocity, pressure, density and temperature of a gas. Shock tubes provide controlled transient flows with uniform velocity and thermodynamic state for test times Δt which are typically a few milliseconds. With stronger shock waves and at lower initial pressures the test times become usually shorter. Shock tube experiments have the advantage that the gas velocity and the thermodynamic state can be chosen conveniently in a wide range. Different types of shock tubes and instrumentation of shock tubes have been described in monographs [300-303] and in many scientific journals. For the investigation of aerosols, dispersions, or sprays the shock tube has the advantage that only small amounts of material are needed. This characteristic of shock tubes may be essential for investigations with particles, which are expensive or consist of poisonous material. Shock waves and relaxation phenomena in gas flows carrying small liquid or solid particles have been described in the literature [66, 304].

The key parameter for describing the thermodynamic state behind the shock is the shock Mach number, which is defined by

$$Ma_s = \frac{v_s}{c_1} , \tag{5.2}$$

where c_1 is the velocity of sound in the undisturbed gas ahead of the shock wave. For ideal gases the pressure and density ratio across the shock wave are given by the relations

$$\frac{p_2}{p_1} = \frac{2\kappa Ma_s^2 - (\kappa - 1)}{\kappa + 1} , \qquad \frac{\varrho_2}{\varrho_1} = \frac{(\kappa + 1)Ma_s^2}{(\kappa - 1)Ma_s^2 + 2} , \tag{5.3}$$

where κ is the ratio of the specific heats. The temperature ratio T_2/T_1 is obtained from the equation of state $(p_2/p_1) = (\varrho_2/\varrho_1)(T_2/T_1)$. For the flow Mach number of the test gas one has

$$Ma_2^2 = \frac{(\kappa - 1)(Ma_s^2 - 1) + (\kappa + 1)}{2\kappa(Ma_s^2 - 1) + (\kappa + 1)} . \tag{5.4}$$

Equations 5.2-5.4 describe the change of the thermodynamic state produced by the shock wave, when the effect of the particle on the gas flow is negligible, as in investigations of single droplets, droplet streams, or sprays with very low droplet concentrations.

When a shock wave travels through a spray with higher droplet concentration, the gas will experience a rapid change of the pressure, density, temperature, and velocity according to Eqs. 5.2-5.4. The velocity and the temperature of the droplet cannot follow instantaneously these changes. By transfer of momentum and heat between droplet and gas the final equilibrium is established in a zone of finite thickness, the so-called relaxation zone. The corresponding time is of the order of the relaxation times τ_v and τ_T defined in Sect. 1.9. This final equilibrium state of the droplet-gas mixture can be obtained from Eqs. 5.2-5.4, when c_1 and κ are replaced by the corresponding properties of the mixture [66].

Shock tubes have been used to study for example drag coefficients and fragmentation of droplets, as well as evaporation, ignition, and ignition delay.

5.2.4 Droplet Deformation

In computational simulations of droplets dispersed in a gaseous or liquid phase it is often assumed that the droplets maintain spherical shapes. Experimental and theoretical investigations indicate, that the motion of droplets in liquid or gaseous media is often associated with complicated deformations of the droplet shape. Effects of droplet deformation on mass, momentum, and heat transfer rates between droplets and surrounding fluid are important for many practical applications. Different modes of droplet behavior have been discussed by Valentine et al. [305] for liquid-liquid systems. Gelbard et al. simulated hydrodynamic effects on binary collisions of droplets in experiments with solid spheres settling in a viscous liquid. The interaction between two spheres was detected by recording the acoustic signals produced by the impact [306]. Numerical simulations have been used to predict transient shapes of droplets dispersed in a gaseous medium [307, 308]. The circulation and the thermodynamic state within the droplet can be predicted with such models.

5.2.5 Droplet Fragmentation by Shock Waves

The fragmentation of liquid droplets, resulting from their sudden exposure to a high-velocity gas stream, has many important applications in the fields of aerodynamics and propulsion. For example the phenomenon of supersonic rain erosion, which is caused by the impingement of rain droplets at high relative speeds on exterior aircraft surfaces, can greatly be alleviated through proper aerodynamic design. A reduction in the damage sustained from impacting droplets is achieved by designing a body whose detached shock is sufficiently far removed to allow for drop shattering in the region separating the shock from the body surface [309].

With regard to propulsion, the rate of mixing can be greatly enhanced by virtue of the fragmentation process of the fuel droplets. As a result of droplet breakup, burning rates are obtainable which are higher than those possible under the conditions of low velocity and forced convection, where no disintegration occurs. The phenomenon of two-phase detonation is pertinent to liquid-propellant rocket motor combustion instability [310].

Droplet breakup in high speed gas flows plays an important role in different engineering applications, which include supersonic flight in rain, instabilities in rocket engine combustion, propagation of detonation waves in two-phase systems or atomization of liquid metals. For these applications it is important to know the breakup time and breakup distance [311, 312].

The fragmentation of water droplets caused by sudden exposure to a high velocity air stream has important applications in the field of missile and heat shield design. Rain erosion by the impinging of rain drops on the exterior surfaces of missiles causes severe damage to ablative heat shields and may result in structural failure of the missile. The effects of erosion damage can perhaps be minimized by judicious selection of the heat shield material to

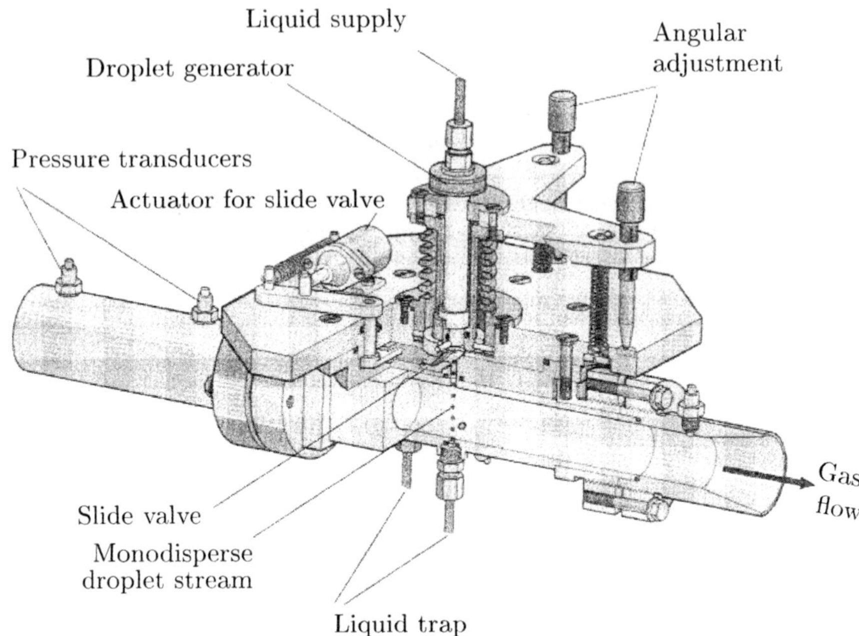

Fig. 5.11. Setup for injection of monodisperse droplet stream into test section of shock tube. After passage of the shock front the droplet stream is exposed to an external flow with velocity v_2 and temperature T_2. The resulting flow Mach number is Ma_2. A shadowgram of the resulting flow is shown in Fig. 5.12

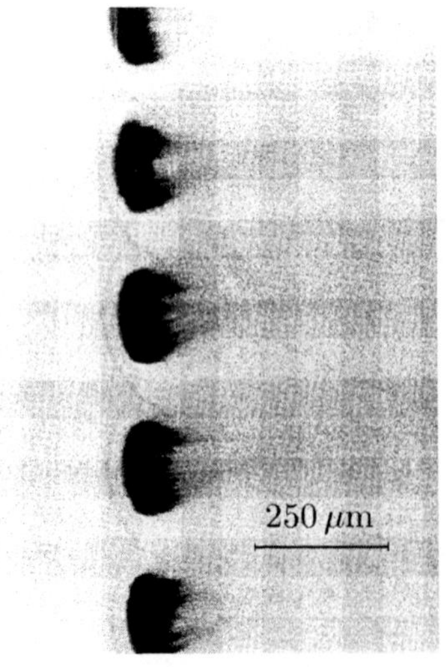

Fig. 5.12. Shadowgram of iso-propanol droplets with the initial diameter $130\,\mu$m taken $2.9\,\mu$s after passage of shock front with shock Mach number $Ma_s = 4.0$ in air with the initial pressure 70 Torr. The velocity of the shock front was $v_s = 1360\,\text{m/s}$, the conditions behind the shock $v_2 = 1060\,\text{m/s}$, $T_2 = 1150\,\text{K}$, $p_2 = 1.7\,\text{bar}$, and $Ma_2 = 1.65$. The droplet Reynolds number was $Re = 2000$ and the Weber number $We = 4000$. A system of detached shock waves in front of the deformed droplets is clearly visible. The photo is an enlargement of frame (e) in Fig. 5.13 (from W. Klenk and N. Widdecke, ITLR)

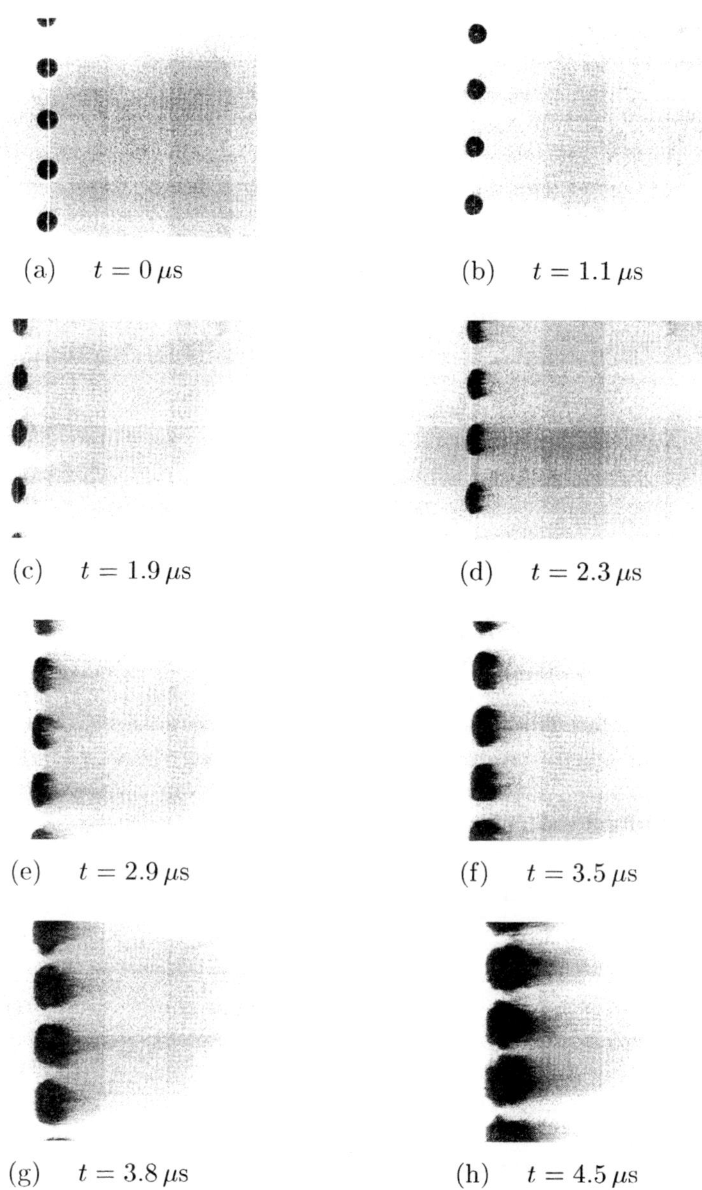

Fig. 5.13. Series of shadowgrams taken at different times after passage of the shock front. The droplet and flow parameters are the same as in Fig. 5.12 (from W. Klenk and N. Widdecke, ITLR)

withstand drop impact or by suitable design of the missile. In this latter approach, the missile nose cap may be designed to generate a shock wave of sufficient strength to cause drop fragmentation before impingement upon the heat shield. Either of these approaches for minimizing rain erosion damage requires knowledge of rain drop mass and shape as a function of time [312].

Different researchers have used shock tubes to study the interaction between shock waves and droplets [309-313]. Some results for the interaction of shock waves and droplets obtained by Widdecke and Klenk will be presented in the following [313, 314]. The setup used by these authors for generating the droplets and injecting them into the test section of the shock tube is shown in Fig. 5.11. The droplets enter the test section through a small aperture in the shock tube wall. Before an experiment this aperture is blocked with a special slide valve in order to avoid accumulation of the droplet liquid and its evaporation in the test section.

The droplets are observed with a microprojection technique. The inner surfaces of the glass windows used for the observation technique have the same curvature as the shock tube. The optical distortion caused by the cylindrical shape of the windows was compensated with an additional specially designed cylindrical lens. The light source was a short duration flash triggered by the approaching shock. One photograph is obtained in each experiment.

A series of shadowgrams taken at different times after passage of the shock wave are shown in Fig. 5.13. Just after passage of the shock wave at $t = 0$ the droplets are not yet deformed. The white line on the photos represents the initial position of the droplet centers. After a few microseconds the droplets are deformed and begin to move downstream. Small particles are stripped off and micromist forms. The breakup of droplets has been compared with the breakup of liquid jets [315].

5.2.6 Ignition Delay Times of Droplets in Shock Tubes

Kauffman and Nicholls have shown in an extensive experimental study that fuel droplets surrounded by an oxidizing atmosphere in the low pressure section of a shock tube may be ignited by the incident shock wave [316]. The experimental conditions involved droplets having diameters between approximately 1 and 2 mm and shock Mach numbers between $Ma_s = 3$ and $Ma_s = 5$. The presence of combustion can be recognized by a region of luminosity behind the incident shock wave in the wake of the fuel droplet. The measured ignition delay time apparently includes the time needed to remove material from the droplet and the evaporation delay time.

Droplets can be ignited in shock tubes by the incident or the reflected shock wave. Different shock tube investigations of the ignition of single droplets, droplet arrays and of sprays have been presented in the literature. Advantages and disadvantages of different techniques for measuring ignition delay times have been discussed by Miyasaka and Mizutani [317]. These authors used an ultrasonic atomizer for introducing a droplet cloud into the low pressure section of a conventional pressure driven shock tube. In these experiments the light emitted after ignition of the droplets was detected with photo-transistor devices. The ignition delay measured in these experiments was only a fraction of the values obtained with conventional techniques for example in Diesel engines or electrical furnaces.

5.3 Droplet Interaction with Solid and Liquid Surfaces

Many applications of droplet systems such as spray painting, injection of droplets into combustion chambers, impact of cloud droplets on airplane wings, ink-jet printing, soil erosion by rain involve motion of droplets towards solid walls. Impaction, collection, and deposition of liquid and solid particles in two-phase flows have been discussed for different body shapes in the literature [318, 319]. The rate of mass, momentum and heat transfer has been studied for different targets and flow regimes. Review articles describing fluid dynamic phenomena of droplet impact on cold solid and liquid surfaces have been published recently by Rein [320, 321]. In many other technical applications the droplet impact on heated surfaces is of interest. Examples are spray cooling of metals in steel industry, fire suppression with sprinkler systems, vaporization of fuel droplets in combustion chambers. When the wall temperature is above the Leidenfrost temperature a vapor layer exist between droplet and wall.

Table 5.1. Definition of important dimensionless parameters. The quantities l, h, d represent different characteristic lengths, f a characteristic frequency, η the viscosity, and g the gravitational acceleration

$We \equiv \frac{\varrho v^2 l}{\sigma}$	$Re \equiv \frac{\varrho v l}{\eta}$
$Bo \equiv \frac{\varrho g h^2}{\sigma}$	$Oh \equiv \frac{\eta}{\varrho \sigma l}$
$Fr \equiv \frac{v^2 d}{g}$	$Ca \equiv \frac{\eta v}{\sigma}$
$St \equiv \frac{\eta v}{\varrho g l^2}$	$Str \equiv \frac{f l}{v}$

Table 5.2. Relations between dimensionless parameters

$Ca = \frac{We}{Re}$	$Oh = \frac{\sqrt{We}}{Re}$
$Bo = \frac{We}{Fr}$	$St = \frac{Fr}{Re}$

The fluid dynamic phenomena occurring when droplets impinge on solid or liquid surfaces are very complicated and depend on many different parameters such as surface tension, viscosity, density and temperature of the droplet liquid, diameter, impact angle and velocity of the droplet, surface roughness and temperature, physical and chemical properties of the wall. For the impact on liquid films the film thickness is an important parameter. Effects of capillary waves and shock waves, electrical charges, deformations of the impinging droplets due to oscillations will not be considered in the following. The different phenomena are usually described with nondimensional numbers. In many situations the Weber number We, the Reynolds number Re, the Ohnesorge number Oh, the Bond number Bo, the Froude number Fr, the capillary number Ca, the Stokes number St, and the Strouhal number Str are the most important dimensionless parameters.

5.3.1 Cold Solid Surfaces

When a liquid droplet with negligible kinetic energy comes in contact with a flat solid wall the free energy of the system will show a tendency to establish a minimum. Two distinct phenomena may be observed, when the droplet has attained its equilibrium state, partial wetting with a finite contact angle ($\theta > 0$) or complete wetting with vanishing contact angle ($\theta \to 0$). The corresponding shapes of the liquid are illustrated in Fig. 5.14. Mechanical equilibrium is described by Eq. 1.10 in the form

$$\sigma_{13} - \sigma_{23} - \sigma_{12} \cos\theta = 0 \ . \tag{5.5}$$

The contact angle θ is obviously defined in terms of the interface tensions σ_{13}, σ_{23}, and σ_{12}. The quantity σ_{12} can be determined in a system containing only the liquid and its vapor. The contact angle θ depends therefore on the difference $\sigma_{13} - \sigma_{23}$. One has to distinguish advancing and receding contact angles [322]. The influence of the curvature of the interface between liquid and vapor near the contact has been discussed in Ref. [323]. Complete wetting occurs when $\theta \to 0$ or $\cos\theta \to 1$. In this special case Eq. 1.10 has the form

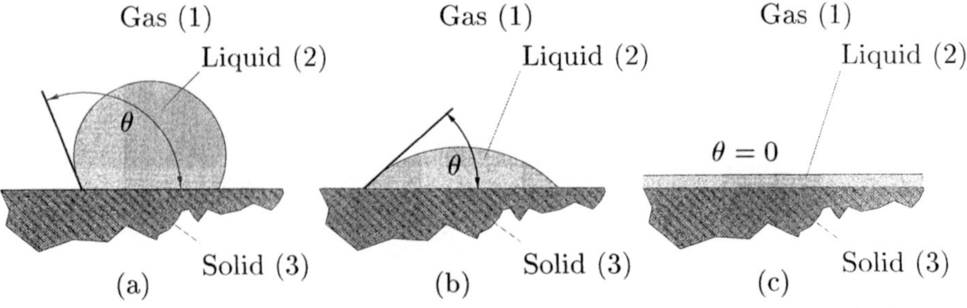

Fig. 5.14. Droplet on flat horizontal surface with partial wetting (a), (b), and complete wetting (c). In the case of complete wetting one has $\theta \to 0$

$$\sigma_{13} - \sigma_{23} - \sigma_{12} = 0 \ . \tag{5.6}$$

In nonequilibrium situations Eq. 5.6 does not hold. As a measure of spreadability the spreading coefficient

$$S = \sigma_{13} - \sigma_{23} - \sigma_{12} \tag{5.7}$$

has been introduced. It has been emphasized in the literature that contaminants may have dramatic effects on contact angles and wetting [8, 323].

Complicated dynamic phenomena are observed when droplets impact with finite velocity on solid surfaces [324]. Immediately after the impact a thin film forms on the wall around the droplet. This film has of course a circular shape when the preimpact velocity is vertical to the wall. During the radial expansion of the film liquid propagates outwards in radial direction and mass is accumulated at the periphery forming a liquid ring. Due to the viscous flow in the thin film kinetic energy will be dissipated. When the film has reached its maximum radius recoiling starts. The recoil of the liquid may lead to a separation of the droplet from the wall. It should be mentioned that the dynamic contact angle must change from advancing to receding.

The described flow has been treated numerically in Ref. [322] and compared with experiments. The disturbances occurring usually along the rim of the film can easily be observed when ink from a fountain pen is dropped on a sheet of paper. The described deformation may lead to bouncing or regular reflection of the droplets. Review articles discussing all these phenomena have been published by Rein [320, 321].

At higher impact velocities splashing occurs. The threshold for splashing depends on the Weber number and the properties of the wall, such as roughness. Different correlations describing the threshold have been proposed [320].

5.3.2 Cold Liquid Surfaces

Some authors have studied the droplet impact on solid walls which were covered with a thin liquid film [325]. Levin and Hobbs come to the conclusion that there is little difference between the phenomena observed with water droplets splashing into thin water layers and splashing on a solid surface [326]. At small Weber numbers coalescence or floating on the liquid surface is observed, at higher Weber numbers bouncing. Splashing droplets produce a crater and a crown forms. Some numerical results for splashing droplets will be presented in Sect. 5.3.5.

5.3.3 Heated Surfaces

The evaporation of liquid droplets on heated surfaces is of great practical interest for many industrial applications, such as fuel evaporation in internal combustion engines, spray cooling in steel industries, sprinkler systems for fire suppression, cooling of electronic components and emergency spray cooling of nuclear reactors.

The different regimes of evaporation and boiling are well known from pool boiling [327]. The heat flux characteristics, which depend essentially on the degree of superheat provided by the heated wall, are depicted in Fig. 5.15. At small values of T_w the liquid is heated by natural convection. When nucleate boiling starts, the boiling curve becomes steeper until the heat flux goes through a maximum. Between the maximum and minimum of the heat flux boiling becomes instable.

The thermodynamic phenomena occurring during the interaction of liquid droplets with high temperature solid walls has been studied in a great number of theoretical and experimental investigations [268]. Most of these studies have been conducted with water as droplet fluid at temperatures around or much higher than the Leidenfrost temperature.

Michiyoshi and Makino studied experimentally the evaporation characteristics of small water droplets on heated plates of various materials at surface temperatures between 80°C and 450°C using small thermocouples to determine the surface temperature of the wall as a function of time [328]. These authors determined the so-called boiling curve for a single water droplet. In the nucleate boiling regime the heat flux is found to be independent of the wall material whereas in the transition regime the heat flux depends on properties of the wall.

The so-called minimum temperature T_{min} belongs to the minimum of the heat flux on the well known pool-boiling curve in Fig. 5.15. Nucleate or transition boiling prevails, when the wall temperature is below the minimum temperature. Film boiling will occur, when the wall temperature is greater than the minimum temperature.

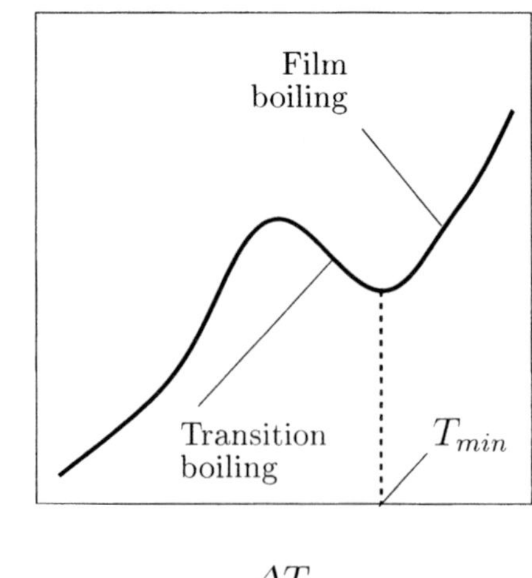

Fig. 5.15. Schematic representation of the well known boiling regimes for pool boiling. The heat flux $\dot{q}$ is depicted as a function of $\Delta T = T_w - T_s$ where T_w represents the wall temperature and T_s the saturation temperature of the liquid. The minimum of the heat flux corresponds to the Leidenfrost point. The relation between the temperature T_{min} and the Leidenfrost temperature has been discussed by Baumeister

The Leidenfrost temperature T_L is associated with single liquid droplets in contact with a hot wall as shown qualitatively in Fig. 5.16. For determining the Leidenfrost temperature experimentally the measured total evaporation time of a droplet t_{evap} is plotted as a function of the wall temperature T_w. The prediction of the Leidenfrost and the minimum temperature has been discussed by Baumeister and Simon [329].

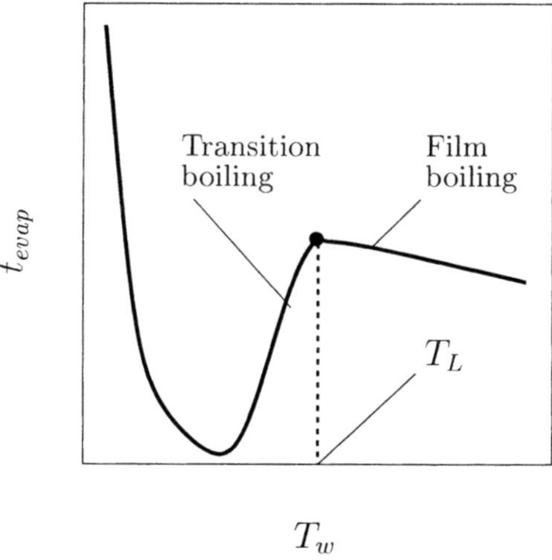

Fig. 5.16. Schematic representation of total evaporation time t_{evap} of liquid droplets in contact with a hot surface or wall. Experimentally this evaporation time curve is determined by placing different droplets of equal volume and equal initial temperature on a hot surface. The Leidenfrost temperature T_L is defined as indicated in the figure at the local maximum of t_{evap}

The physical phenomena, which are observed, when liquid droplets impact and evaporate on hot surfaces, is of interest in a number of areas of engineering such as spray evaporators or cooling of hot metal ingots. For high temperatures around the Leidenfrost temperature these questions have received considerable attention in the literature. The interaction of droplets below the Leidenfrost temperature has not been studied to the same degree.

In experimental studies it is often assumed that the thermal resistance in the wall is negligible. This assumption leads to the isothermal wall condition. When the heat capacity of the wall cannot be neglected the isothermal wall condition does not hold. A significant reduction in wall temperature of up to 30 K for droplet sizes of 600 μm have been reported in Ref. [330]. Similar results have been found by Baumeister [329].

Using high-speed photography the droplet behavior on heated walls has been studied by Makino and Michiyoshi [331]. The heat transfer during the so-called contact period has been discussed by the same researchers.

Experimental Setup. The interaction between monodisperse droplet streams and a hot wall has been studied in great detail by Karl [332] for droplet sizes ranging from 70 to 260 μm. The experimental setup used for these investigations is shown in Fig. 5.17. The droplets are generated with

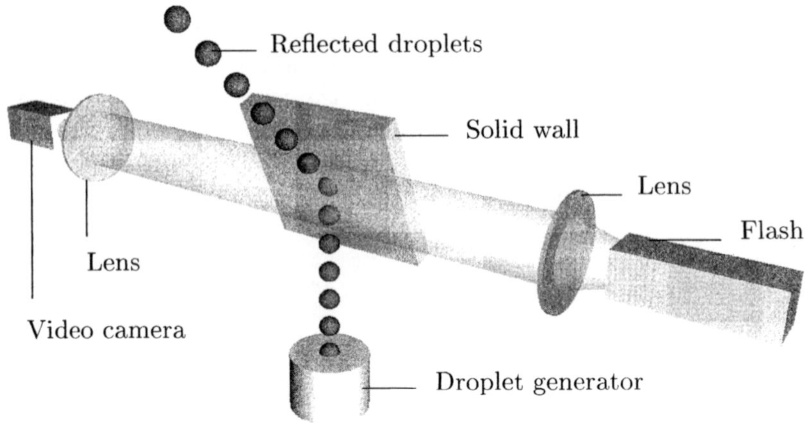

Fig. 5.17. Schematic view of experimental setup for the investigation of droplets impacting on heated solid walls. The droplets are illuminated with a flash light and recorded as shadows by the video camera

the droplet stream generator described in Sect. 2.3. The temperature of the wall, which is heated electrically, is determined with a thermocouple below the surface of the wall. Chromium plated copper and steel were used as wall materials.

The incident droplets move upwards before they impinge on the hot wall. The impinging angle of the droplet streams can be varied continuously over a wide range. All experiments were performed well above the Leidenfrost temperature. Depending on the initial conditions the droplet deformation during the interaction process results in nearly perfect reflection of the droplets, formation of secondary droplets or splashing. These phenomena occur during a time ranging typically from few microseconds to a few hundred microseconds.

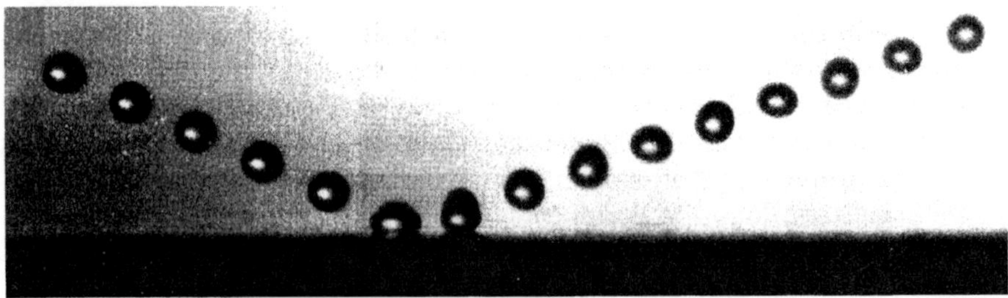

Fig. 5.18. Nearly specular reflection of ethanol droplet stream impinging with velocity $v_1 = 3\,\mathrm{m/s}$ on solid wall. The impact angle is $\beta_1 = 25°$, the droplet diameter $d = 140\,\mu\mathrm{m}$, and the temperature of the wall $T_w = 280°\mathrm{C}$. Definitions of important parameters are given in Fig. 5.19 (from A. Karl, ITLR)

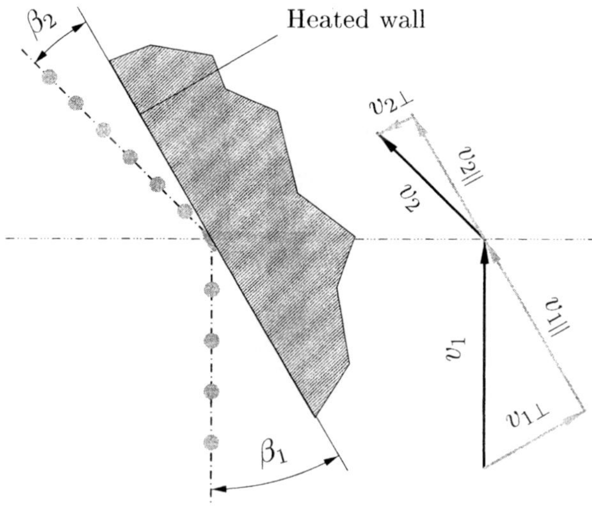

Fig. 5.19. Droplet stream impinging on solid wall. Definitions of the impinging angle β_1, reflection angle β_2 and of the different velocity components for the impinging and reflected droplet streams

With monodisperse droplet streams the impact phenomena are repeated periodically. The periodicity allows a special observation technique which uses an ordinary CCD-video camera in combination with a flash light with a repetition rate of 50 Hz. The short duration of the light flashes of approximately 200 ns guarantees sharp pictures of the moving droplets and the highly dynamic interaction phenomena during and immediately after the impact of the droplets (see Sect. 4.4).

With the optical arrangement of video camera and flash light shown in Fig. 5.17 shadow images of the droplets are obtained. An example for nearly perfect reflection is shown in Fig. 5.18. The definition of impact parameters and velocities are given in Fig. 5.19.

Loss of Momentum. An important parameter for the reflection of liquid droplets by a solid wall is the loss of momentum, which is represented in the dimensionless form

$$L = \frac{mv_{1\perp} - mv_{2\perp}}{mv_{1\perp}} = 1 - \frac{v_{2\perp}}{v_{1\perp}} = 1 - r_n \ , \tag{5.8}$$

where r_n is the ratio of the normal velocities $v_{2\perp}$ and $v_{1\perp}$.

In the experiments the velocity ratio r_n is determined by measuring the normal velocities $v_{1\perp}$ and $v_{2\perp}$ before and after the interaction with the heated wall. All experiments by Karl have been performed for impinging angles $\beta_1 < 50°$. Most results reported in the literature are for impinging angles $\beta_1 \approx 90°$. Experimental results for the loss of normal momentum are represented in Fig. 5.20.

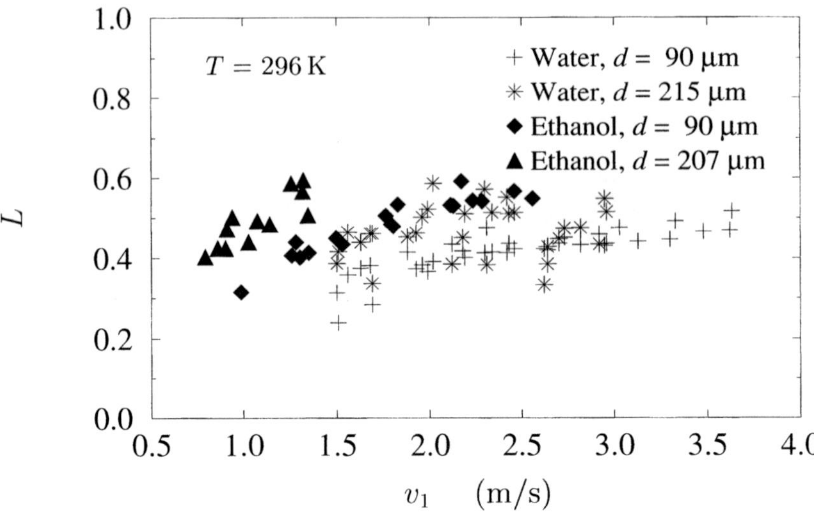

Fig. 5.20. Loss of momentum L normal to the heated wall for water and ethanol droplets with two different droplet diameters in dependence of the impact velocity v_1. The wall temperature is $T_W = 548\,\mathrm{K}$ for ethanol and $T_W = 598\,\mathrm{K}$ for water (from A. Karl, ITLR)

Droplet Deformation. During the droplet wall interaction above the Leidenfrost temperature the droplet floats on a vapor cushion. Thus the liquid has no direct contact with the wall. The corresponding heat transfer regime of the droplet wall system is called by some authors nonwetting regime [333].

Immediately after the impact the droplet liquid starts to propagate in radial direction on the vapor cushion. When the impact velocity is not too high, this spreading motion is stopped by surface tension and the droplet reaches a state of maximum deformation. In the following receding motion the fluid starts to flow radially inwards. This results in the formation of an elongated liquid complex moving away from the wall and lift off. After separation from the wall the liquid mass begins to oscillate until viscosity damps the motion and finally a spherical droplet travels away from the wall.

The resulting deformation with important parameters are shown schematically in Fig. 5.21. The diameter $d(t)$ represents the outer diameter of the deformed droplet and $d_K(t)$ the circular area above the vapor cushion. The deformation of the droplet is described by the dimensionless parameter $r_d = d(t)/d$, where d is the diameter of the undeformed droplet before the impact with the wall. Experimental results for the droplet deformation r_d are shown in Fig. 5.22 as a function of time for different droplet liquids and sizes.

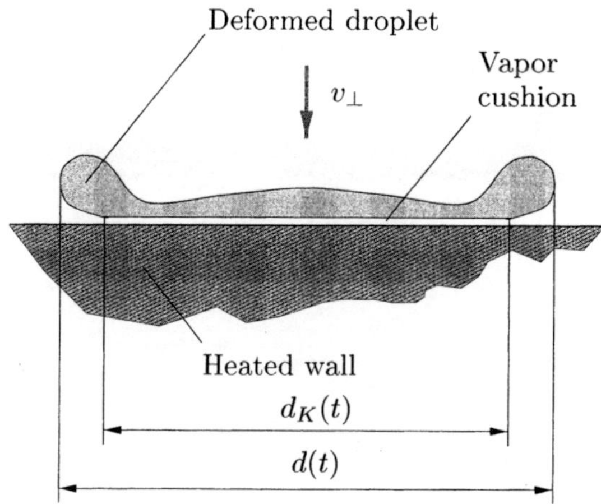

Fig. 5.21. Deformation of droplet after impinging with normal velocity $v_\perp$ on heated solid wall. The diameter of the vapor cushion is $d_K(t)$ and the outer diameter $d(t)$

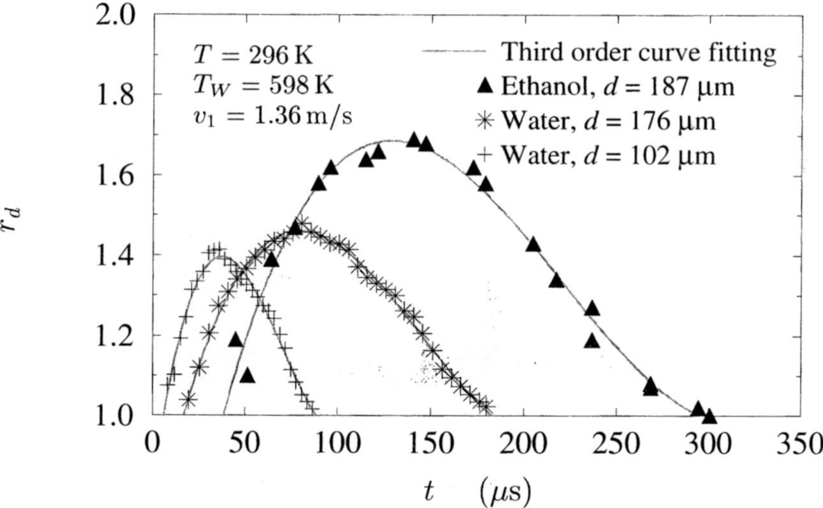

Fig. 5.22. Influence of droplet diameter and droplet liquid on the droplet deformation $r_d = d(t)/d$ for the impact velocity $v_1 = 1.36\,\text{m/s}$ (from A. Karl, ITLR)

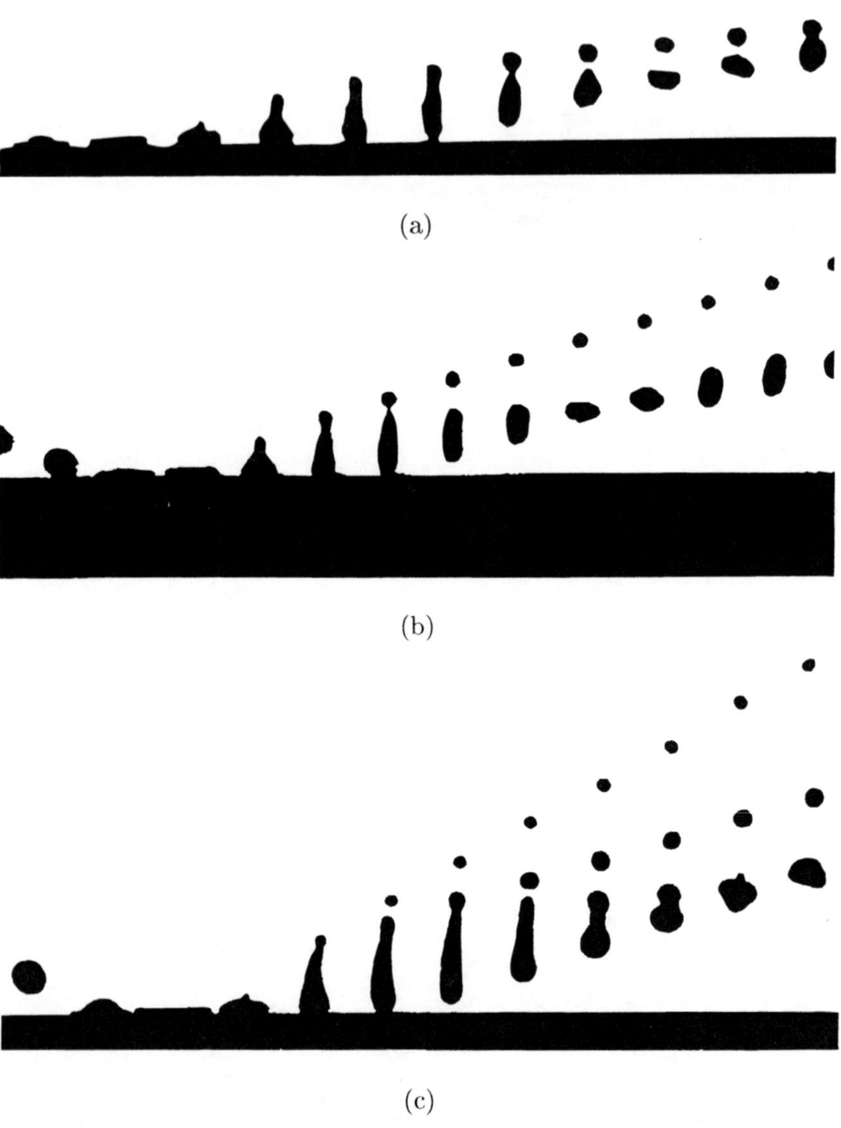

Fig. 5.23. Three different types of secondary droplet formation during interaction of small droplets with a hot wall. For all cases the wall temperature is well above the Leidenfrost temperature. The impact energy increases from (a) to (c). All images have been converted into binary images for the ease of presentation (from A. Karl, ITLR)

Formation of Satellite Droplets. It has been reported by different authors in the literature that upon droplet impingement on hot walls above the Leidenfrost temperature a transition occurs from ideally rebounding droplets, which remain intact, to droplets that break up [334, 333]. Different types of droplet wall interactions are shown in Fig. 5.23, where the droplets move from left to right. These prints have been obtained by digitizing video recordings and transforming them to shadow pictures by an image processing program.

The case shown in Fig. 5.23 (a) represents conditions for the beginning of secondary droplet formation. During the interaction with the wall the droplet is strongly deformed, which results in large velocities inside the reflected droplet. The surface tension is not sufficient to maintain a closed surface and a small droplet forms. This disintegration process occurs, when the droplet liquid is in a contracting motion. As a consequence the formed droplets have a relative velocity towards each other and coalesce after traveling a short distance. So a stream of single droplets departing from the wall is observed finally.

When the impact velocity is increased by a small amount the secondary droplets remain permanently separated, as shown in Fig. 5.23 (b). Here the disintegration takes place when the droplet liquid is still in a stretching motion, so the formed droplets move away from each other and cannot coalesce. The secondary droplet is usually smaller than the primary droplet, which exhibits strong oscillations.

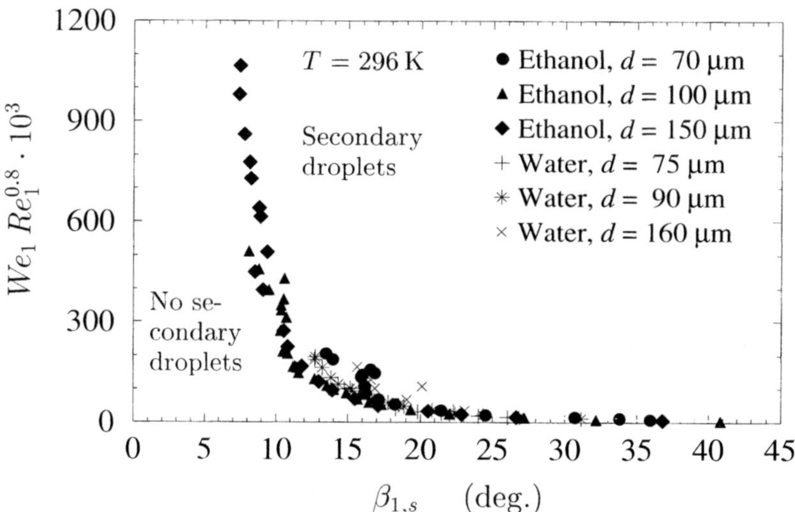

Fig. 5.24. Parameter $We_1 \cdot Re_1^{0.8}$ of impinging droplet stream for secondary droplet formation in dependence of angle $\beta_{1,s}$. Presented are results for ethanol and water droplets with different diameters. The wall temperature is $T_W = 548$ K for ethanol and $T_W = 598$ K for water (from A. Karl, ITLR)

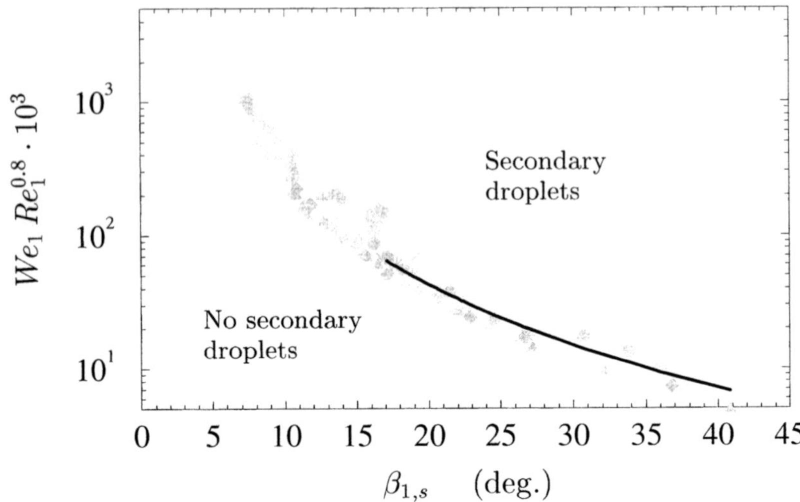

Fig. 5.25. Parameter $We_1 \cdot Re_1^{0.8}$ of impinging droplet stream for secondary droplet formation in dependence of angle $\beta_{1,s}$. Shown are the experimental results together with the correlation $We_1\, Re_1^{0.8} = 9.988 \cdot 10^7 \beta_{1,s}^{-2.598}$, which is represented by the solid line (from A. Karl, ITLR)

For even larger impact velocities this behavior is repeated and a second disintegration occurs during the stretching motion as shown in Fig. 5.23 (c), where three stable droplet streams are formed by the reflection processes. By further increasing the impact velocity even more droplet streams may form until the energy for the so-called splashing is reached. The droplet size increases from the first secondary droplet to the primary droplet.

For the following analysis of the interaction phenomena the case shown in Fig. 5.23 (a) is considered as transition to secondary droplet formation. For the experimental determination of the transition point the impinging angle β_1 is increased in small steps at constant impinging velocity until secondary droplets are observed. This procedure is repeated for different droplet diameters and velocities in order to determine the conditions for the transition to secondary droplet formation.

For the representation of the results the Weber number $We = \varrho v_1^2 d/\sigma$ is introduced, with the velocity v_1 before the impingement and the droplet diameter d as characteristic length.

Influence of Surface Roughness. It has been shown by Karl that the minimum impinging angle for the formation of secondary droplets described in the last section is related directly to the friction at the wall, which depends on the surface roughness. [332] In technical applications the surface roughness is described with the so-called average depth of roughness R_z. This quantity is related to the maximum depth of surface roughness over a defined length.

In the presented investigations the values of R_z ranged from $0.41\mu m$ for chromium-plated copper walls to $2.4\mu m$ for steel walls. The smaller value, which represents a very smooth surface, can easily be obtained by polishing the surface. The larger value is usually obtained by grinding the surface. With some effort this value can also be obtained directly by machining on a lathe or a milling machine.

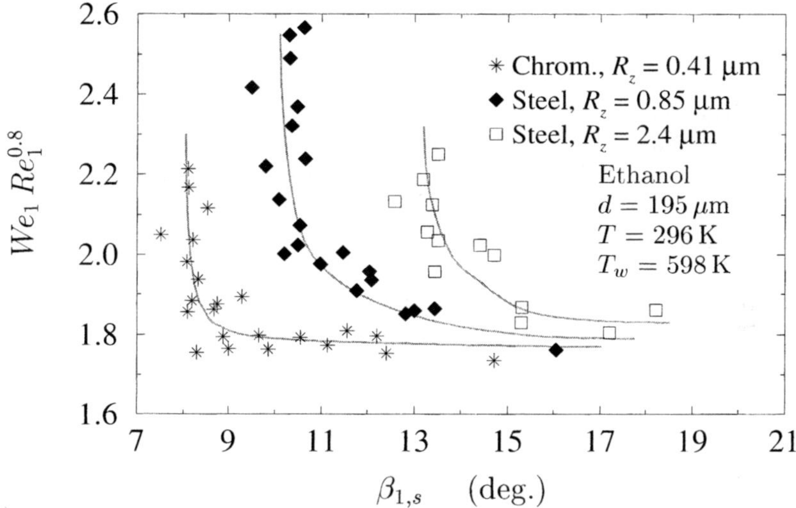

Fig. 5.26. Experimental results for the onset of secondary droplet formation at three different values for the surface roughness R_z (from A. Karl, ITLR)

The experimental results of Ref. [332] show, that the general behavior during the reflection process is not altered. However, an essential influence on secondary droplet formation is observed, especially at small impinging angles β_1, as can be seen in Fig. 5.26, where the onset of secondary droplet formation is shown for three values of the surface roughness R_z. The solid lines separate the regimes of ideal or specular reflection and secondary droplet formation for three different surfaces. Above and to the right of the shown solid lines secondary droplets are formed, whereas to the left and below the curves secondary droplets do not occur. The experimental results reveal the transition between these two regimes, which will be discussed in the following for chromium-plated copper.

In the case of chromium plated copper the normal impinging velocity needed for the formation of secondary droplets is constant and independent of the impinging angle β_1 in the approximate range $\beta_1 > 8°$. This corresponds to constant values of the parameter $We_n \cdot Re_n^{0.8}$, as shown in Fig. 5.26, which means that a certain minimum amount of impact energy is necessary for the formation of a secondary droplet. Approaching the impinging angle 8° a steep increase of the velocity needed for secondary droplet formation is observed. Below impinging angles $\beta_1 \approx 8°$ secondary droplets are not formed, even for higher impact energies.

The described behavior can be explained as follows: when the impinging angle is decreased along the transition limit for the formation of secondary droplets a strong increase of the velocity tangential to the wall $v_{1\|} = v_{1\perp}/\tan\beta_1$ occurs at constant normal impinging velocity. With increasing tangential velocity the velocity gradient normal to the wall within the vapor cushion increases also. As shown in Ref. [335] the interaction time is independent of the normal impinging velocity in the investigated parameter range. For higher tangential velocities this results in greater distances traveled by the droplet on the vapor cushion along the heated wall. Both effects cause an increase of energy dissipation. Thus for smaller impinging angles there is not sufficient kinetic impact energy for forming a secondary droplet.

As the dissipation at the hot wall depends on the surface roughness, this minimum impinging angle for the formation of secondary droplets should also depend on the surface roughness. This is confirmed in Fig. 5.26 for three different values of the surface roughness. All other parameters like droplet diameter, wall temperature and droplet liquid are held constant for this investigation. For all three values of the surface roughness the same behavior is observed: a constant region for larger impinging angles and with decreasing impinging angle a sudden increase of the impinging velocity, which is needed for the formation of secondary droplets.

This minimum impinging angle increases with increasing surface roughness, which confirms the explanation given above. For ethanol droplets with the diameter 195 μm the minimum angles are approximately 8°, 10° and 13° when the surface roughness has the values 0.41 μm, 0.85 μm and 2.4 μm, respectively.

The described behavior has been studied with different droplet diameters. For all investigated droplet diameters the behavior is similar: at constant surface roughness the minimum impinging angle decreases for increasing droplet diameter. This means, that the minimum impinging angle depends not only on the surface roughness but also on the droplet diameter. Therefore, all results for the minimum impinging angle have been represented in Refs. [332, 336] as a function of the ratio of the depth of roughness and the droplet diameter R_z/d. This parameter seems to be the controlling factor at least for a given droplet liquid. It can be shown that a reasonable correlation exists for most of the experimental data points. The minimum impinging angle can be described with the empirical relation

$$\beta_{min} = 45.7 \cdot \left(\frac{R_z}{d}\right)^{0.272} \tag{5.9}$$

in the region $0.002 \leq R_z/d \leq 0.015$.

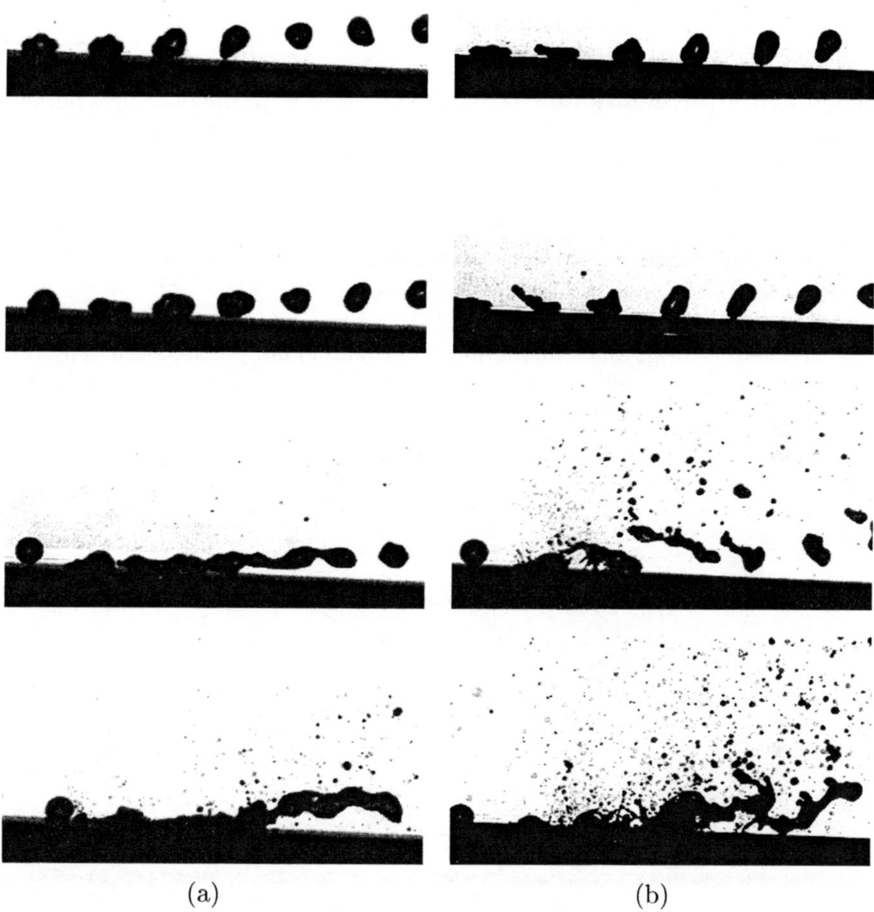

Fig. 5.27. Impact of droplets on hot wall (a) for pure water and (b) for binary mixture of water and 50 vol% ethanol. The droplet diameter is 195 μm in all cases. For water the impact velocity is 6.90 m/s and the wall temperatures are 290°C, 250°C, 240°C, 210°C from top to bottom, for the mixture these parameters are 7.22 m/s and 230°C, 220°C, 200°C, 180°C. For the highest temperatures, which are above the Leidenfrost temperature, regular reflection without secondary droplet formation is observed for water and for the mixture. At the lower temperatures vigorous disintegration occurs after the impact (from A. Karl and A. Sommerer, ITLR)

Results close to the Leidenfrost temperature are shown in Fig. 5.27 for pure water (a) and for a mixture of water and ethanol (b). In both cases the temperature decreases from top to bottom. Both frames at the top show practically ideal reflection of the droplets. With decreasing temperature the droplets disrupt during the impact. It can be observed, that the number of small droplets is larger for the mixture.

Observation perpendicular to the wall. The experimental and numerical studies of the present section reveal that droplet wall interaction processes are rather complex phenomena, which depend on a great number of different parameters. All experiments described so far provide droplet observations in the plane which contains the droplet velocities before and after the collision. Additional experiments have been performed to observe the lateral extension of the contact area between droplet and wall.

A heated glass wall has been used in a series of experiments, which allows to observe the impact phenomena perpendicular through the wall. The observation technique shown in Fig. 5.17 has been slightly modified for this purpose to allow simultaneous observations in two directions with two video

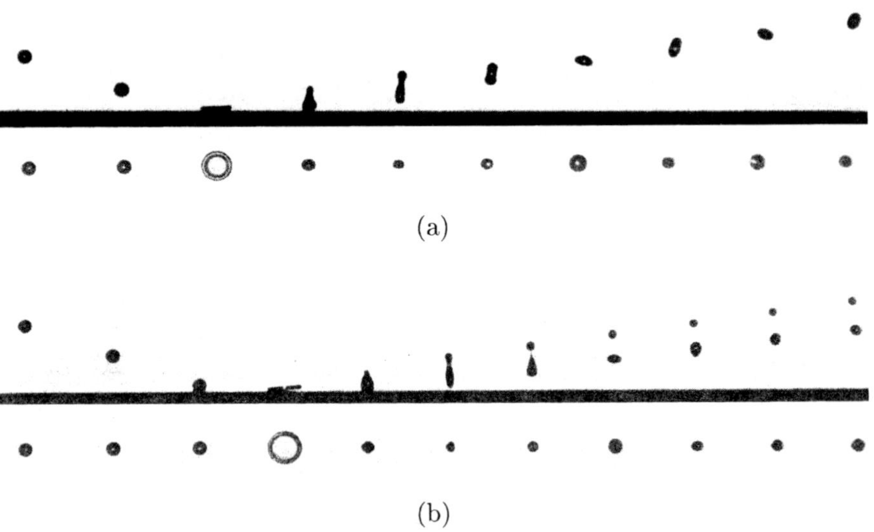

Fig. 5.28. Ethanol droplets impinging at the angle $\beta_1 \approx 20°$ on heated glass wall above the Leidenfrost temperature. The droplets are observed simultaneously from the side and perpendicular through the glass wall. The droplets move from left to right. For (a) the droplet velocity is $v_1 = 8.8\,\mathrm{m/s}$ and the droplet diameter $d = 160\,\mu\mathrm{m}$. In (b) the droplet velocity is $v_1 = 12.2\,\mathrm{m/s}$ and $d = 174\,\mu\mathrm{m}$. The droplet in (a) do not disintegrate, in (b) secondary droplets form (from A. Karl and J. Wolber, ITLR)

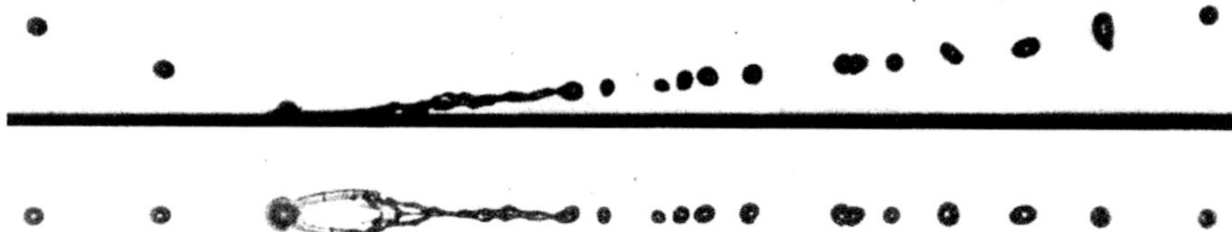

Fig. 5.29. Ethanol droplets impinging on heated glass wall close to, but below the Leidenfrost temperature. The droplets are observed simultaneously from the side and perpendicular through the glass wall. The droplets move from left to right. The droplet velocity is $v_1 = 13.3\,\mathrm{m/s}$, the droplet diameter $d = 155\,\mu\mathrm{m}$ and the impinging angle $\beta_1 \approx 20°$. The reflection process appears irregular (from A. Karl and J. Wolber, ITLR)

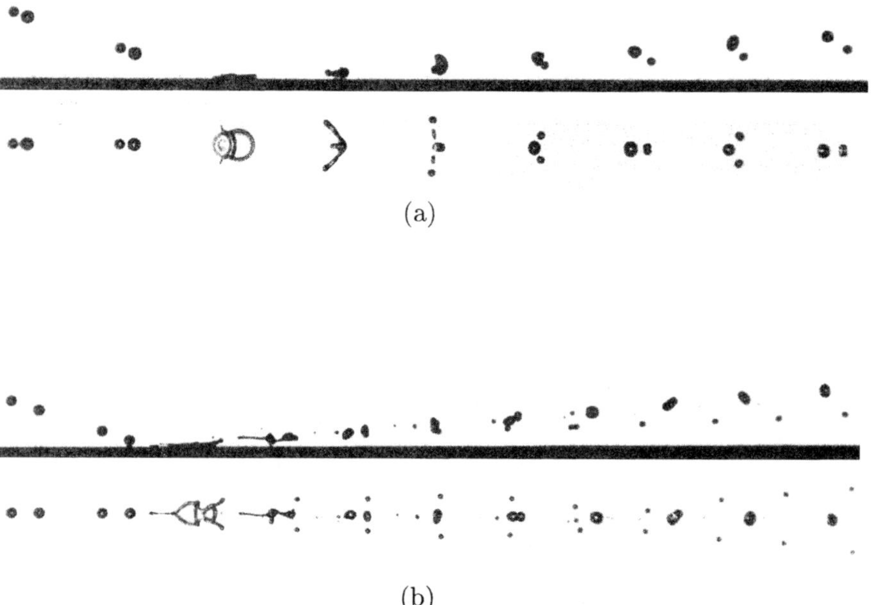

(a)

(b)

Fig. 5.30. Pairs of ethanol droplets impinging on heated glass wall above the Leidenfrost temperature. The droplets are observed from the side and perpendicular through the glass wall. The droplets move from left to right. The droplet velocity is $v_1 = 12.2\,\mathrm{m/s}$, and the impinging angle $\beta_1 \approx 20°$. In (a) a droplet with the diameter $d \approx 160\,\mu\mathrm{m}$ is followed by a smaller droplet. In (b) two droplets with approximately the same diameter impinge on the wall. In both cases the droplets interact with each other during the impinging process. The outcome after the reflection is quite different (from A. Karl and J. Wolber, ITLR)

cameras. The first direction is the same as in Fig. 5.17, the second direction is perpendicular through the glass wall. The light of the flash is split, in order to provide the light for both observations. Some results obtained with this observation technique are shown in Figs. 5.28-5.30.

5.3.4 Wetting of Solid Walls During Droplet Impact

Various examples of droplet-wall interactions have been described in the previous sections. However, open is the question if, or under which conditions the wall is wetted by the droplet liquid during the interaction process. In the following an experimental method will be described, which allows to decide if the wall is wetted by impinging droplets. This method uses total reflection of light. The well known phenomenon of total reflection is illustrated in Fig. 5.31. An optical arrangement for studying the wetting of a glass wall by impinging droplets is shown in Fig. 5.32.

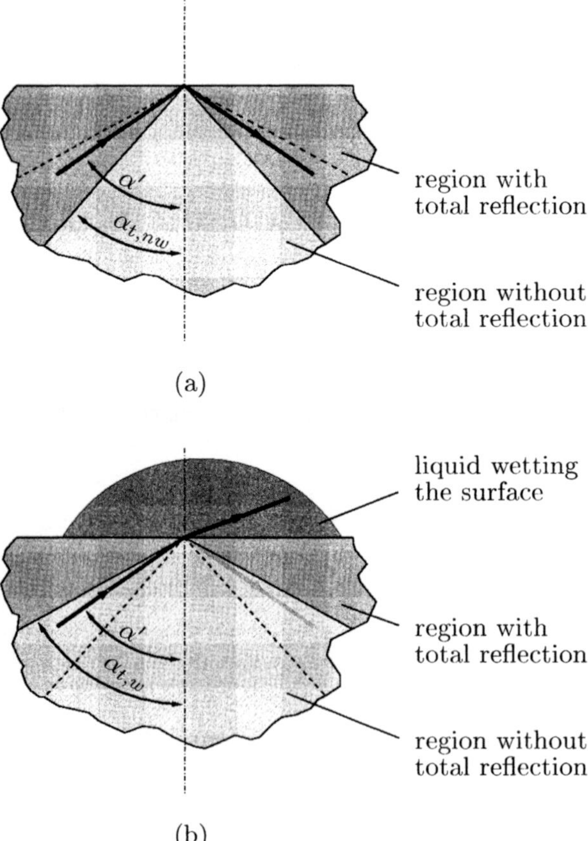

Fig. 5.31. Behavior of a light ray illuminating a glass surface from the inner side. Shown is the ray path (a) in the case of a nonwetted surface and (b) in the case of a wetted surface.

Total reflection can be observed, when light passes from a medium of higher optical density into another medium of lower optical density. In the optical system described here, the glass of the prism represents the medium with the higher optical density. The medium with the lower optical density is air, when wetting does not occur, and droplet liquid, when the prism is wetted, as shown in Fig. 5.31. The indices of refraction are denoted by n_{prism}, n_{air}, and n_{liq} with $n_{pris} > n_{liq}$ and $n_{air} \approx 1$. A light ray is totally reflected, when the angle of incidence α' is larger than the critical angle [97]. Without wetting the critical angle α_t can be expressed by the refractive indices of glass and air; in this case one has $\alpha_{t,nw} = \sin^{-1}(1/n_{prism})$. In the case with wetting the relation $\alpha_{t,w} = \sin^{-1}(n_{liq}/n_{prism})$ is valid. As the refractive index is a function of the wavelength of the light, the critical angle depends also on the wavelength. As white light is used for illumination in the experiments, it follows, that a certain range of critical angles has to be taken into account. In Fig. 5.31 the path of a light ray incident (a) on a nonwetted surface and (b) on a wetted surface is shown. In both cases one has for the angle α' of the incident ray $\alpha_{t,w} > \alpha' > \alpha_{t,nw}$. The light ray is therefore totally reflected in the nonwetted case shown in (a). In the figure the critical angle $\alpha_{t,w}$ is indicated by a dashed line. In addition the regions, in which total reflection can be observed, are shown by a darker gray shade. The behavior of the same light ray for a wetted surface is shown in Fig. 5.31(b). As α' is lower than $\alpha_{t,w}$, only a fraction of the incident light is reflected, most of the light leaves the glass prism. The critical angle $\alpha_{t,nw}$ is indicated again by a dashed line in the figure and the regions of angles α', for which total reflection is observed, are marked by a darker gray shade.

In the experiments the illuminated inner side of the surface is observed in order to detect the wetted regions. In nonwetted regions the incident light is reflected almost completely, and can be detected. However, the wetted regions, in which only a small part of the incident light is reflected at the inner side of the glass surface, appear darker. It has to be emphasized, that the angle of incidence α' of the illuminating light has to be between the critical angle for the nonwetted surface $\alpha_{t,nw}$ and the critical angle for the wetted surface $\alpha_{t,w}$. As the critical angle for the wetted case depends on the refractive index of the liquid, it is possible to distinguish different liquids.

The experimental setup depicted in Fig. 5.32 allows the observation of the impacting droplets as shadows and the detection of the wetted regions simultaneously [337]. This technique has been described in Sect. 5.3.3 and in Ref. [338]. The traces of the light rays for the shadowgraphs and for the detection of wetting are depicted schematically in Fig. 5.32. Shown are the prism with the surface for the droplet impact and a beam-splitter.

The rays for the shadowgraphs, here indicated in black, are perpendicular to the impact surface and enter the prism from outside. They are reflected twice within the prism, pass the beam splitter, and are imaged by a lens system on a camera. The droplets are observed as dark shadows, when they

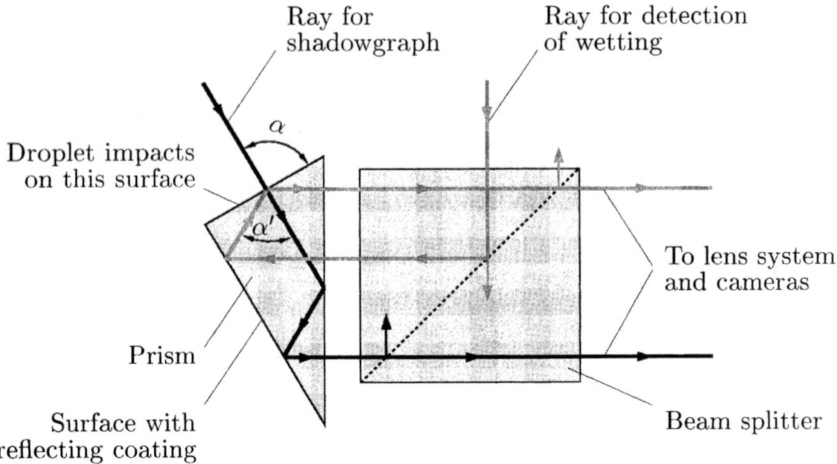

Fig. 5.32. Illustration of ray paths for shadowgraphs and for detection of wetting. Shown are the prism and the beam splitter

block incident light rays. The rays for detecting the wetting of the surface, indicated in dark gray, are reflected by the beam splitter towards the prism; they are reflected in the prism by the surface with a reflecting coating, before they illuminate the surface, which is impacted by the droplets. This surface reflects the rays totally when it is nonwetted. The lens system images the inner side of the wall on a second camera. When the surface is wetted, only part of the light is reflected at the glass surface, the remaining part is refracted and enters the droplet liquid. For a complete understanding of the results, the interaction of the refracted light ray with the liquid-air interface has to be taken into account. If the prism is wetted by a film of constant thickness the light rays entering the liquid are reflected totally at the film surface; therefore only the border lines of the liquid film appear darker.

A schematic view of the whole experimental setup is shown in Fig. 5.33. The prism and the beam splitter, shown in Fig. 5.32, can be recognized. The droplets of a monodisperse droplet stream impact on the surface of the prism. The equally spaced monodisperse droplets are produced with a droplet stream generator. Another possibility is to produce the droplets with a droplet on demand generator with a repetition rate higher than the frequency of the video signal. In this case the influence of neighboring droplets in the droplet stream can be reduced. Even impacts of single droplets can be studied.

With the combination of a flash lamp and a video system the impact of the droplets can be observed perpendicular to the wall with high time resolution. The principle of this illumination and observation technique has been described in Sect. 4.4. The light of the flash lamp is split by a beam splitter. One part of the light is used to illuminate the inner side of the surface in order to detect wetting, the other part of the light is used to produce the shadows of the droplets. The surface is imaged by a lens system. One image is obtained with the rays for the shadowgraph. The other image is obtained with the rays illuminating the inner side of the surface. Both images are detected by separate video cameras. As the path length of the rays from the surface to the lens is not the same for the two directions, the magnification of the images is different.

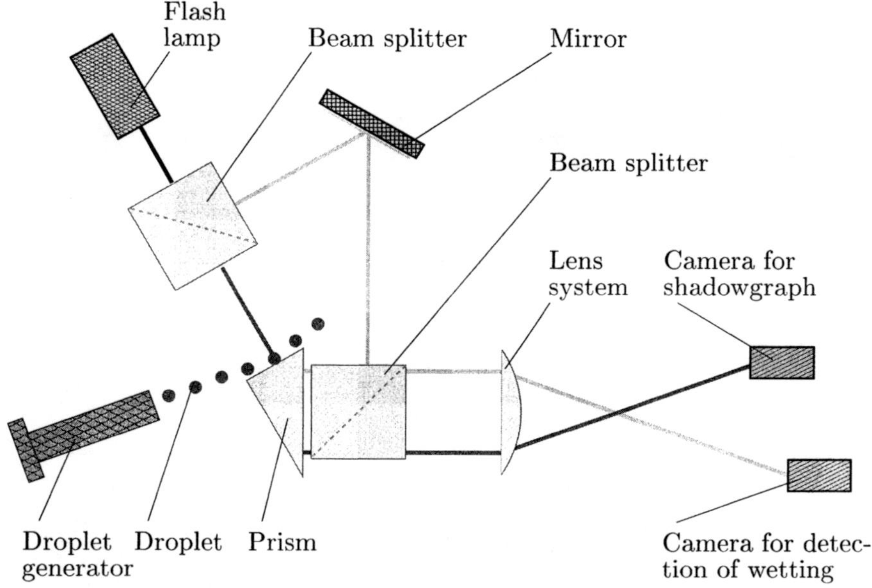

Fig. 5.33. Experimental setup for simultaneous observation of impacting droplets and wetted regions on the wall

In the experiments performed at ITLR the setup of Fig. 5.33 has been extended, to allow the observation of the impacting droplets additionally from the side, as shown in Figs. 5.17 and 5.18. Thus the impact angle β_1 can be determined. For the illumination in this case the light of the flash lamp is split once more. A second lens system and a third video camera are used for imaging. This extension is not depicted in Fig. 5.32.

A set of pictures from the three cameras is shown in Fig. 5.34. Shown is a monodisperse, equally spaced ethanol droplet stream impinging on the smooth glass surface of the prism. Although the surface has ambient temperature, the droplets are reflected regularly, since the angle between droplet stream and surface is very shallow, as can be seen from Fig. 5.34(a). In (b) the shadows of the droplets, which come from the left hand side, can be seen in the direction perpendicular to the wall. A liquid film is observed near the impact of the droplets. Figure 5.34(c) shows the light, which is totally reflected at the inner side of the surface. Only the border of the liquid film and

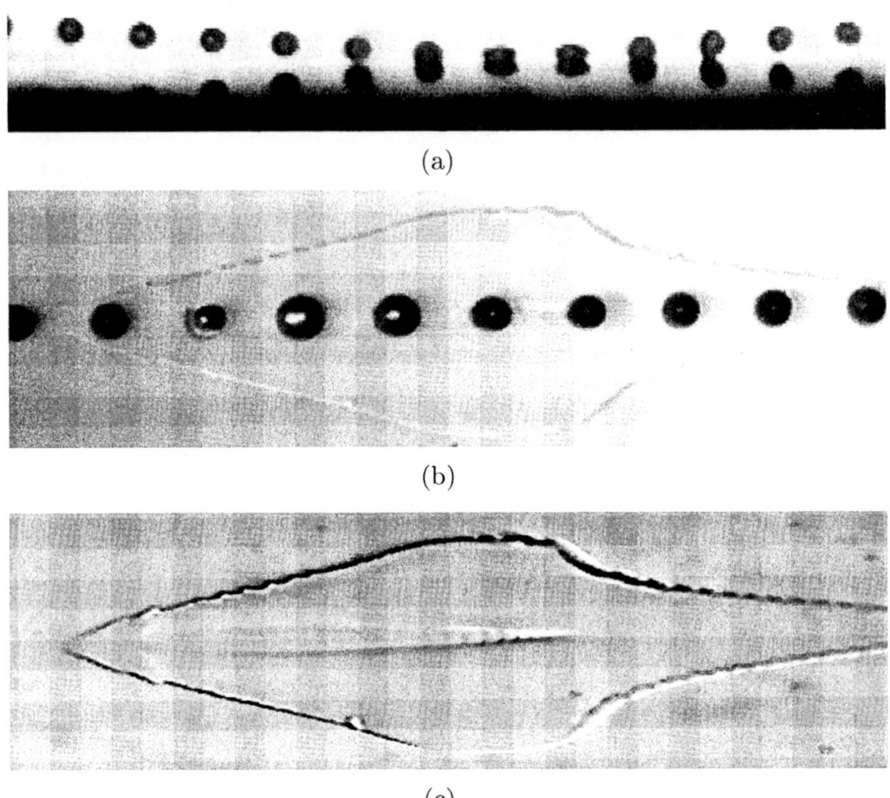

Fig. 5.34. Impact of a monodisperse droplet stream on glass wall at ambient temperature shown from the side (a). The light passing through the wall shows the droplets together with a liquid film (b). The totally reflected light (c) reveals only the liquid film. The diameter of the droplets was $d = 150\,\mu$m, their spacing $s = 360\,\mu$m and their velocity $v = 7\,$m/s. The impact angle of the impinging droplets was $\beta_1 \approx 5°$

the locations where the film changes its thickness can the recognized on the image. At the locations where the obviously thin film is parallel to the glass surface the light is totally reflected at the interface between the liquid and the ambient air, which results in the same brightness on the image as without film. The droplets themselves cannot be seen in (c) even at the locations on the film where the film is disturbed by approaching droplets. The only effect produced by the droplets is the trace in the middle of the film. This observations indicate that there is no liquid bridge between the droplets and the film. The film may be caused by condensation of vapor on the wall. Depending on the temperature of the wall and the vapor concentration in the region of the wall, where the droplets impact the liquid film may decrease until it has disappeared or the film may grow until a balance between the amount of vapor condensing on the wall close to the impact region and the amount of vapor evaporating from the film is reached. It has been found, that cooling of the wall supports growing of the film. The film becomes visible, when the frequency of the droplet impacts is high enough because only small amounts of vapor can be delivered by one droplet.

5.3.5 Comparison with Numerical Results

Confidence in the experimental results described in the present section can be strongly improved by numerical simulations of the corresponding processes, provided that experimental and numerical results show good qualitative and quantitative agreement. On the other hand, any quantitative difference may reveal errors in the measurements or may indicate weaknesses in the physical, mathematical or numerical model underlying the simulations, thus showing the way to possible improvements in both the experimental and the numerical methods.

When it can be assumed, that the simulation gives correct results, numerical calculations provide detailed information about distributions of pressure, velocity or temperature, quantities which are usually not easily accessible to experiments. Two completely different numerical methods have been developed at ITLR. The first method is based on the Navier-Stokes equations of the continuum theory, the second code is based on the Boltzmann equation of of the kinetic theory of gases.

Navier-Stokes method. Liquid and gas may be considered as continua with constant density and viscosity, which are separated by a sharp interface with surface tension. Thus in principle one may describe each phase by the Navier-Stokes equations for incompressible single phase flow, supplemented by appropriate jump conditions at the interface. This method however is very unwieldy, because there are three sets of equations, each defined only in parts of the time-space domain under consideration. In the so-called volume-of-fluid method the Navier-Stokes equations for incompressible fluids are solved with variable density and viscosity including free interfaces with surface tension. One has in usual notation

$$\frac{\partial(\varrho \boldsymbol{u})}{\partial t} + \nabla \cdot [(\varrho \boldsymbol{u}) \otimes \boldsymbol{u}] = -\nabla p + \nabla \cdot \eta[\nabla \boldsymbol{u} + (\nabla \boldsymbol{u})^T] + \nabla \cdot \mathsf{T} \; , \qquad (5.10)$$

$$\nabla \cdot \boldsymbol{u} = 0 \; . \qquad (5.11)$$

The above set of equations is defined at every time in the liquid phase, in the gas phase and at the interface. The difference between these equations and the Navier-Stokes equations for single phase flow is due to the last term in the momentum equation, which accounts for surface tension according to

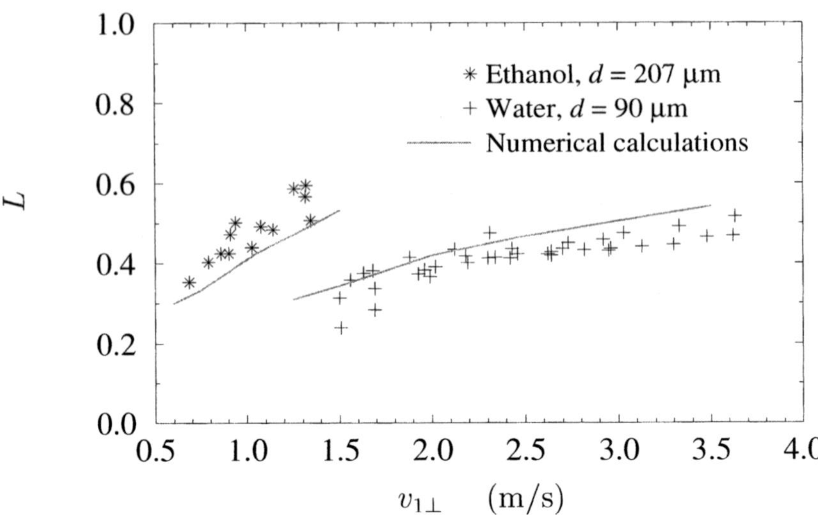

Fig. 5.35. Comparison of experimental and numerical results from Navier-Stokes calculations for the loss of momentum L normal to the wall for two different droplet liquids. The wall temperature for the experimental results is $T_W = 548\,\mathrm{K}$ for ethanol and $T_w = 598\,\mathrm{K}$ for water. The numerical results are represented by solid lines

the conservative model of Lafaurie et al. [339], where T is the capillary stress tensor. It should be mentioned, that these equations have to be interpreted in the weak sense, because density, viscosity and pressure vary discontinuously at the interface.

The distribution of the two phases in the time-space domain is described by the the volume fraction α of the liquid. In the volume-of-fluid method [340] density and viscosity are related to α by

$$\varrho = \varrho_{gas} + (\varrho_{liq} - \varrho_{gas})\alpha \,, \tag{5.12}$$

$$\eta = \eta_{gas} + (\eta_{liq} - \eta_{gas})\alpha \,. \tag{5.13}$$

Advection of the liquid volume, and thus of the discontinuity, is governed by the transport equation

$$\frac{\partial \alpha}{\partial t} + \nabla \cdot (\boldsymbol{u}\alpha) = 0 \,. \tag{5.14}$$

In the framework of the volume-of-fluid method all flow discontinuities are captured in a straightforward way using conservative finite volume discretizations for Eqs. 5.10 to 5.14. Specifically, a second-order conservative Godunov projection method on a staggered grid has been implemented on a parallel supercomputer. Details of this numerical method have been described in the literature [341].

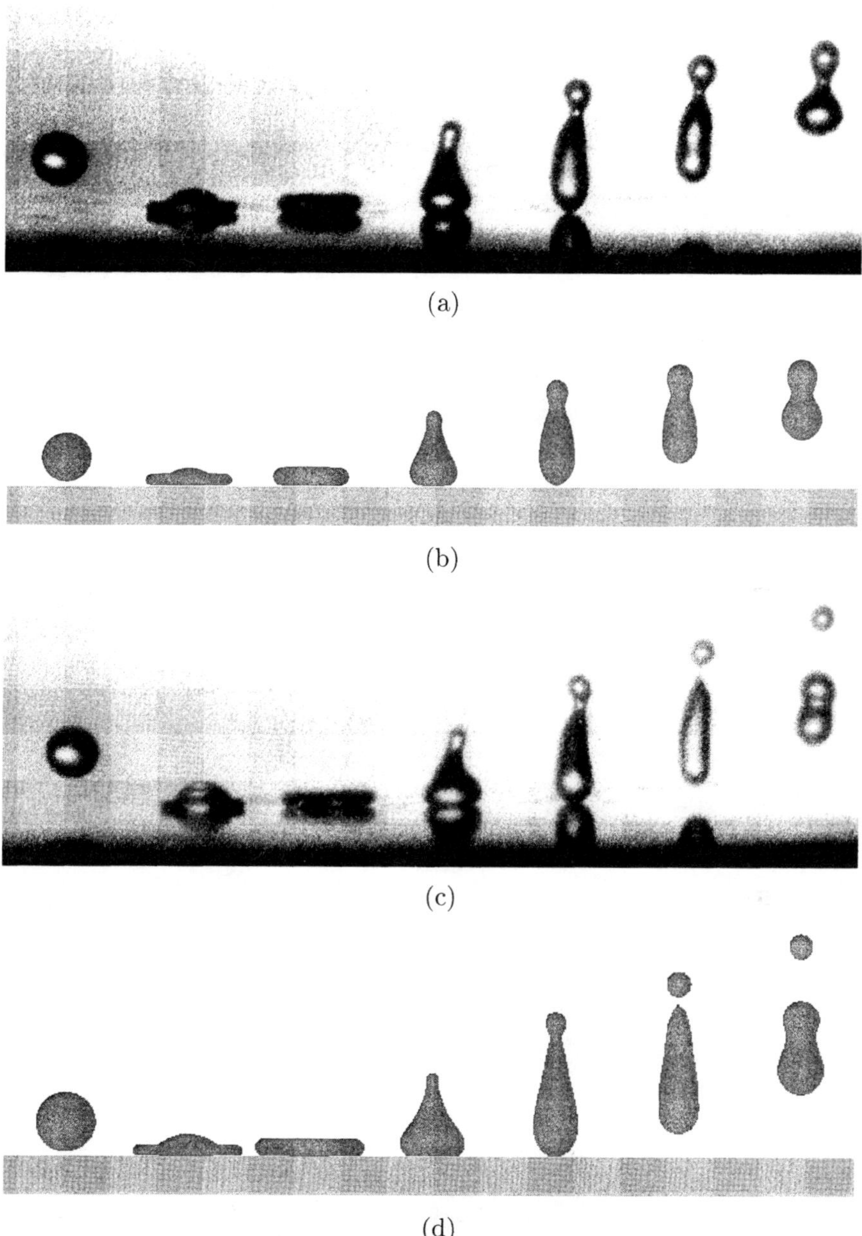

Fig. 5.36. Experimental and numerical results for the wall impact of ethanol droplets with diameter $d = 140\,\mu$m. In (a), (b) with the velocity $v_1 = 5.3\,$m/s formation of satellite droplets does not occur. In (c), (d) with $v_1 = 5.8\,$m/s one satellite droplet forms. The impact angle was $\beta_1 = 25°$ and the wall temperature was $T_w = 280°$C. The numerical results have been obtained with the volume-of-fluid method

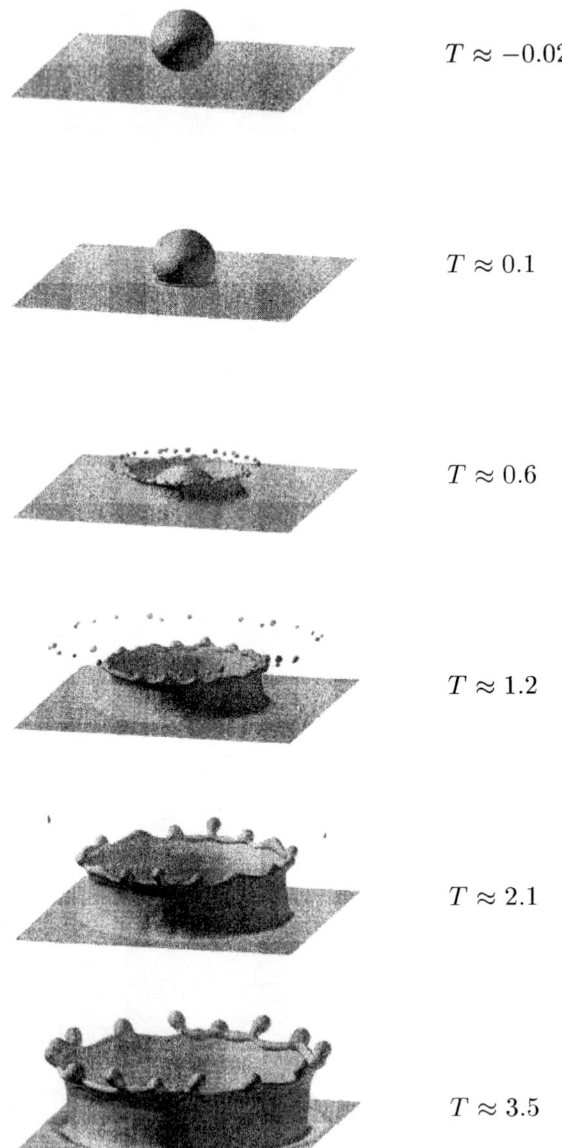

Fig. 5.37. Disintegration of the splashing lamella after impact of a droplet on a thin liquid film with $We = 250$, $H = 0.12$, $Re \approx 10000$. A 320^3-grid has been used for the simulation (from M. Rieber, ITLR)

A comparison of numerical and experimental results for the loss of momentum is shown in Fig. 5.35. In this figure the quantity L defined in equation 5.8 is shown as a function of the impact velocity $v_{1\perp}$ for different droplet diameters and liquids.

In Figs. 5.36 experimental and numerical results for the droplet deformation during impingement on a hot wall are shown. Figure 5.36(a) and (b) shows the transition to satellite formation, whereas in Fig. 5.36(c) and (d) one secondary droplet is formed after the impact [342].

The impact of a single drop on a liquid film is another challenging CFD problem because the effects of incompressibility, inertia and surface tension interact in a strong way, while liquid structures of very different length scales develop. Thus very fine grids are needed to resolve small liquid structures. Famous photographs, which give an idea of the fascinating phenomenon of droplet impact are due to Edgerton and Killian [343], while numerical simulations have been performed by Rieber et al. [344]. The impact parameters are characterized by the Weber number $We = \varrho dv^2/\sigma$, the Reynolds number $Re = \varrho dv/\eta$, the time parameter $T = tv/d$, and by the dimensionless parameter $H = h/d$, which characterizes the film thickness.

At low impact energies, i.e. for low values of the Weber number, the droplet coalesces with the liquid film without splashing. The threshold for splashing, which depends on We, Re, and H has been investigated by Cossali et al. [325]. Figure 5.37 shows the simulation of a droplet impact with $We = 250$ and $H = 0.12$, which results in splashing in agreement with experimental observations. Only one quarter of the visualized splash was really simulated in the numerical calculations. A high resolution simulation with a spatial grid with 320^3 cells and using 2553 time integration steps resulted in a run time of 4 h 30 min on 250 processors of the parallel computer Cray T3E/512-900. Due to the high resolution of the problem, the formation of a characteristic crown with finger ejecting small droplets, the so-called splashing, is in good agreement with experimental observations.

Lattice Boltzmann method. In this method the fluid is represented by discrete particles. The evolution of the particle distribution function f is described with the Boltzmann equation in the form

$$\frac{\partial f}{\partial t} + \xi_i \frac{\partial f}{\partial x_i} = \frac{1}{\tau}(f^{(0)} - f) \,. \tag{5.15}$$

The collision term on the right hand side has been replaced by Krook's model with the local Maxwellian distribution function $f^{(0)}$ and the relaxation time τ. This Krook's approximation has often been used to solve problems in the kinetic theory of gases [345]. The finite-difference approximation of this equation

$$\frac{f_j(\boldsymbol{x} + \boldsymbol{\xi}_j \cdot \delta t, t + \delta t) - f_j(\boldsymbol{x}, t)}{\delta t} = \frac{f_j^{(0)}(\boldsymbol{x}, t) - f_j(\boldsymbol{x}, t)}{\tau} \tag{5.16}$$

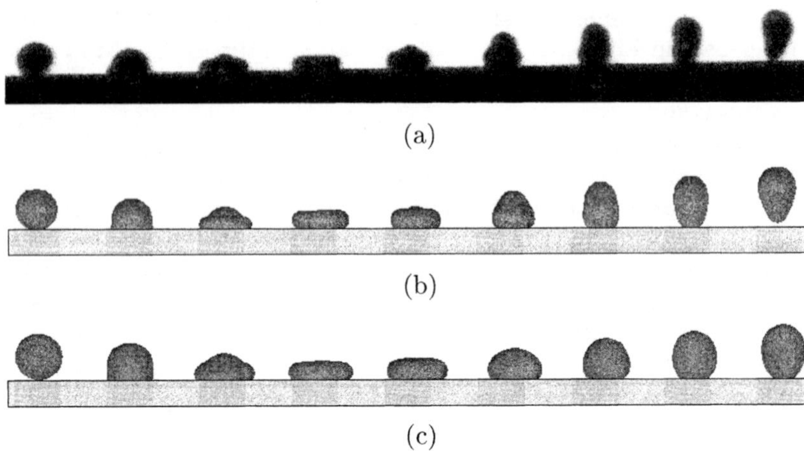

Fig. 5.38. Experimental result for droplet-wall interaction (a) in comparison with Navier-Stokes simulation (b) and lattice Boltzmann simulation (c). In the experiment the droplet liquid was ethanol, the droplet diameter 185 μm and the normal velocity $v_{1\perp} = 0.86\,\mathrm{m/s}$. The Navier-Stokes simulation has been obtained with the same droplet diameter and velocity and the properties of ethanol. These data result in $We_n = 5$ and $Re_n = 104$. The lattice Boltzmann solution is for $We_n = 5$ and $Re_n = 146$

has been used for the numerical simulation of two-phase flows with free surfaces [346]. It has been shown, that realistic values can be obtained for physical parameters, like surface tension and viscosity, when this model is used [347]. By extending this method to three dimensions the interaction between droplets and solid walls has been simulated. An example is shown in Fig. 5.38 together with a Navier-Stokes simulation and an experimental result [348].

5.4 Droplet-Droplet Interaction

The present section deals with binary encounters between droplets. During an encounter a droplet may experience very rapid changes of its velocity and size. As a consequence the velocity and size distribution of sprays may alter. The size distribution has an essential influence for example on the evaporation and ignition behavior of fuel sprays in combustion chambers. Interaction phenomena between droplets represent on the other hand fascinating phenomena of fundamental interest in other fields of science. Collisions between droplets and solid particles, which are important for scavenging [349], will not be considered in the following

The colliding droplets will be considered in the following as perfect spheres. Droplet deformations due to shape oscillations or aerodynamic forces are assumed to be negligible during the interaction.

Binary droplet collisions have been investigated experimentally by many researchers. Rayleigh studied collisions between droplets, which were provided by two crossing liquid jets, which were resolved into two regular droplet streams by applying an acoustic field [52]. A survey of experiments devoted to the interaction of water droplets relevant to cloud physics has been presented by Abbott [350]. Detailed investigations of different collision parameters have been presented for water [19, 150, 351-354], and hydrocarbons [355, 356] as droplet liquid. Other investigators have studied the coalescence of curved water surfaces [357], coalescence and separation in droplet collisions [358, 359] or included collisions of droplets with different diameters [356] or different liquids [360].

A review article on droplet collisions has been published by Orme [361]. Experimental studies of the mechanism of coalescence have been performed by Berg et al. [362]. Different numerical simulations of droplet collisions have been presented in the last years. These calculations were based on the Navier-Stokes equations [363] or on lattice gas and lattice Boltzmann simulations [364, 365]. Slow coalescence occurs in sintering processes [366-369].

5.4.1 The dynamics and geometry of a binary encounter

In the absence of external forces the mass-center of the two droplets will move at constant velocity throughout the encounter. Two droplets having the equal diameter d and initial velocities $\boldsymbol{v}_1$ and $\boldsymbol{v}_2$ are shown in Fig. 5.39. From conservation of angular momentum it follows that the colliding droplets

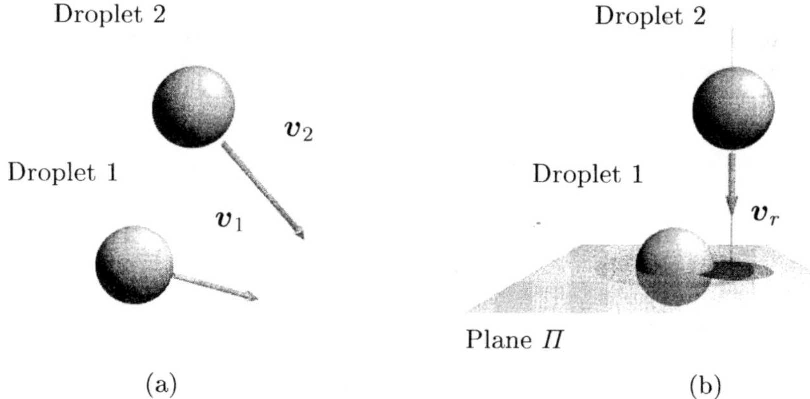

Fig. 5.39. Encounter of droplet 1 and droplet 2 in different coordinate systems. In the laboratory system (a) the initial velocities of the droplets are $\boldsymbol{v}_1$ and $\boldsymbol{v}_2$. In (b) it is illustrated how the encounter is observed in a coordinate system moving with the velocity $\boldsymbol{v}_1$; in this system droplet 1 is at rest and droplet 2 moves with the relative velocity $\boldsymbol{v}_r = \boldsymbol{v}_2 - \boldsymbol{v}_1$. The velocity $\boldsymbol{v}_r$ is vertical to the plane Π, which contains the center of droplet 1

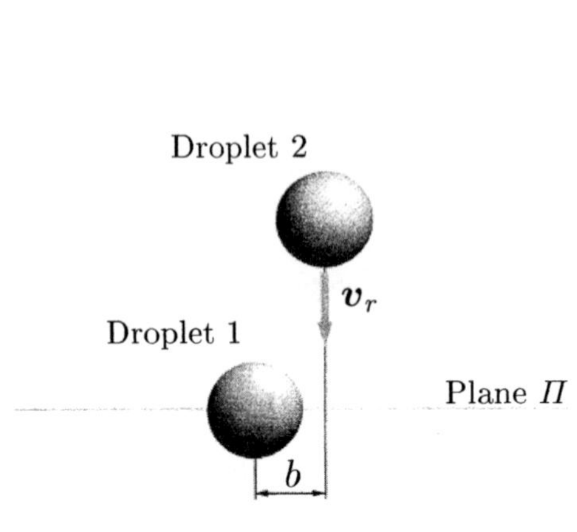

Fig. 5.40. From Fig. 5.39 it follows that the description of the binary encounter is only complete, when the coordinates of the intersection point of $\boldsymbol{v}_r$ and Π are known. Usually an angle ϵ and a radius b are introduced as polar coordinates in plane Π around the center of droplet 1. Most of the experimental results described in the following sections have been obtained at constant ϵ. Thus only the radius b is introduced, which is often called collision parameter

are always located in a plane of unvarying orientation, but moving with the velocity of the center of mass. For a convenient description of the encounter the relative velocity $\boldsymbol{v}_r = \boldsymbol{v}_2 - \boldsymbol{v}_1$ is introduced. For an observer moving with relative velocity $\boldsymbol{v}_1$ droplet 1 is at rest and droplet 2 moves with the velocity $\boldsymbol{v}_r$. This transformation is illustrated in Figs. 5.39 and 5.40.

5.4.2 Experimental Setup and Results

The coalescence or noncoalescence of droplets has been studied by different researchers by methods, which mechanically constrain the droplets. In these experiments droplets supported on two wires or capillary tubes were made to collide. In the experiments performed by Schneider et al. the droplets were not mechanically constraint [150]. These authors used streams of uniform droplets, which were formed launching a disturbance of defined wavelength onto a jet of liquid. The disturbance was produced with a piezoelectric transducer. The droplets could be charged electrically. The charge of a single droplet in this stream could be changed by applying a voltage pulse to the charging electrodes. These droplets undergo a different deflection between the deflecting electrodes and can be separated from the main droplet stream, as described in Sect. 3.3.3. For studying collision processes two equal droplets were produced by operating two of the described devices at right angles. Photographs of colliding and coalescing droplet pairs were obtained in the experiments with water droplets having diameters of 192 μm and 158 μm and the velocities 3.3 m/s and 1.1 m/s. Similar techniques have been used in Ref. [19] to study the influence of the collision parameter b for an extended diameter and velocity range. Results were presented in a collision parameter-impact velocity diagram.

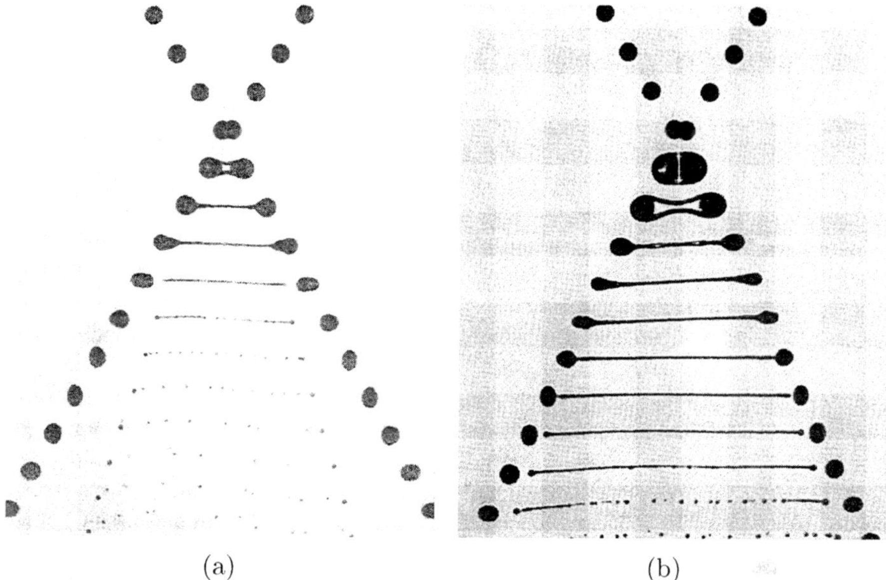

Fig. 5.41. Photograph of two colliding droplet streams of iso-propanol with the impact parameter $b = 0.68$ in (a) and impact parameter $b = 0.56$ in (b). In both cases the droplet stream generators have been excited with the frequency $f_{dis} = 48.63\,\mathrm{kHz}$. The droplet radius was $r = 48\,\mu\mathrm{m}$, the droplet velocity $v = 10.7\,\mathrm{m/s}$. The Weber number was $We = 401$ and the Reynolds number $Re = 354$ for the collision process

Brazier-Smith et al. studied the formation of satellite droplets during collisions between water droplets in order to explain increased rain fall in certain clouds [354]. A histogram for the size of satellite droplets produced by collisions is presented in this paper. The same authors report in a later paper on the conditions under which colliding droplets of equal size will coalesce, separate or bounce [352]. Results are formulated using the Weber number. Arkhipov et al. studied the outcome of droplet interactions for a large Weber number range of $0.1 < We < 120$ [370].

Experiments with droplet stream generators have the advantage, that all phenomena occur strictly periodically and rather simple observation techniques can be used [199]. With two such droplet generators periodical droplet collisions between two droplet streams have been studied at ITLR. In some of these investigations an argon ion laser has been used in combination with the droplet stroboscope technique of self-illumination described in Sect. 4.3 [356]. In other experiments a Nanolite flash triggered with the excitation frequency of the droplet stream generator was used. Two photos of colliding droplets obtained with this technique are shown in Fig. 5.41. The number of satellite droplets formed by disintegration of the ligament between the droplets af-

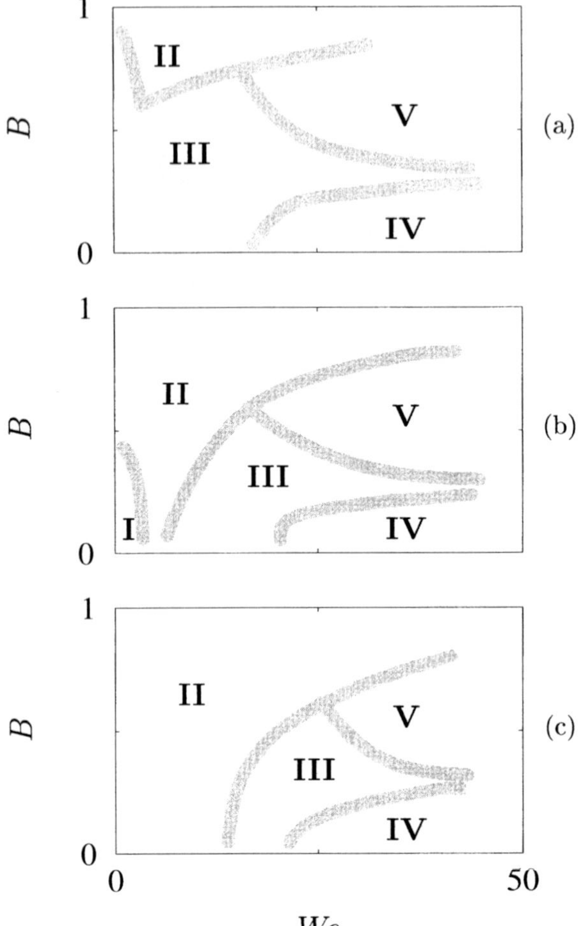

Fig. 5.42. Qualitative representation of different outcomes of binary droplet collisions in We, B-diagram according to the findings of Qian and Law. Here We is the Weber number associated with the droplet diameter d and $B = b/d$ a dimensionless representation of the impact parameter. One has $B = 0$ for head-on collisions and $B = 1$ for grazing collisions. The pressure increases from (a) to (c). In regimes I and III coalescence is observed, in regime II bouncing, in IV and V separation

ter the collision is rather large. Central and noncentral collisions have been studied with this technique. Ashgriz and Poo [353] used droplet streams to study colliding water droplets in the Weber number range of 1 to 100. They used a high-speed video camera in black and white and a conventional color video camera for recording collisions of differently colored droplets. Jiang et al. studied collisions of water and n-alcane droplets using droplets streams generated with the ink-jet technique [355]. The same technique for the generation of colliding droplet streams was used by Quian and Law to study different collision regimes in We, B-diagrams [358]. The entire apparatus was housed in a chamber whose pressure could be varied between 0.1 and 20 bar. The image acquisition system allowed a fine resolution of the impact phenomena needed for distinguishing the different collision regimes. Results of these investigations are presented qualitatively in Fig. 5.42.

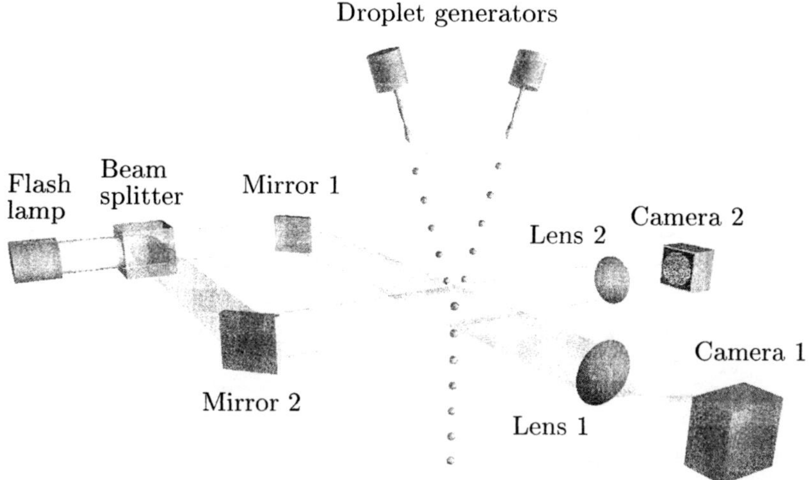

Fig. 5.43. Experimental setup for the observation of colliding droplets from two directions, which are perpendicular to each other

In the experiments described so far the droplet collisions were observed from one direction, usually perpendicular to the plane containing the initial velocities of both initial droplet streams. For a correct interpretation of the transient three-dimensional structures immediately after a collision it is desirable to have photos from different directions. An experimental setup for observing colliding droplet streams from two directions perpendicular to each other is depicted in Fig. 5.43. Some results are shown in Fig. 5.44 for different impact parameter B. In this figure the Weber number was constant $We \approx 76$, while the impact parameter B has been varied. Different situations are observed after the collision. In (a) a smaller satellite droplet and two larger droplets of approximately the same size form. For a smaller impact parameter in (b) this satellite droplet disappears. A further decrease of the impact parameter leads to coalescence of the colliding droplets (c), whereas for a head-on collision with $B = 0$ the colliding droplet separate again.

The same experimental setup has been used to study high energy head-on collision of droplets with much higher We numbers than in the examples shown in Fig. 5.44. These high We numbers have been obtained in the experiments mainly by increasing size and velocity of the droplets and by increasing the angle between the droplet streams to 67°. The initial spacing between neighboring droplets has been enlarged using the modulation technique described in Sect. 2.3. A collision regime with new effects has been found, which is not presented in Fig. 5.42. The photos of two colliding droplet streams in Fig. 5.45 show the transition between reflective separation and a mode, which is similar to splashing of droplets impacting on a liquid film, which has been shown in Fig. 5.37. The droplet liquid was iso-propanol with the

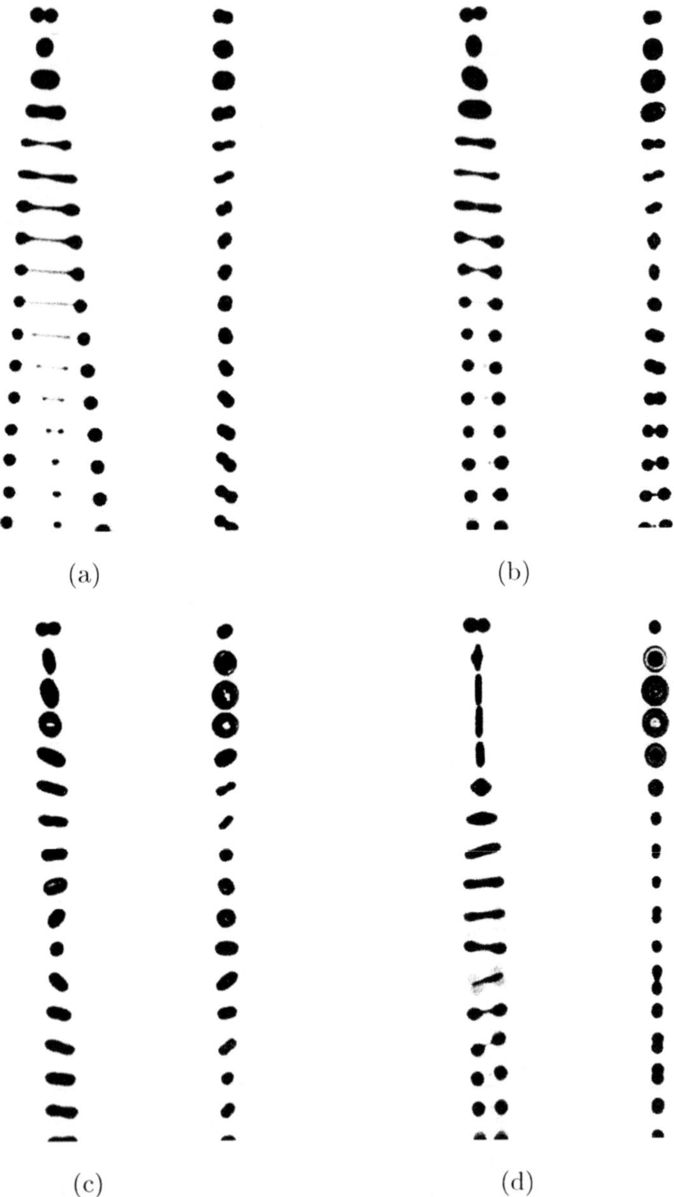

Fig. 5.44. Binary droplet collisions observed from two directions perpendicular to each other using the setup of Fig. 5.43. The droplet diameter was $d \approx 98\,\mu\text{m}$ with an inter-droplet distance in a stream of $s \approx 197\,\mu\text{m}$ for all collisions shown, the velocity was $v \approx 9.5\,\text{m/s}$. The impact parameter $B = b/d$ was $B = 0.738$ in case (a), $B = 0.643$ in case (b), $B = 0.6$ in case (c), and $B = 0$ in case (d)

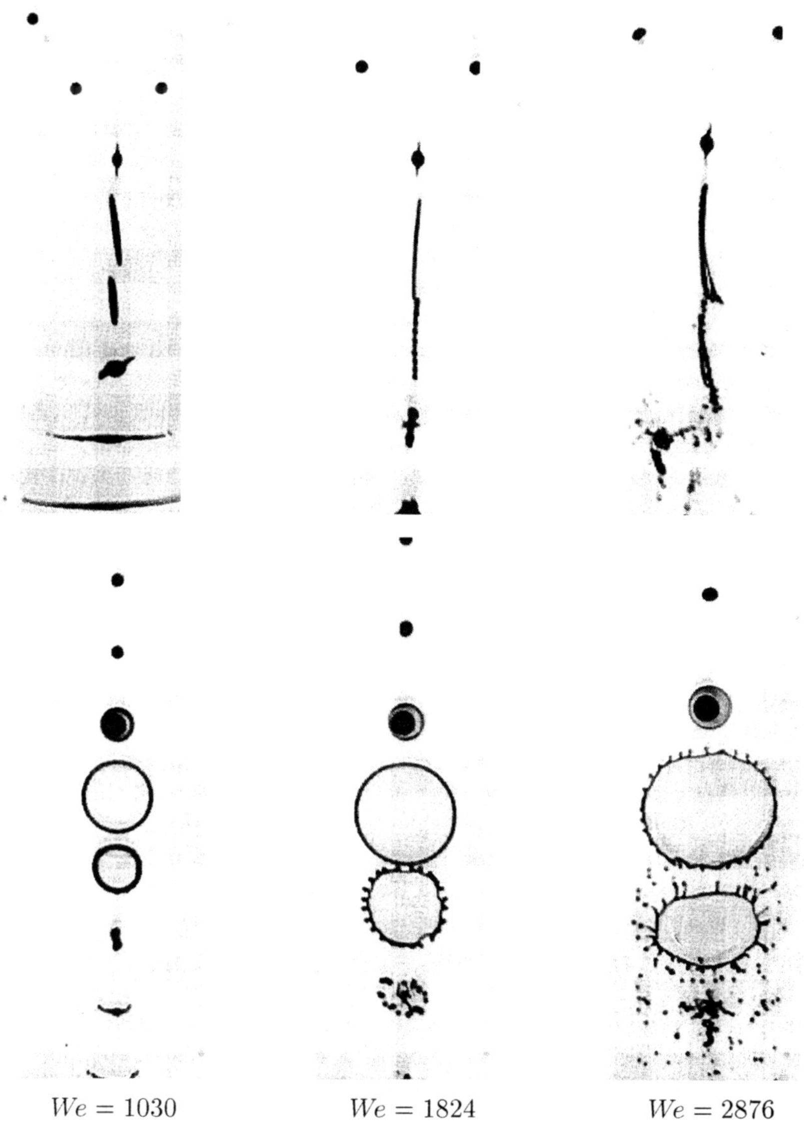

Fig. 5.45. Transition between reflective separation (left) and splashing (right). Photographs of two colliding droplet streams at three different values of the Weber number. The upper row shows the plane of the trajectories of the droplet streams, the lower row shows the corresponding perpendicular view. The parameters of the collision are from left to right $v_r = 12.8\,\mathrm{m/s}$; $16.4\,\mathrm{m/s}$; $19.9\,\mathrm{m/s}$, $d = 170.5\,\mu\mathrm{m}$; $184\,\mu\mathrm{m}$; $197\,\mu\mathrm{m}$, $We = 1030$; 1824; 2876, and $Re = 759$; 1050; 1364.

density $\varrho = 800\,\mathrm{kg/m^3}$, the surface tension $\sigma = 0.0217\,\mathrm{N/m}$, and the viscosity $\eta = 0.0023\,\mathrm{kg/(ms)}$.

In Fig. 5.45 high energy droplet collisions for three different Weber numbers are shown. In the upper row of the figure the collision process is shown in the plane of the initial velocities of the droplet streams. In the lower row the same situation is presented in the corresponding perpendicular view. As can be seen, the initial phase of the collision process looks similar for all three cases shown. Immediately after the impact a disc with a liquid sphere in the center forms. After a short time the disc transforms into a thin liquid film with a thicker rim. Later phases of the collision process depend strongly on the Weber number. For the lower Weber number shown on the left hand side of the figure the thickness of the rim starts to increase, whereas the diameter of the disc decreases. After the collapse of the disc a stretched filament is spreading in the plane shown in the upper photograph. This filament, which may disintegrate into several droplets, is typical for the so-called reflective separation described in Ref. [353]. At higher Weber numbers the rim of the disc shows growing disturbances, which form first nodes. This can already be observed in the center column for the Weber number $We = 1824$. Then liquid fingers form in radial direction, which finally disintegrate into many tiny droplets, as can be seen in the column on the right hand side for the Weber number $We = 2876$. There the overall result after disintegration is a ring of secondary droplets around a droplet cloud originating from the collapse of the remaining liquid disc in the center. This is typical for splashing, which is observed for instance in collisions between droplets and thin liquid films on solid walls.

Numerical calculations of binary droplet collisions at higher energy show the same effects described above, as growing disturbances with the formation of fingers and disintegration into droplets. These calculations have been performed with the Navier-Stokes method described in Sect. 5.3.5. More details of these experiments may be found in Ref. [371].

5.4.3 Numerical Results for Binary Droplet Collisions

The numerical methods described in Sect. 5.3.5 have been used to simulate binary droplet collisions. The outcome of a droplet collision depends on a rather large number of parameters. For a dimensionless representation the following dimensionless parameters have been introduced: the geometrical parameter $B = b/d$, the Weber number $We = \varrho d v^2/\sigma$, the Reynolds number $Re = \varrho d v/\eta$, and the time parameter $T = vt/d$. In these definitions d represents the droplet diameter. Figure 5.46 shows a comparison of a lattice Boltzmann simulation with an experimental result, which has been obtained with camera 1 in the experimental setup of Fig. 5.43. The results of the numerical simulation, especially the formation of the satellite droplet, are very satisfying.

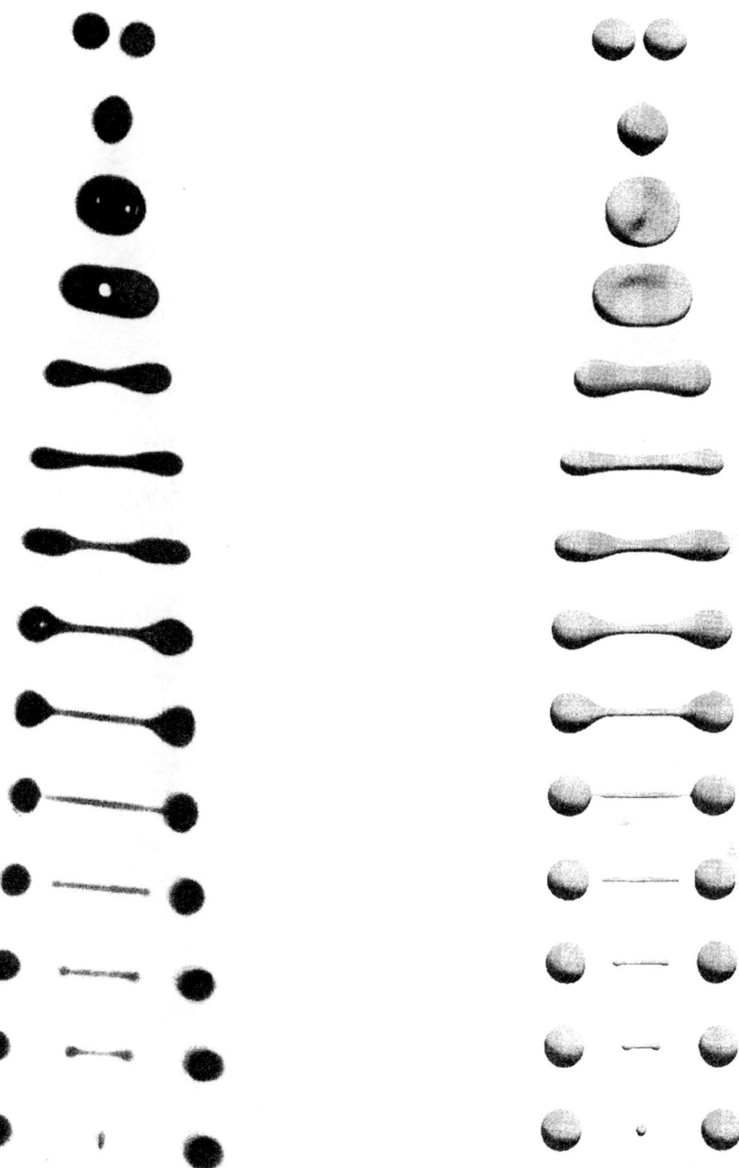

Fig. 5.46. Comparison of experimental result (left) and lattice Boltzmann simulation (right) of binary droplet collision with $We = 106$, $B = 0.5$, and $Re = 100$. The colliding droplets have the same size. The agreement between the droplet shapes obtained by experiment and simulation is very good (from M. Schelkle, ITLR)

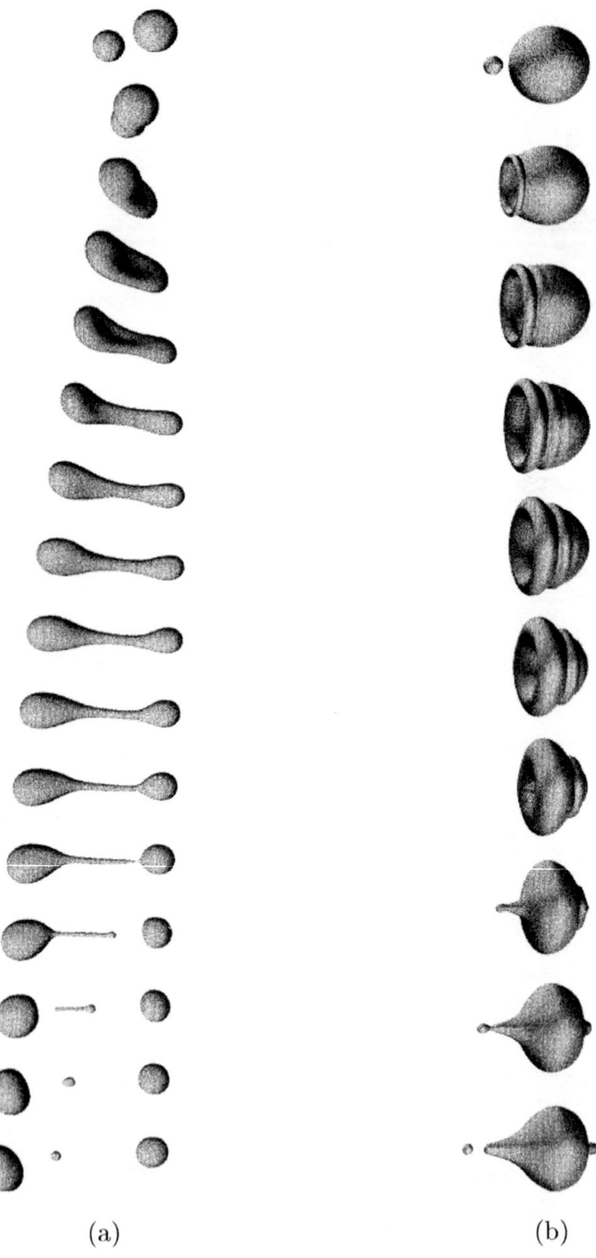

Fig. 5.47. Numerical simulations of droplet collisions for droplets with different diameters. Lattice Boltzmann simulations (a) with $We = 81$, $Re = 75$, $B = 0.5714$, $D = d_1/d_2 = 3/4$. Navier-Stokes simulations (b) with $We = 180$, $Re = 3600$, $B = 0$, $D = d_1/d_2 = 1/9$ (from M. Rieber and M. Schelkle, ITLR)

When the diameters of the two colliding droplets are not equal, the additional parameter $D = d_1/d_2$ is introduced. Figure 5.47 shows two simulations of collisions between droplets with different diameters.

6. Experiments to Study Phase Transition Processes

6.1 Introduction

Processes with droplets in technical and in natural systems involve different phase transitions. The evaporation of fuel droplets is essential for heterogeneous combustion. Freezing and melting as well as condensation and evaporation are important for cloud droplets in the atmosphere of the Earth. Sublimation, the phase transition between solid and vapor, is observed for instance with frozen water droplets, or when ice crystals grow on frozen droplets. Another kind of phase transition may occur in the case of solutions. If a droplet consisting of a solution evaporates a solid particle or in many cases a crystal remains at last.

Phase transition processes in droplets can be studied with different experimental setups. Some of these will be presented in this chapter. Described are the principles of the experimental arrangements, which may vary in detail from application to application. Optical or acoustical levitation allows to study the behavior of single droplets. Different configurations of lenses and sensors are used to observe the scattered light of levitated droplets, or droplets moving within droplet streams. Optical arrangements to determine size and shape, index of refraction and temperature of the droplets will be presented.

Most of the experimental results presented in this chapter have been obtained with setups, which were the same as or similar to the ones described in Chap. 4. Evaporation processes of droplets were studied on droplets within monodisperse droplet streams and on single droplets levitated optically or acoustically.

Closely related to evaporation is droplet combustion. Basic phenomena can be observed in burning droplet streams. As the evaporation rate, in this case called burning rate, is high, even the influence of neighboring droplets can be studied. Droplet heating and flame propagation are interesting phenomena observed in spray combustion processes. Disruption or microexplosion of droplets may occur, when burning droplets consist of fuel mixtures.

The phase transition from liquid to solid, commonly called freezing, has been studied on optically levitated water droplets. Before the droplets freeze they are supercooled to temperatures below 0°C. The frozen droplet sublimates under subsaturated conditions, whereas crystals grow on the frozen droplet under supersaturated conditions.

At very high ambient pressure the critical point of the droplet liquid may be reached. A distinction between liquid and gaseous phase is no longer possible. Droplets do not exist any more as the surface tension becomes zero. At high pressures sophisticated experimental setups are necessary.

All experiments shown in this chapter are examples to give an impression of the behavior of droplets under different ambient conditions. The experiments should be suggestions for investigations of phase transition processes.

6.2 Experimental Setups

6.2.1 Arrangements and Sensors to Detect Scattered Light

Optical measurement techniques, which use light scattered by droplets to determine droplet properties, have become very popular because they are practically nonintrusive. Depending on the purpose of the measurement and the physical principle of the measurement technique, different optical arrangements with different types of detectors or sensors are used to detect the scattered light. These arrangements are employed in different experimental setups to study for instance droplet streams or levitated droplets. Three principal types of sensors will be described below in more detail. In the descriptions given here it is supposed, that the behavior of lenses used is according to Eq. 4.1.

In many cases the output signal of the sensor is used to determine the intensity of the incident light. With such sensors, like photomultipliers or different types of photodiodes, no spatial light intensity distribution can be obtained; therefore these sensors may be called point sensors, to distinguish them from linear or planar sensors.

Lenses are often used for the collection of light scattered within a given solid angle, as the area of the light-sensitive sensor surface is in many cases rather small. The light collected by the lens is focused on the sensor. The solid angle is determined by the distance s_l between the lens and the droplet, and by the aperture of the lens. Only if s_l is larger than the focal length f_l of the lens a focus is obtained at the distance s_l' from the lens, which is given by Equ. 4.1. Since s_l is very large in comparison with the droplet radius the scattered light is supposed to come from the center of the droplet. The application of lenses allows large distances between droplet and sensor. Point sensors are used for instance in LDV, PDV, L2F velocimetry, briefly described in Sect. 4, or in a rainbow measurement techniques according to van Beeck [372].

Linear sensors like linear CCD cameras or CCD arrays detect light along a line. Linear CCD sensors consist of small light sensitive elements so-called pixels arranged in a line. Measurements with a linear sensor can be compared directly with calculations of the intensity distribution of the scattered light, as presented in Sect. 1.13; then, however, the extension of the pixels perpendicular to the array line should be relatively small.

The intensity distribution of the scattered light in the far-field can be detected directly without a lens, when the distance between sensor and droplet is large enough. Lenses may be applied for different purposes. When lens and sensor are arranged in the same way as described above, the near-field of the scattered light is observed and the intensity maxima which belong to glare points, which have been described in Sect. 1.13.2, can be detected in the forward scattered light. Glare points allow to determine the droplet size and refractive index, as described in Sects. 4.6.5 and 4.7.3. The intensity distribution of the far-field may be obtained if a larger distance s_l' between sensor and lens is chosen. In this case it is possible to put an orifice in the plane, where the scattered light is focused. This spatial filtering with an orifice allows to pass only light coming directly from the droplet; disturbances due to other light sources near the measurement volume like ignition sources or reflections from parts of the mechanical setup are avoided. In a common arrangement the distance s_l between lens and the droplet equals the focal length f_l of the lens. In this case the scattered light behind the lens is parallel and the sensor can be positioned in any place.

For the determination of the rainbow position the special arrangement sketched in Fig. 6.1 is suitable [105]. The sensor is positioned exactly in the

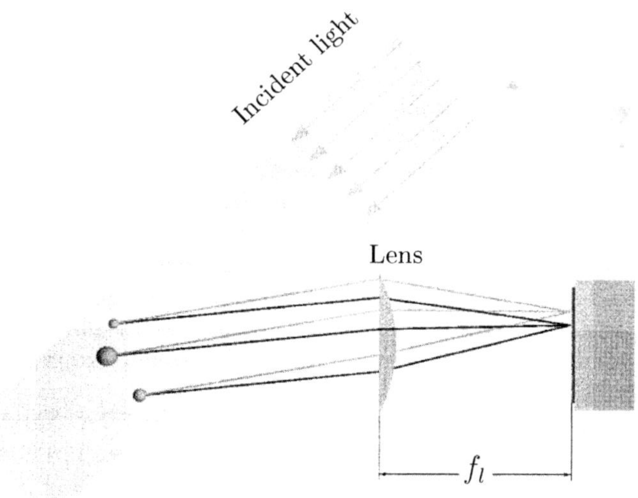

Fig. 6.1. Optical arrangement to study the scattered light in the rainbow region

focal plane of the lens. With this so-called Fourier lens the intensity detected by the sensor is practically independent of the position of the droplets in the measurement volume. This can be seen from the rays plotted in Fig. 6.1. Therefore the result obtained for the rainbow position does not change, when the droplet moves.

A special linear sensor is the position-sensing-diode, called PSD sensor. The output of a PSD sensor is not the intensity distribution along the sensor, but the first moment of this intensity distribution. In addition the total light intensity incident on the sensor surface can be obtained. The use of such sensors is one way to detect the position of the rainbow [105]. Normally PSD sensors are used to detect the position of a light spot on the sensor. When the light scattered by a droplet is focused on the sensor, the droplet position can be determined in the direction parallel to the sensor line. To detect the position in two directions planar PSD sensors are available.

Planar sensors are used for imaging purposes as described in Sects. 4.2 to 4.4. The sensor may be a common photographic film or the CCD chip of a video camera with pixels arranged in a plane. A three-dimensional impression of the droplets may be obtained by diffuse illumination. All arrangements of lens and sensor described for linear sensors can be applied. Planar sensors can be employed for both, for near-field observations to obtain the glare points as well as for far-field observation in the forward direction for size measurements as described in Sect. 4.6.3.

6.2.2 Optically Levitated Droplets

Optical levitation and stabilization of droplets is based on radiation pressure forces as described in Sect. 1.13.6. This technique allows to study processes in systems consisting of single droplets, as described in Sect. 3.4.9. For different purposes like evaporation, condensation, or freezing studies distinct experimental configurations are necessary. Main features of the setup concern the generation and trapping of the droplets, the manipulation of the levitated droplets, the choice of the ambient conditions of the droplets, and the application of different measurement techniques. An example of a possible configuration is shown in Fig. 6.2 schematically. In most cases the droplet has to be levitated within an observation chamber to allow a control of the ambient conditions and to prevent that the trapped droplet is lost due to external disturbances.

In the case shown in Fig. 6.2 the dimensions of the chamber are large in comparison with the droplet size. For a droplet with a diameter of approximately hundred micrometer the distance to the walls of the observation chamber is more than hundred droplet diameters, the vertical length of the chamber is more than thousand droplet diameters. At least one of the walls of the chamber is made of glass or has appropriate windows to allow optical access for the observation of the droplet. The laser beam for levitation and stabilization is directed vertically upwards via a mirror into the observation

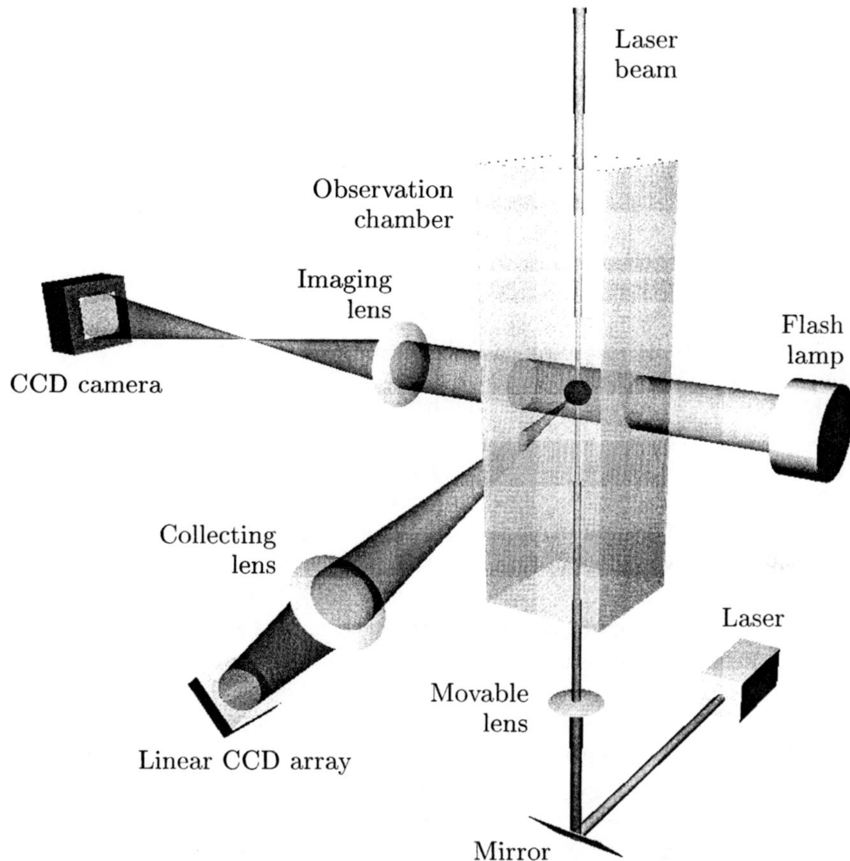

Fig. 6.2. Experimental arrangement for studying optically levitated droplets. The position of the droplet in the observation chamber can be determined with the movable lens. Shadows of the droplet are obtained with a CCD camera, whereas a linear CCD array is used to record the intensity distribution of the laser light scattered in the backward direction

chamber. The movable lens focuses the beam; the droplet is stabilized near the focal plane of this lens. The position of the droplet is closely related to the position of the lens. The position of the droplet can be manipulated by moving the lens.

High intensities of the laser light exist at the location of the droplet, as the laser beam is focused. Therefore it should be checked if the absorption of the droplet liquid is so small, that droplet heating can be neglected. On the other hand, absorption of the laser light by the droplet liquid allows to increase the temperature of the droplet. A more detailed description of droplet heating may be found in Ref. [95].

After trapping the droplet is brought in the desired observating position with a vertical movement of the lens. In the observation position the shape of the droplet and the scattered light can be observed. In the setup depicted in Fig. 6.2 the droplet is imaged by a lens and a CCD video camera. For the illumination a flash light has been used to obtain shadows of the droplets. This technique has been described in Sect. 4.4. The scattered light in the backward hemisphere is observed with a collecting lens and a linear CCD array or a PSD sensor, in order to record the light scattered in the region of the first rainbow. The arrangement of the collecting lens and the sensor is according to Fig. 6.1. Various other observation techniques are possible. The polarization behavior of the droplet liquid during freezing can be studied for instance with the optical setup shown in Fig. 6.44.

Phase transition processes of the droplet liquid depend essentially on the ambient conditions in the observation chamber. The temperature in the observation chamber can be varied by heating or cooling the walls. Sections of the chamber, whose temperature cannot be controlled, have to be thermally insulated. In addition the outside of the windows has to be flushed with dry air in order to prevent condensation at very low temperatures.

Defined temperature gradients along the vertical axis of the chamber can be achieved by heating or cooling different sections of the observation chamber. Well defined variations of the droplet's ambient temperature with time can be attained by moving the droplet along the axis.

The droplet can be bathed in a gas stream, which is directed vertically downwards with a velocity of a few centimeter per second. This gas stream can be dry or humidified. When a humidified gas stream is cooled while flowing through the observation chamber, supersaturated conditions can be obtained. In this case the inner side of the windows may have to be heated in order to prevent condensation or icing. With temperature gradients inside the chamber one should be aware of natural convection effects.

An essential problem of optical levitation experiments is the method for injecting the droplets into the observation chamber and trapping them there. Different techniques have been described. One possibility is to spray a droplet cloud into the chamber using for instance an ultrasonic droplet generator. This droplet generator produces droplets with low initial velocities. In the observation chamber the droplets fall downwards. Droplets which move along the axis of the laser beam may be trapped if their size is appropriate with respect to the power and the shape of the laser beam. However, more than one droplet may be trapped. These additional droplets are stabilized not along the beam axis. They are stabilized in the diffraction maximum of the diffraction pattern of the droplet, which has the lowest vertical position [173]. The probability to trap more than one droplet can be reduced if the droplets are sprayed first into a larger chamber above the observation chamber. A small gate is then opened to let only a small number of droplets fall into the observation chamber [173].

Another technique for injecting the droplets into the observation chamber uses a droplet on demand generator. This allows to bring only one single droplet into the observation chamber. The droplet is injected from the side of the chamber as depicted for example in Fig. 6.44. When the injected droplet has traveled its stop distance, it must fall exactly along the axis of the laser beam. Therefore the droplet generator has to be adjusted precisely.

The location, where the droplets are brought into the chamber is normally above the position, at which the droplet is stabilized. At the beginning of an experiment this plane is not necessarily identical with the observation position. Depending on the purpose of the experiment the observation position may be above or below the position in which the droplet is levitated first. If this first position is above the observation position it will be difficult to obtain dry conditions of the gas stream due to sticking of droplets on the walls of the chamber especially if a droplet cloud is sprayed in the chamber. This can be avoided, when a droplet on demand generator is used carefully.

The freezing experiments described in Sect. 6.5 have been performed with such a configuration. To study evaporation the droplets have been brought into the chamber below the observation position and then moved upwards with the lens until the observation position is reached. In this position the droplets are bathed in a dry air stream. If the evaporation rate is high the droplets should be levitated as close as possible to the observation position. This is important especially for multicomponent droplets to conserve the desired composition for the beginning of the measurement.

This description of various arrangements for optical levitation of droplets reveals a high flexibility of this technique. The setup can be applied for very different purposes in investigations of single droplets.

6.2.3 Acoustically Levitated Droplets

Acoustic levitation is, like optical levitation, a technique for studying individual droplets. The droplet size range, however, for which levitation is possible, is shifted to larger droplets. It is possible to levitate droplets with diameters up to a few millimeters. It should be emphasized, that very large droplets do not remain spherical. Photographs of the setup consisting of a vibrating plate at the bottom and a reflector at the top are shown in Figs. 3.9 and 3.10 of Sect. 3.4.10. In this section the physical principle of acoustical levitation has been described in more detail. From Fig. 3.10 it can be seen, that more than one droplet can be levitated simultaneously. Even arrays of droplets can be levitated, in which the droplets are aligned along the axes of the levitator. To study the influence of different ambient gases the levitator can be mounted in a chamber.

The distance between the vibrating plate and reflector has to be adjusted very precisely for a given frequency to obtain a standing wave system. This distance between the plates should be large enough to allow good optical access and to bring a droplet in the levitation position between the plates. For larger distances more pressure nodes and therefore more vertical positions, in which a droplet can be levitated, are obtained, as can be seen from Fig. 3.10. Naturally the distance cannot be varied continuously for a given excitation frequency; only steps of $\lambda/2$ are possible. Various arrangements to photograph the droplets are possible. The configuration allows the application of different optical measurement techniques.

The choice of the technique used for injecting the droplets into the space between vibrating plate and reflector, is important. A special method is illustrated in Fig. 3.10. A fine spray, which may be generated for example with an ultrasonic droplet generator, is sprayed into the space between the plates. In such sprays the relative velocities between the droplets are low. The fine droplets are levitated in horizontal node planes of the system of standing sound waves. This can be seen in Fig. 3.10(a). Then the droplets begin to coagulate to form larger droplets, as shown in Fig. 3.10(b). Another method is to bring a single droplet, which is suspended on a thin fiber, in the levitation position. It is possible to use a dropper instead of the fiber.

With a droplet on demand generator small droplets can be injected into the levitator. It is possible to feed a larger droplet already formed and levitated. In Fig. 5.6 of Sect. 5.2.2 the trajectories of these small droplets in the acoustic field are visible. Coagulation occurs, when the small droplets collide with the large droplet. The trace of the small droplet is visible due to continuous illumination. The droplets themselves can be recognized due to illumination by a flash lamp using the setup described in Sect. 4.4. With this method the initial size of the levitated droplet can be adjusted very precisely. This technique allows for instance to dope the large droplet with small amounts of special liquids like surfactants.

6.2.4 Droplet Streams

Experiments with monodisperse droplet streams take advantage of the regularity of the droplets and the periodicity of the observed phenomena, as described in Sect. 3.3.1. For imaging of the droplets the droplet stroboscope technique from Sect. 4.3 may be used instead of a flash lamp. For the detection of the scattered light sensing and evaluation within an extremely short time are not necessarily needed. Due to the high regularity of the processes the sensors can detect a superposition of the scattered light of many droplets instead of detecting the scattered light of a single droplet, which traverses the measurement volume.

An example of a setup is shown schematically in Fig. 6.3. The detectors are arranged in a horizontal plane, which is perpendicular to the droplet stream. The distance of this horizontal plane from the orifice of the droplet generator determines the observation position of the droplets and can be changed easily with an adjustment of the vertical position of the droplet generator. The droplet stream is illuminated continuously with a laser beam. The direction of polarization of the monochromatic laser is perpendicular to

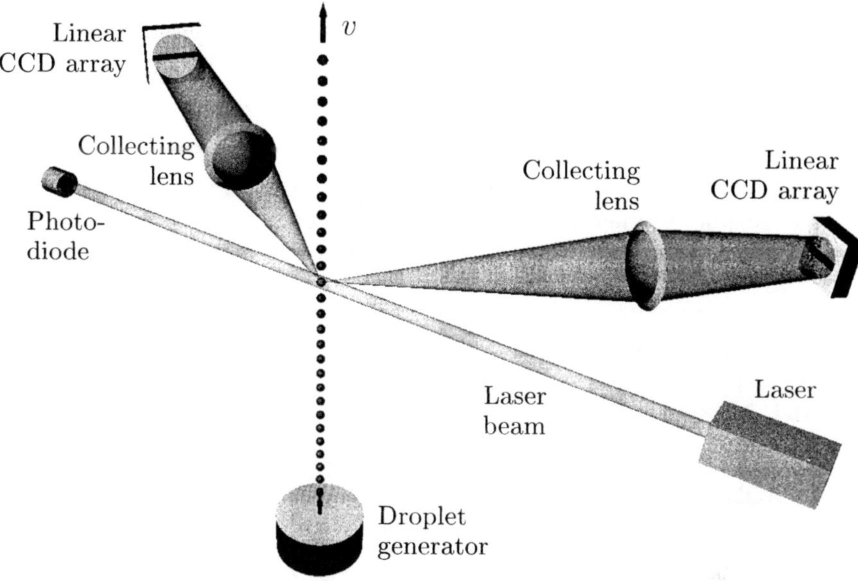

Fig. 6.3. Example of experimental setup for experiments with monodisperse droplet streams. Here linear CCD arrays are used for detecting the distribution of the scattered light

the horizontal plane in order to detect the intensity distribution of the first rainbow. Depending on the purpose of the measurement the beam should have an appropriate diameter, with respect to droplet size and spacing. The adequate beam diameter is obtained by focusing the laser beam with a lens, which is not shown in the figure. The initial diameter of the laser beam and the focal length of the lens determine the waist of the beam [108].

The laser light is detected by a photodiode as shown in Fig. 6.3. Every time a droplet enters or leaves the laser beam the output signal of the photodiode changes. This allows to control the regularity of the droplet stream and to check if the disintegration process is completed. In Fig. 6.3 the scattered light is observed with a linear CCD camera and a lens in the forward direction at scattering angles $\theta < 90°$. With this arrangement the droplet size can be obtained as described in Sect. 4.6.3. A second similar arrangement is available to detect the scattered light in the backward hemisphere for scattering angles $\theta > 90°$ in the region of the first rainbow. This arrangement of lens and sensor is used for the detection of the rainbow position. According to the description in Sect. 6.2.1 the sensor is placed in the focal plane of the lens.

The described setup is also suitable for investigations on droplet arrays consisting of several parallel droplet streams, as described in Sect. 3.3.2. Measurements are then performed at different distances h for each droplet stream. To change between the neighboring streams the droplet generator can be adjusted in a plane perpendicular to the streams. Measurements are possible even when the droplet streams are surrounded by a flame, as can be seen from Figs. 6.27 or 6.33.

The setup described in Fig. 6.3 is an example. For different purposes adaptations are necessary. For instance two lasers of different colors may be necessary if different diameters of the laser beam are suitable. For instance for size measurements according to Sect. 4.6.3 another beam diameter is appropriate than for velocity measurements according to Ref. [207].

6.3 Evaporation

6.3.1 Measurements on Single Droplets

Many of the experiments described in the following have been performed with optically or acoustically levitated droplets. Optically levitated droplets can be observed during a long part of their life time, when they have been levitated, stabilized, and moved in the observation position. However, due to this inevitable initial procedure it is difficult to study the initial phase of evaporation just after generation of the droplet. For the investigation of the initial phase of evaporation monodisperse droplet streams are more suitable. Results for the evaporation of droplet streams will be presented in Sect. 6.3.2.

Droplets of different initial size have been levitated optically and it is shown in the next section, that the evaporation process can be described for most of the lifetime by the d^2-law, which has been introduced in Sect. 1.12. Different droplet liquids show usually different evaporation rates. This can be seen directly in diagrams when the influence of the initial radius is eliminated according to the d^2-law. Rates of the change of radius were measured directly using partial wave resonances, as described in Sect. 4.10. For the description of evaporating multicomponent droplets the d^2-law is not or only partly appropriate. Measurements of droplet size and size change rates during the whole evaporation process are therefore important for multicomponent droplets. Under supersaturated ambient conditions condensation of vapor on the levitated droplets has been observed.

Different Initial Droplet Sizes. Experimental results for the evaporation behavior of optically levitated n-hexadecane droplets of different initial size are shown in Fig. 6.4. The experiments have been performed at room temperature. The droplets were bathed in a vertical downward directed air stream to avoid accumulation of the vapor in the observation chamber. With this example it can be shown, that the influence of different initial sizes can be eliminated, when the d^2-law describes the evaporation process with sufficient accuracy. In this case one has

$$r^2 = r_0^2 - \beta_v t \, , \tag{6.1}$$

where r is the momentary and r_0 the initial droplet radius. The evaporation

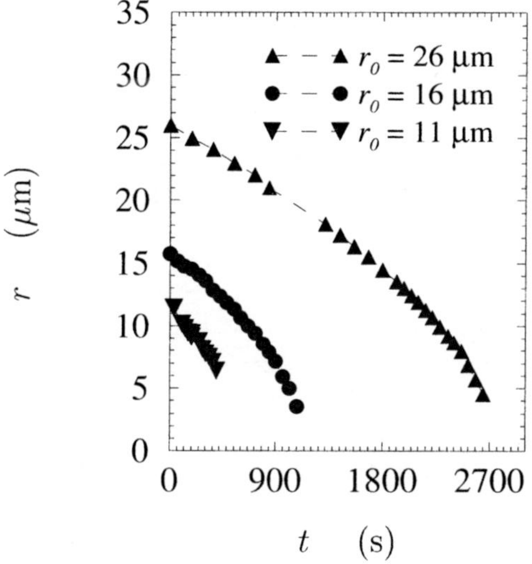

Fig. 6.4. Droplet radius as a function of time for three droplets of n-hexadecane with different initial radii r_0

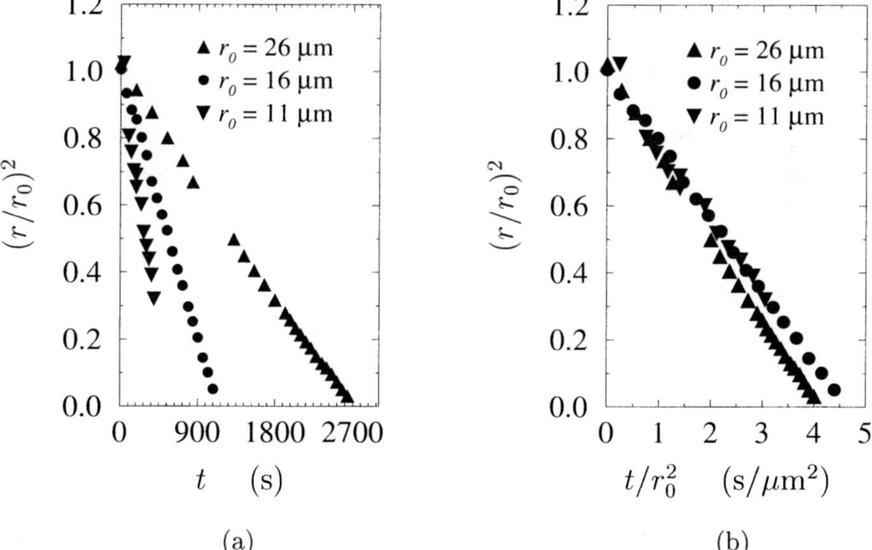

Fig. 6.5. Evolution of $(r/r_0)^2$ in (a) as a function of time t and in (b) as a function of t/r_0^2. The data are from n-hexadecane droplets with different initial radii

coefficient β_v depends on the thermodynamic properties of the droplet liquid and on conditions of the ambiance of the droplets like temperature, velocity, or distance to neighboring droplets. The d^2-law does not account for the initial phase of evaporation, when the droplet temperature changes with time or is not uniform [373]. However, this initial part of evaporation has not been resolved in the described experiment. The measurements shown in Fig. 6.4 can therefore be described by the d^2-law without too large errors. When Eq. 6.1 is divided by the square of the initial radius, one obtains

$$\frac{r^2}{r_0^2} = 1 - \beta_v \frac{t}{r_0^2} \ . \tag{6.2}$$

The plot r^2/r_0^2 as a function of time t results in straight lines, as can be seen from Fig. 6.5(a), where the data of Fig. 6.4 has been used. From a linear regression for each measurement one obtains the slope of the curves, which according to Eq. 6.2 is β_v/r_0^2. When, however, $(r/r_0)^2$ is plotted against t/r_0^2, the negative slope of the curves represents directly the evaporation coefficient β_v. A corresponding diagram is shown in Fig. 6.5(b).

All measurements show approximately the same evaporation coefficient β_v. As can be seen, this representation of evaporation is independent of the initial droplet radius. Small differences between the measurements may be caused by slightly different ambient conditions, for instance a different flow rate or temperature of the gas stream. The evaporation coefficient β_v characterizes the evaporation process for a given pure droplet liquid under given ambient conditions.

Different Droplet Liquids. Different liquids with different vapor pressures result in different evaporation coefficients β_v under the same ambient conditions. Results for three hydrocarbons are shown in Fig. 6.6. Here the influence of the initial radius has again been eliminated. The initial radius was in the range of $16\,\mu\text{m} \leq r_0 \leq 27\,\mu\text{m}$. The different behavior of the three liquids is clearly pronounced by a different value of the evaporation coefficient β_v.

The change of droplet radius with time dr/dt is obtained from Eq. 6.1. By differentiating this equation one finds

$$\frac{dr}{dt} = -\frac{\beta_v}{2r} \ . \tag{6.3}$$

From Eq. 6.3 it can be seen, that for a constant evaporation coefficient β_v the change of radius per time dr/dt depends only on the momentary radius r. Using the method based on partial wave resonances, which has been described in Sect. 4.10, the rate dr/dt can be measured directly. In order to determine the evaporation coefficient β_v it is sufficient according to Eq. 6.3, when the rate dr/dt and the droplet radius r are measured at the same time.

With a PSD sensor, which detects the scattered light in the rainbow region, dr/dt can be determined quasi continuously. The output signal of the PSD sensor is a measure for the rainbow position and according to the description in Sect. 4.10, the rate of radius change per time can be determined.

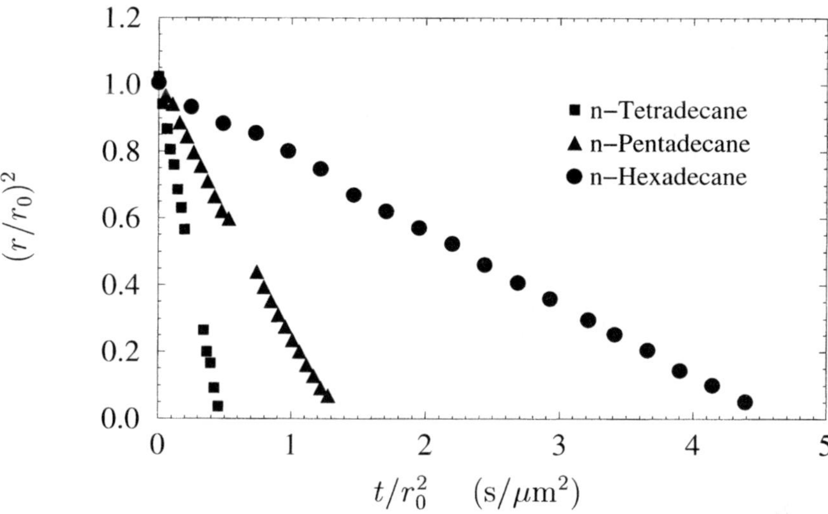

Fig. 6.6. Variation of $(r/r_0)^2$ with t/r_0^2 for three droplets, consisting of n-tetradecane, n-pentadecane, and n-hexadecane

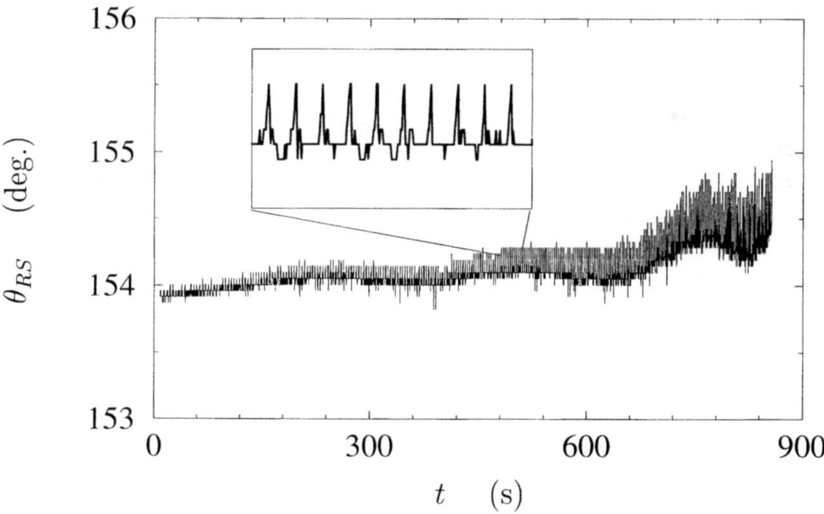

Fig. 6.7. Output signal of PSD sensor for evaporating n-pentadecane droplet. Recorded is the region of the first rainbow. This signal, which is proportional to the first moment of the intensity distribution, is interpreted as rainbow position θ_{rs}

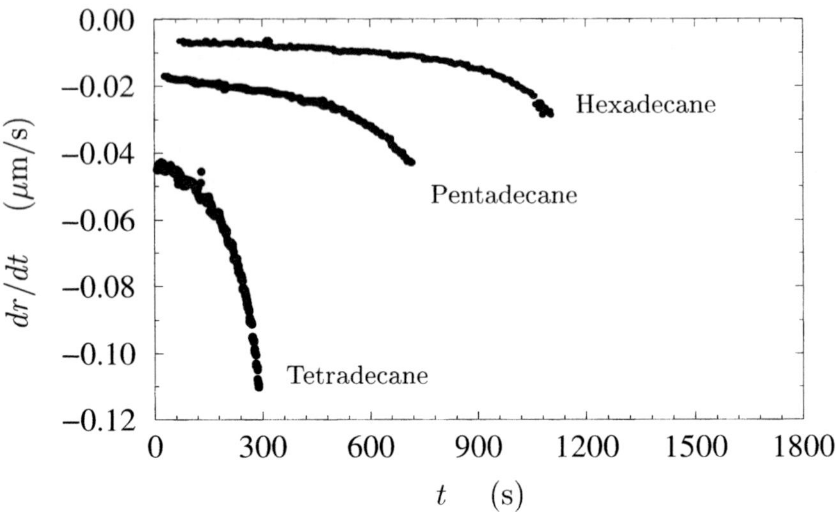

Fig. 6.8. Size change rate as a function of time for three droplets consisting of n-tetradecane, n-pentadecane, and n-hexadecane

The output signal of a PSD sensor is shown in Fig. 6.7 for an optically levitated evaporating n-pentadecane droplet. The output signal of this sensor is proportional to the first moment of the light intensity distribution on the sensitive sensor surface, as described in Sect. 4.6.5. Oscillations of the first moment caused by partial wave resonances can be seen clearly in the enlarged view. From the frequency of these oscillations the rate dr/dt is obtained. This rate is shown in Fig. 6.8 for three hydrocarbon droplets, the corresponding evolution of radius in Fig. 6.6. The amount of liquid, which evaporates per time, can be determined easily, when the droplet radius is measured at least once during the evaporation process.

The influence of different ambient temperatures on the evaporation rate is presented in Refs. [174, 374]. There the evaporation of supercooled water droplets has been studied. For experiments at low temperatures a setup, as sketched in Fig. 6.44, can be employed. In this setup the walls of the observation chamber can be cooled.

In all these measurements pure liquids have been used and the ambient conditions of the droplets were held constant during the evaporation process. This results in a constant evaporation coefficient β_v, the evaporation process can be described with the d^2-law except in the initial phase. The described techniques can be used to study evaporation processes, even when the evaporation rate does not remain constant. Transient evaporation may be due to changes of the ambient conditions or droplet composition during the evaporation of multicomponent droplets.

Multicomponent Droplets. In many technical applications droplets consist of a mixture of two or more pure liquids; this is especially pronounced for fuel droplets used in technical combustion processes. Experimental investigations of evaporation or combustion of multicomponent droplets have been performed by different workers. The evaporation of sessile droplets consisting of binary mixtures was studied by W.-J. Yang et al. [375]. Randolph et al. studied burning droplets consisting of n-hexadecane and one other n-alcane, which was n-tetradecane, n-dodecane, and n-decane. The droplets moved through a flat flame burner and its post combustion zone. These investigators determined the droplet size by imaging the droplets. The mixture ratio during the evaporation and burning process in the hot environment was measured with sampling probes at different locations [376].

The combustion of droplets consisting of n-heptane/n-hexadecane mixtures has been studied by J. C. Yang and Avedisian at low gravity conditions in a drop tower facility [377]. The droplets, which were produced with a droplet on demand generator, were imaged with a high speed camera, in order to determine the size during the combustion process.

The evaporation of commercial fuels, as heavy Diesel fuel, light Diesel fuel, gasoline, and kerosine, has been studied by Elkotb et al. at elevated temperatures [378]. Also binary mixtures of these fuels have been studied. It has to be mentioned, that the commercial fuels are themselves mixtures of different liquids. In this study it was found, that the evaporation coefficient β_v changes with time.

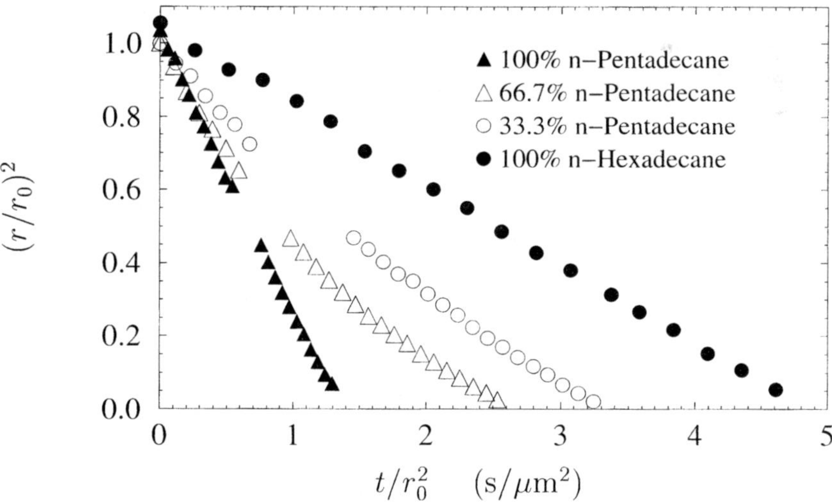

Fig. 6.9. Variation of $(r/r_0)^2$ with t/r_0^2 for four droplets, consisting of two pure liquids and of binary mixtures of these liquids

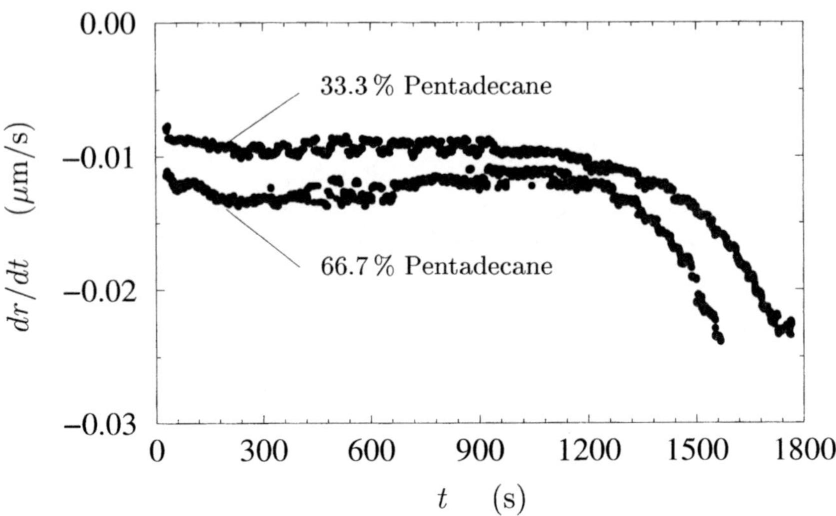

Fig. 6.10. Size change rate as a function of time for two droplets consisting of binary mixtures of pentadecane and hexadecane with different mixture ratio

Here, as an example, results from measurements on optically levitated droplets consisting of binary mixtures of pure liquids will be presented [379]. In Figs. 6.9 and 6.10 some results for mixtures of n-hexadecane and n-pentadecane are shown. The evolution of $(r/r_0)^2$ during the evaporation process is presented in Fig. 6.9. The size has been measured with the interference method described in Sect. 4.6.3. The plot shows $(r/r_0)^2$ as a function of t/r_0^2. For pure hexadecane and pure pentadecane a linear behavior is obtained, as expected. For mixtures of these liquids nonlinear behavior is observed. The slope of the curves, which represents the evaporation coefficient β_v changes with time. Therefore only momentary values of β_v are obtained. In the beginning of the evaporation process the momentary evaporation rate is in between the values found for the pure liquids. Towards the end of the evaporation process the slope is approximately the same as for the less volatile component n-hexadecane. The rate of the size change dr/dt as a function of time is shown in Fig. 6.10 for two mixtures. The data has been obtained with the technique based on the evaluation of partial wave resonances mentioned above. The observed behavior is quite different from the case of pure liquids presented in Fig. 6.8.

The system of water and sulfuric acid is an example for mixtures with very different volatility of the components [380]. Results for droplets with initially 2 vol.% sulfuric acid will be presented. In these experiments a setup similar to the one shown in Fig. 6.44 has been used. The ambient temperature in the observation chamber has been decreased below 0°C. The size of the droplets

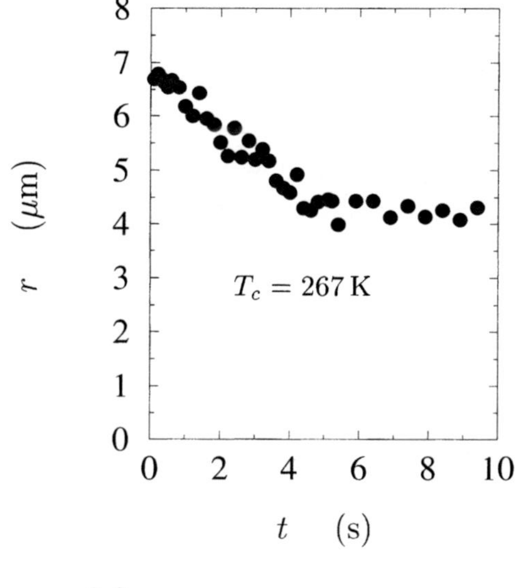

Fig. 6.11. Droplet radius as a function of time for a droplet consisting of water and initially 2 vol.% sulfuric acid

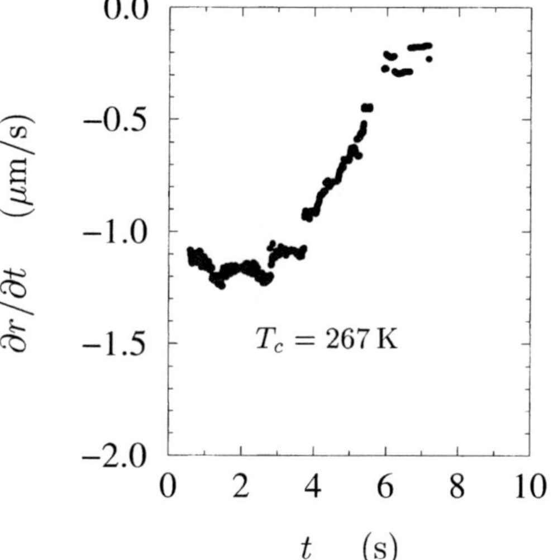

Fig. 6.12. Size change rate as a function of time for a droplet consisting of water and initially 2 vol.% sulfuric acid. It was assumed, that the refractive index of the mixture was constant $n = 1.33$

has been determined with the video observation technique of Sect. 4.4. A result is shown in Fig. 6.11. In the beginning of the evaporation process the droplet radius decreases until a droplet with a high concentration of sulfuric acid of low volatility is obtained. Within the time range of the present experiment a further decrease of the droplet radius is not observed. The same behavior is observed, when the rate dr/dt is determined and plotted as a function of time, as shown in Fig. 6.12. The rate dr/dt tends to zero when the concentration of sulfuric acid increases. These results have been obtained from oscillations in the vertical droplet position caused by partial

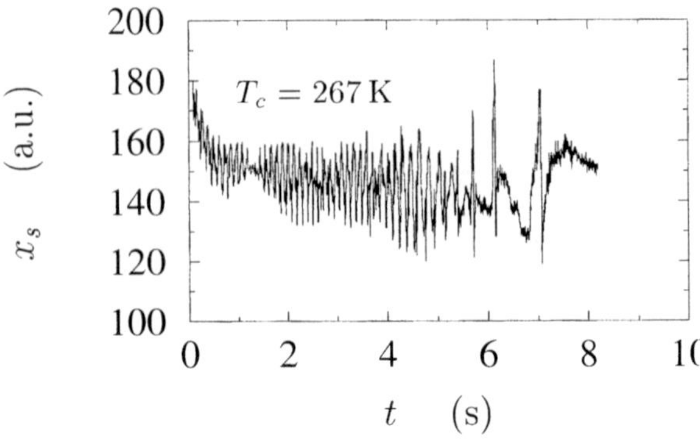

Fig. 6.13. Output signal of PSD-sensor as a function of time for a droplet consisting of water and initially 2 vol.% sulfuric acid

wave resonances. The corresponding output signal of the PSD sensor is shown in Fig. 6.13. It can be seen, that the oscillation frequency decreases with time until very slow oscillations are observed.

Condensing droplet. The experimental setup described in the previous section in connection with evaporating droplets, has been used in another experiment to study condensation processes [380]. For obtaining condensation it is necessary to have supersaturation at the position of the droplet. For this purpose the levitated droplets were bathed in an air stream, which could be humidified. Temperature gradients along the axis of the observation chamber have to be large enough to obtain supersaturation at the location of the levitated droplet. It cannot be avoided, that vapor condenses on the walls and windows of the the chamber. As the geometrical dimensions of the observation chamber are large in comparison with the droplet size, va-

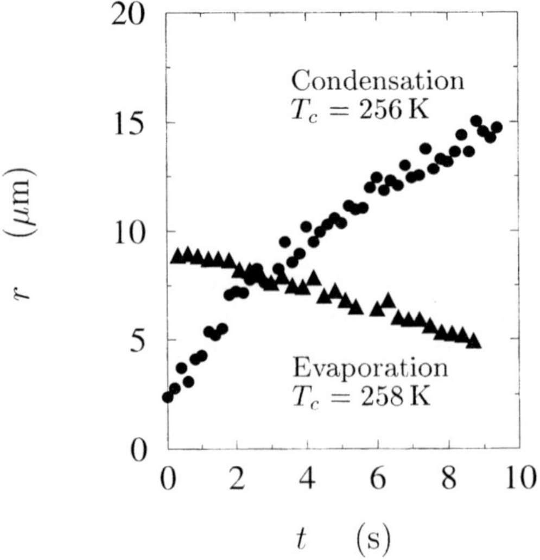

Fig. 6.14. Droplet radius as a function of time for an evaporating and for a condensing water droplet

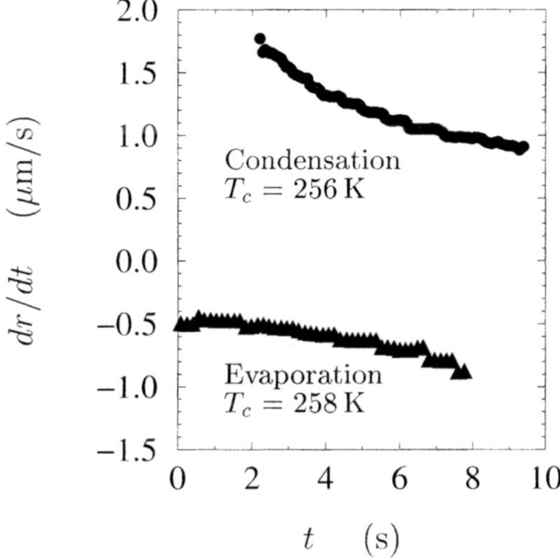

Fig. 6.15. Size change rate as a function of time for an evaporating and for a condensing water droplet

por condenses on the droplet, despite the reduction of the supersaturation by the wall effects. The described experimental arrangement has been used to study evaporation and condensation of supercooled water droplets. The droplet size has been determined from video observations. Results for a condensing and for an evaporating droplet are shown in Fig. 6.14. The rates dr/dt, which have been obtained from the oscillations of the vertical droplet position are shown in Fig. 6.15. It has been found, that the d^2-law is valid for both, condensation and evaporation of pure water droplets. In the case of condensation the evaporation coefficient β_v has negative values, while the rate dr/dt is positive, as shown in Fig. 6.15.

Evaporation of Solutions. In an evaporating droplet, which contains an aqueous solution of a salt, the salt concentration will increase until saturation is reached. With further evaporation the solution within the droplet becomes supersaturated and finally the formation of crystals is observed. The crystals may grow either on nuclei of the own material or on surfaces of foreign material.

Two examples of evaporating droplets consisting of an initially saturated solution of water and sodium chloride at room temperature are shown in Fig. 3.11. The droplets of initially different size are levitated with an acoustic levitator, as described in Sect. 3.4.10. The droplets were brought into the levitator using a thin wire, which was withdrawn as soon as the droplet was levitated. The photographs have been taken at different times during the evaporation process. The time steps are not equal, they were selected to get a good impression of the different stages of the evaporation process. The time increases from top to bottom. The droplets rotate due to inhomogeneities caused by the crystals. Therefore a flash with very short duration had to be used, in order to get sharp photos.

The photographs at the top of Fig. 3.11 have been taken in both cases one minute after levitation. For the photos on the left hand side (a) the droplet liquid was impurified by small crystals on the thin wire. Such impurities may be found often on solid surfaces or the containment of the liquid. Triggered by these particles different crystals start to grow at the droplet surface, as can be seen from the second and third photograph. At last a polycrystalline salt particle remains as can be seen on the photo at the bottom. For the photos on the right hand side (b) it was tried to avoid impurities. It can be seen, that the crystal formation is different in this case. It starts somewhere inside the liquid and more or less regular salt crystals begin to grow. It should be mentioned, that the acoustic field influences the evaporation process and the formation of the crystals. This example shows, that evaporation processes with solutions are influenced by complex parameters like impurities and crystal structure.

Evaporation of electrically charged droplets. Doyle et al. conducted an experimental study of evaporating electrically charged droplets, $60 - 200\,\mu$m in diameter [225]. The charged droplets were produced by applying a voltage of 30 kV to a hypodermic needle on a syringe. By squeezing the syringe a cloud of charged droplets was sprayed into the space between two horizontal metal plates, where they could be suspended with an electrical field. As the droplet evaporate and decrease in size, the electrical charge density on the surface increases, until small highly charged droplets are ejected. The observed phenomena are explained by Rayleigh's stability criterion for charged droplets, which has been described in Sect. 1.4.

6.3.2 Measurements on Droplet Streams

Results presented in this section are obtained from measurements on droplet streams, which were produced with droplet stream generators. Monodisperse droplet streams allow to study evaporation and combustion of droplets in a quasi-stationary system without using high speed measurement techniques [131]. This technique is described in Sect. 3.3.1 and experimental setups, as shown in Fig. 6.3, are used.

These techniques can be used for studying combustion and evaporation of droplets as well. In the case of combustion the ambient temperature is given by the thin flame sheet surrounding the droplets or the droplet streams. The flame temperature is mainly determined by thermodynamic properties of the droplet liquid. In the case of evaporation the ambient temperature far away from the droplet is independent of the droplet properties. It has to be mentioned, that the evaporation and combustion processes of droplets within droplet streams are influenced by the neighboring droplets [381]. A numerical study of droplet stream evaporation has been published by Leiroz and Rangel [382].

In contrast to experimental investigations of optically levitated droplets the initial phase of evaporation and combustion processes is more pronounced in investigations with droplet streams. The liquid jet exiting from a droplet stream generator disintegrates in a very short time and the just formed droplets can be studied as soon as oscillations become negligible. However, after a certain distance the droplet streams become irregular and coalescence between neighboring droplets occurs, as described in Sect. 2.3.

Examples of droplet evaporation at room temperature and normal pressure will be presented in the following. In Fig. 6.16 the evolution of droplet size along monodisperse droplet streams is shown for n-octane and n-pentane. The measurements have been performed with an experimental arrangement similar to the one sketched in Fig. 6.3.

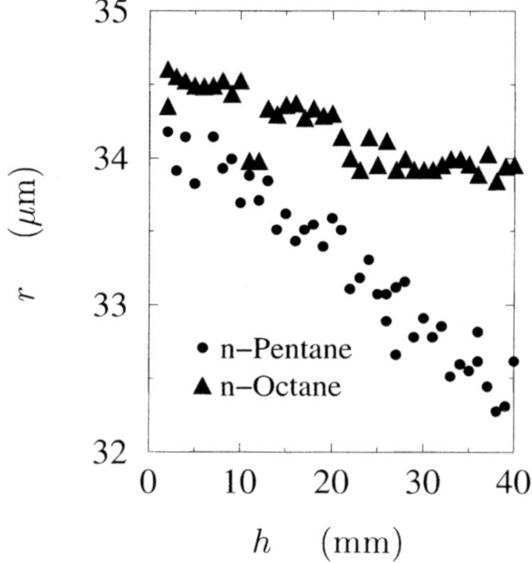

Fig. 6.16. Evolution of droplet radius r along a droplet stream for n-pentane and n-octane

In the experiments described here the droplets pass through a small aperture of a metal plate, which has been placed at a certain distance above the droplet generator. This distance is large enough to neglect oscillations of the droplets passing through the aperture. Between generator and plate the droplet stream has been surrounded with a heatable tube, in order to avoid cooling of the droplets. When the tube is heated appropriately evaporation will be avoided due to the saturated atmosphere within the tube.

The droplet radius has been measured at different distances h above the metal plate using the interference method of Sect. 4.6.3. The initial values of droplet radius, velocity, and temperature are the same for both liquids. Due to the high sensitivity of the interference method even the small radius changes of the n-octane droplets can be detected. The difference between the

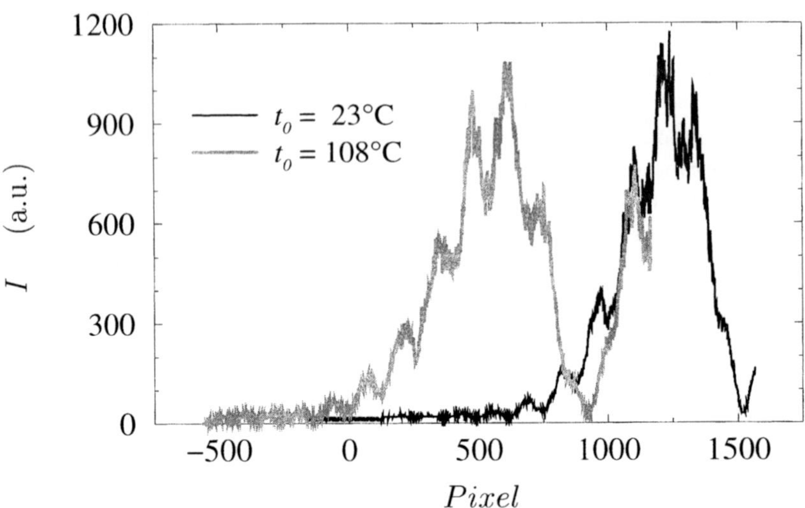

Fig. 6.17. Intensity distribution in the region of the first rainbow for two n-hexadecane droplets with different initial temperatures. The results were obtained with a linear CCD camera at $h = 2\,\text{mm}$ above the horizontal metal plate. Shown are outputs of the camera in arbitrary units as a function of pixel numbers. The location of the optical axis of the measurement unit depicted in Fig. 6.1 is given by pixel number zero

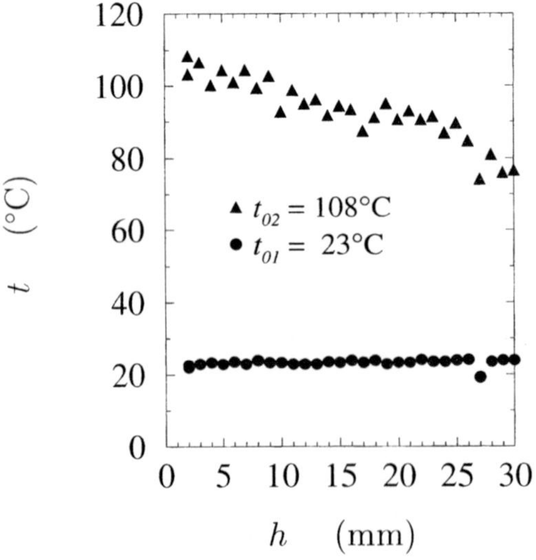

Fig. 6.18. Droplet temperature t as a function of the distance h above the horizontal metal plate for two droplet streams with different initial droplet temperature t_0

highly volatile n-pentane and the less volatile n-octane is pronounced clearly in Fig. 6.16.

Droplet temperatures can be determined using the rainbow method described in Sect. 4.7.4. Two intensity distributions of the scattered light in the region of the first rainbow of n-hexadecane droplets with different temperatures are shown in Fig. 6.17. The different angular positions of the main maximum of the rainbow can be recognized clearly. At higher temperatures the refractive index is lower. As a consequence the main maximum of the first rainbow is found at lower scattering angles.

Measured intensity distributions, as shown in Fig. 6.17, have been evaluated. After calibration the corresponding temperatures have been determined with the correlation technique described in Ref. [105]. The evolution of droplet temperature along droplet streams is shown in Fig. 6.18 for two different initial temperatures of n-hexadecane droplets. No change of droplet temperature is observed, when the droplets have room temperature. The vapor pressure of n-hexadecane at room temperature is very low, cooling effects due to evaporation processes are therefore very small and cannot be detected. In the case of the higher initial droplet temperature the cooling of the droplets by the colder surrounding ambiance can be measured.

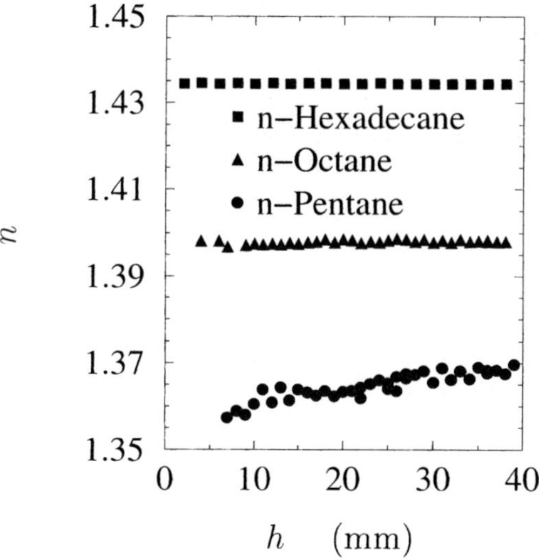

Fig. 6.19. Evolution of the refractive index n of the droplet liquid along a droplet stream for n-pentane, n-octane, and n-hexadecane

The rainbow method can be used in the same experimental setup to measure refractive indices. This allows for example to distinguish between different droplet liquids. The evolution of the refractive index along droplet streams for different liquids is presented in Fig. 6.19. Only data for negligible droplet oscillations are shown, the measurements for n-pentane and n-octane start therefore at a higher distance h above the horizontal metal plate. The liquids with different refractive indices can be distinguished very well. However, for n-pentane droplets the refractive index increases due to cooling effects, which are caused by the much higher evaporation rate. This effect is less pronounced for n-octane. For n-hexadecane droplets changes of the refractive index cannot be detected along the droplet stream.

The evaporation under rarefied conditions has been described in Sect. 1.12.1 theoretically. Experiments on the evaporation within monodisperse droplet streams under rarefied conditions have been presented in Ref. [383]. A comparison of experimental and theoretical results for intermediate Knudsen numbers has been performed by Anders [212].

6.4 Combustion

Combustion of droplets has been described experimentally and theoretically by many authors. For a better interpretation of experimental results it is essential, that comparisons with theoretical predictions can be made. Some theoretical results for single component droplets will be presented in the following. They are based on three theoretical models described for instance by Law [373]. These models differ in the assumptions for the liquid phase inside the droplet.

The first model is the d^2-law, described already in Sect. 1.12. In this model the droplet temperature is assumed to be constant during the whole combustion process; it is below, but close to the boiling temperature of the liquid. For multi-component droplets it is assumed, that the species concentration in the liquid phase remains constant. A more complicated model is the so-called

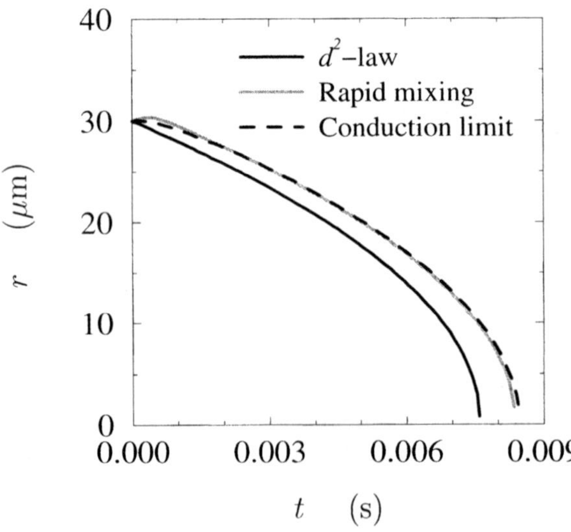

Fig. 6.20. Droplet radius r versus time t for a droplet with the initial radius $r_0 = 30\,\mu$m and the initial temperature $T_0 = 293\,$K. The calculations have been performed with three different theoretical models

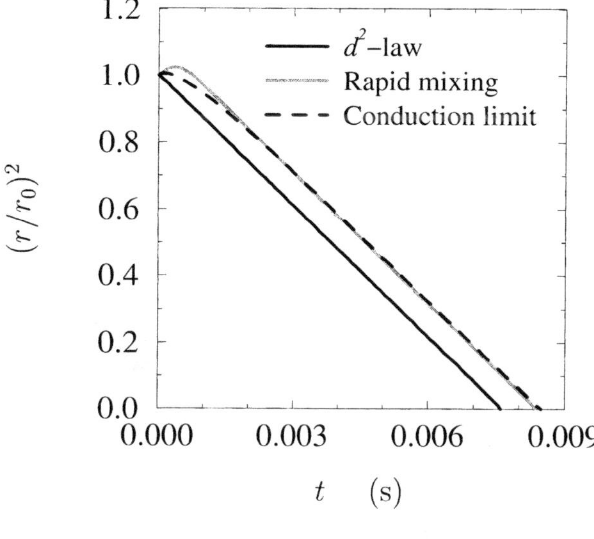

Fig. 6.21. Variation of $(r/r_0)^2$ with time for a droplet with the initial radius $r_0 = 30\,\mu$m and the initial temperature $T_0 = 293\,$K. The calculations have been performed with three different models

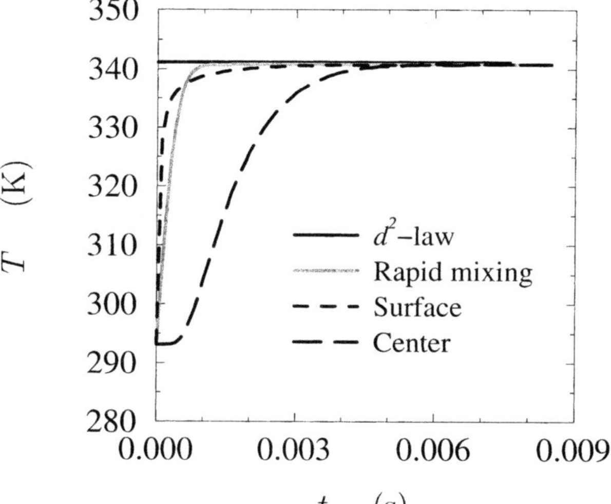

Fig. 6.22. Droplet temperature versus time for a droplet with the initial radius $r_0 = 30\,\mu$m and the initial temperature $T_0 = 293\,$K. The calculations have been performed with three different models. The surface temperature and the center temperature have been obtained with the conduction limit model

rapid-mixing model, in which it is assumed, that transport processes within the droplet are extremely fast, resulting in a zero temperature gradient and a zero concentration gradient in the droplet. The droplet temperature is homogeneous, but varies with time. The third model is the conduction-limit or diffusion-limit model. In this model the transport of energy and species is described by one-dimensional equations within the droplet.

Results for an ethanol droplet with the initial radius $r = 30\,\mu$m will be presented in the following. In Figs. 6.20 and 6.21 the decrease of the droplet radius is shown as a function of time during the combustion process. The

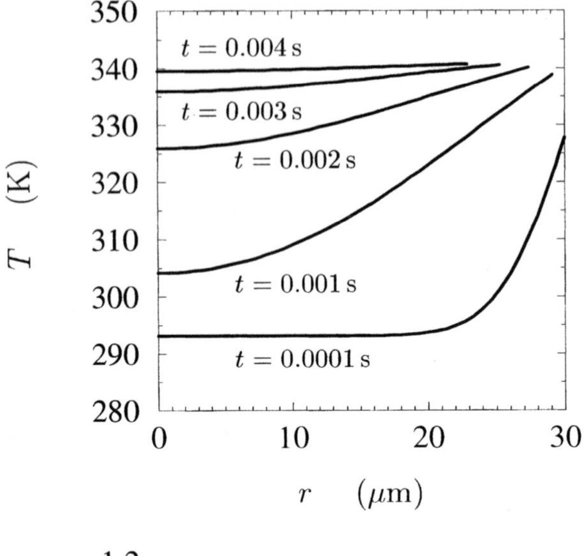

Fig. 6.23. Temperature within the droplet as a function of droplet radius. The results, which have been calculated with the conduction limit model, are shown for a droplet with the initial radius $r_0 = 30\,\mu$m and the initial temperature $T_0 = 293$ K at different times

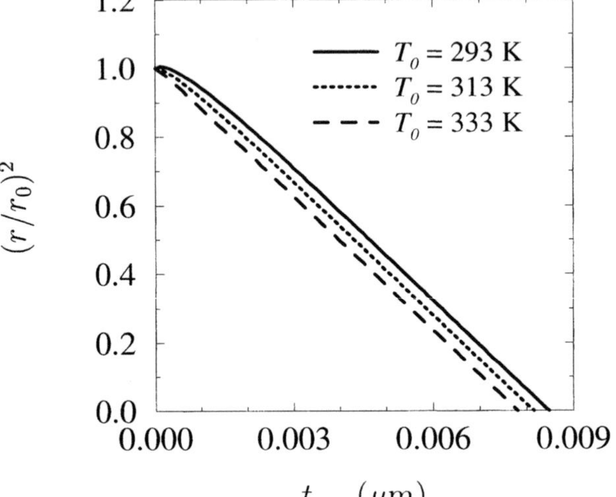

Fig. 6.24. Variation of $(r/r_0)^2$ with time for a droplet with the initial radius $r_0 = 30\,\mu$m and different initial temperatures. The calculations have been performed with the conduction-limit model

initial droplet temperature was $T_0 = 293$ K, which is approximately room temperature. Figure 6.20 shows the radius and Fig. 6.21 the quantity $(r/r_0)^2$ as a function of time for the three theoretical models mentioned above.

In the plot of Fig. 6.21 the d^2-law is represented by a straight line. Both other models show first a small increase of the droplet size, which is due to an increase of droplet temperature. This increase is more pronounced in the rapid-mixing model. After a short while approximately linear behavior is obtained. In this later phase of the combustion process the same slope is found as with the d^2-law and hence the same burning or evaporation coefficient β_v

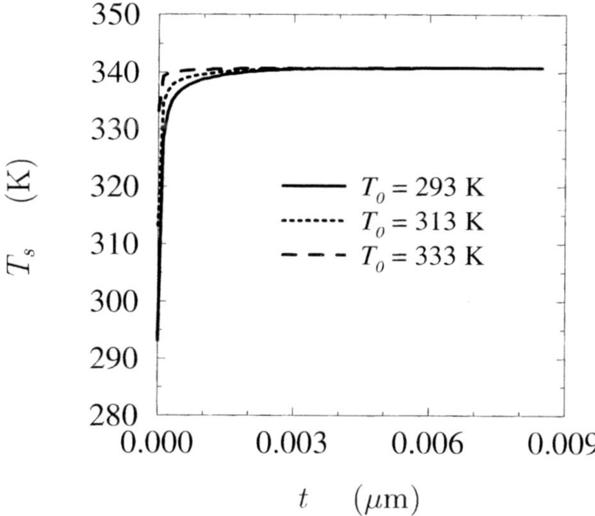

Fig. 6.25. Droplet surface temperature versus time for a droplet with the initial radius $r_0 = 30\,\mu$m and different initial temperatures. The calculations have been performed with the conduction-limit model

is obtained. The same linear behavior is obtained at a later time with the conduction-limit model. The droplet life-time is approximately the same with the rapid-mixing and the conduction-limit model, however, it is larger than for the d^2-law.

The behavior of droplet temperature is shown in Fig. 6.22. According to the d^2-law the droplet temperature is constant of course. After a certain time a constant temperature is reached also with the two other models. This temperature is the same as for the d^2-law. The temperature within the droplet is not homogeneous for the conduction-limit model. Therefore surface and center temperature of the droplet are shown for this model. The droplet center temperature reaches an approximately constant value at a rather late time. The temperature distribution within the droplet obtained with the conduction-limit model is presented in Fig. 6.23 at different times. Large temperature gradients are found especially at the beginning of the combustion process, when large deviations from the d^2-law exist.

The influence of different initial temperatures T_0 is shown in Figs. 6.24 to 6.26. The results presented are obtained with the conduction-limit model. In Fig. 6.24 the quantity $(r/r_0)^2$ is shown as a function of time. It can be seen, that the results resemble more and more the result obtained with the d^2-law, when the initial temperature increases. This effect is due to the fact, that the droplet heating phase becomes shorter with increasing initial temperature, as can be seen from Figs. 6.25 and 6.26. For higher initial temperatures the surface temperature reaches very fast a constant value, as can be seen from Fig. 6.25. The same behavior is found for the center temperature, which is depicted in Fig. 6.26.

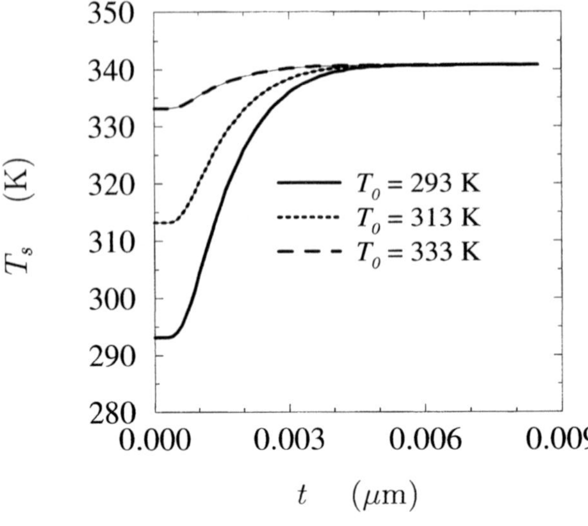

Fig. 6.26. Droplet center temperature versus time for a droplet with the initial radius $r_0 = 30\,\mu$m and different initial temperatures. The calculations have been performed with the conduction-limit model

Extensions beyond these models to multi-component droplets and to droplets with internal circulation are possible, or the whole flow field inside and outside the droplet can be obtained by solving the Navier-Stokes equations including the energy and species transport equations [384]. Experimental investigations on droplet combustion are numerous. Many papers may be found in the publications of the Combustion Institute. Only a few selected papers have been cited here. Some results on the combustion of monodisperse droplet streams will be presented in the following.

6.4.1 Burning Rates

The present section is devoted to the experimental determination of the combustion rate of droplets. The combustion of single droplets consisting of different liquids has been studied by Godsave [385]. In these experiments the droplets were suspended with thin silica fibers. It has been found, that the square of the droplet size decreases linearly, as predicted by the d^2-law. Values for the evaporation coefficients found for different liquids have been tabulated. Hall and Diederichsen studied the combustion of suspended droplets under elevated pressure up to twenty atmospheres [386]. Different authors report on experiments under microgravity conditions as described in Sect. 3.4.11. The droplets can be suspended with thin fibers [387] or they can be free [388].

For technical applications the evaporation of droplets within spray flames is very important. Within a spray the evaporation and burning rate of droplets are influenced by the presence of neighboring droplets. This effect has been studied by different investigators in regular droplet systems. Theoretical as well as experimental investigations have been performed. The combustion behavior of droplets within linear droplet arrays with two or more droplets,

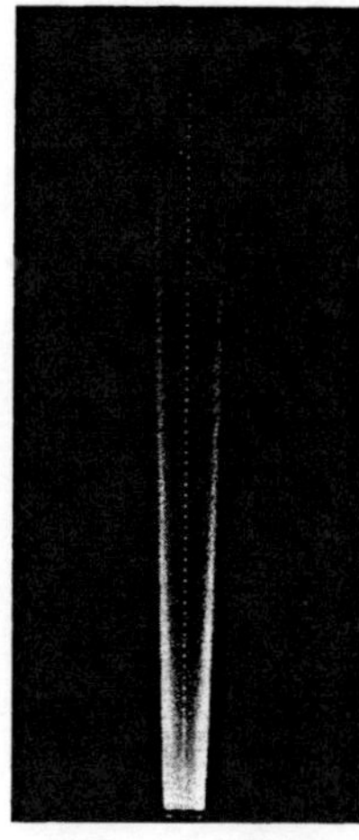

Fig. 6.27. Monodisperse droplet stream surrounded by laminar diffusion flame. The fuel was ethanol. The luminosity of the flame is sufficient for a photo, whereas the droplets have been illuminated with a very short flash. The direction of the flash is such, that glare points can be observed on the droplets

which were free or suspended, have been studied even under microgravity conditions [89, 91, 389-392]. Another regular droplet system is the droplet stream described in Sect. 3.3. In the performed investigations these droplet streams are often monodisperse.

Nuruzzaman et al. studied the combustion of horizontally ejected droplet streams. After ignition a self-supporting flame was obtained [133]. Other investigators studied burning droplet streams, which were directed vertically upwards. In this case the flame can be stabilized by a heated coil. Then the droplets are surrounded by a common quasi-stationary flame front, as can be seen in Fig. 6.27. Individual flame fronts for each droplet may be obtained, when the spacing between neighboring droplets is enlarged [393].

Results obtained for burning droplet streams surrounded by a common diffusion flame, as shown in Fig. 6.27, will be presented in the following. Ethanol at room temperature has been used as fuel. One purpose of these experiments was, to study the influence of neighboring droplets on the burning rate. In Fig. 6.28(a) the droplet radius r is shown as a function of time t for two burning droplet streams with a different initial spacing s_0 between neighboring droplets. The initial droplet size and temperature were in both cases approximately the same. The size measurements have been performed along the droplet streams using the interference method described in Sect. 4.6.3. The residence time of the droplets in the flame region is obtained according to the description in Sect. 3.3.1. The droplets in the stream with the wider spacing evaporate and burn faster than in the case with a smaller spacing s_0.

The data of Fig. 6.28(a) can be presented in a dimensionless form [394]. For this purpose the quantity $(r/r_0)^2$ has been plotted in Fig. 6.28(b) as a function of the dimensionless time parameter $t^* = tD/r_0^2$, where r_0 is the initial droplet radius and $D = 0.135\,\mathrm{cm}^2/\mathrm{s}$ the diffusion coefficient for ethanol in air. The data for each droplet stream have been approximated with a linear

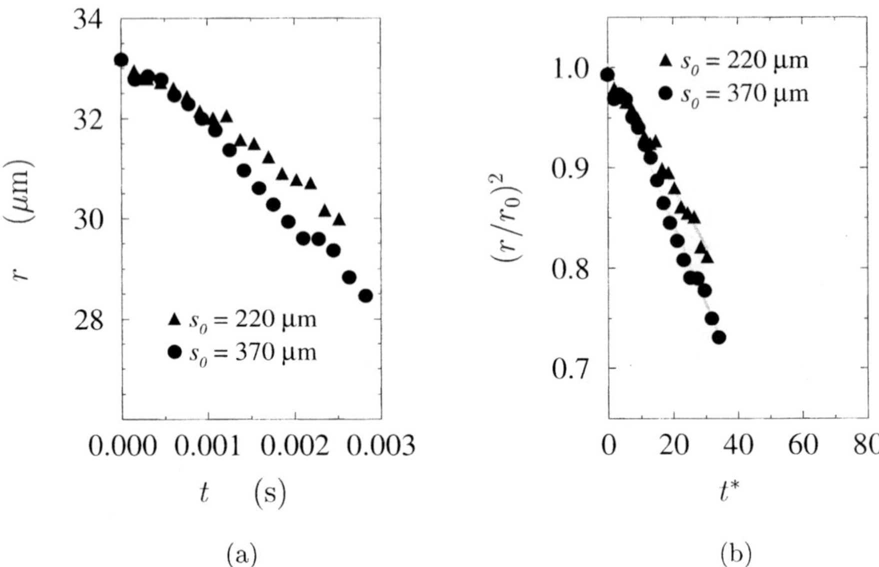

Fig. 6.28. Evolution of droplet radius for two droplet streams with different initial droplet spacing s_0. In (a) the droplet radius r is shown as a function of time t. In (b) the same data is shown in dimensionless form with $(r/r_0)^2$ as a function of the dimensionless time parameter $t^* = tD/r_0^2$

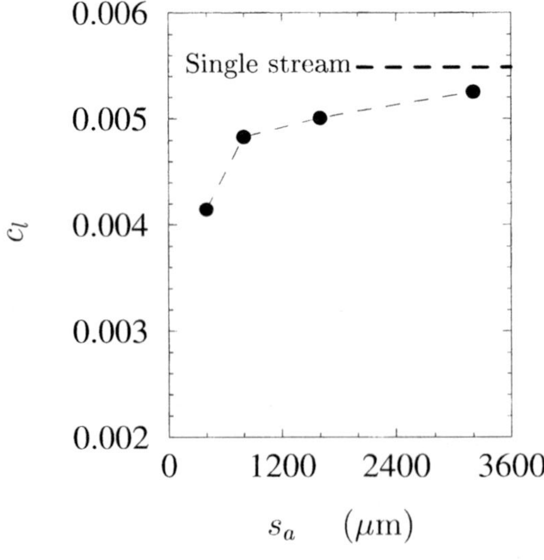

Fig. 6.29. Dimensionless representation of the burning rate for systems of two parallel droplet streams as a function of spacing s_a between neighboring droplet streams. The initial spacing between neighboring droplets of a droplet stream was $s_0 = 181\,\mu$m, the initial droplet radius $r = 30\,\mu$m

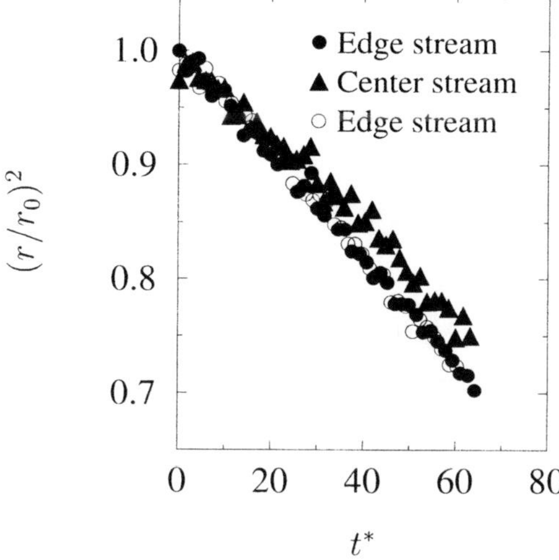

Fig. 6.30. Square of normalized droplet radius as a function of the dimensionless time parameter $t^* = tD/r_0^2$ for droplet streams of a droplet system with three streams and the same initial conditions as for the systems with two streams

regression curve. The negative slope of the regression line can be defined as dimensionless representation of the burning rate of the droplets $c_l = \beta_v/D$. For the wider spacing between the droplets the burning rate is larger. It should be mentioned, that the linear regression does not account for the non-linear behavior during the initial phase in which droplet heating has to be taken into account.

A few experiments have been performed with several parallel monodisperse droplet streams in a plane. The rate c_l has been determined for each droplet stream of these planar droplet arrays in the same way as described above. Results for systems with two parallel droplet streams are depicted in Fig. 6.29. The initial droplet size and temperature were in both cases approximately the same. The techniques for determining the droplet parameters were the same as with single droplet streams. Size measurements have been performed along each droplet stream of the droplet array. It has been found that both streams of an array have practically the same burning rate. The spacing s_a between the droplet streams has been varied. It can be seen from Fig. 6.29, that the rate c_l tends to the value of a single droplet stream with the same initial droplet parameters, when s_a increases.

More complicated planar droplet arrays are obtained, when more than two parallel droplet streams are arranged in a plane. The temporal evolution of the droplet size along the droplet streams is shown in Fig. 6.30 for an array consisting of three streams. The droplets in the streams at the edges burn faster than the droplets in the center stream [395].

Results for planar droplet arrays consisting of two, three, or five streams are presented in Fig. 6.31. The same initial conditions are chosen as for the results described above, in order to allow comparisons. The rate c_l as a func-

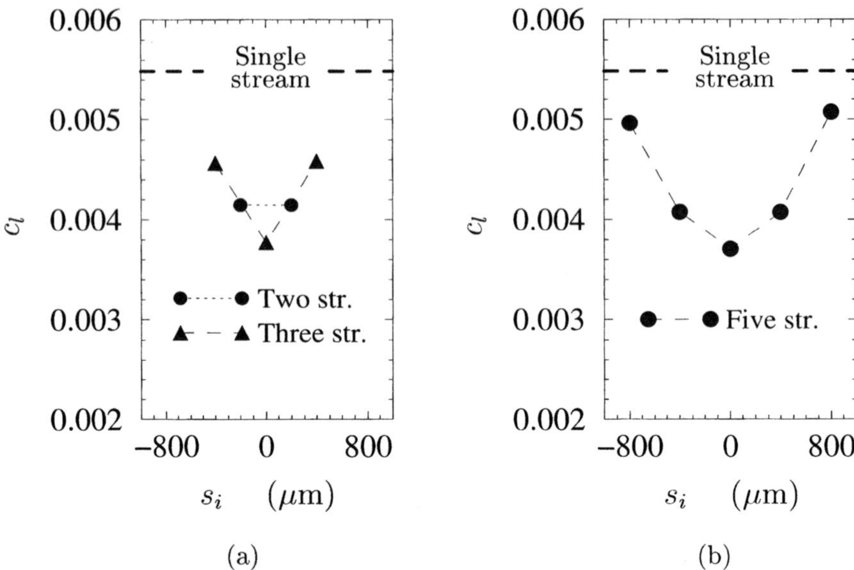

Fig. 6.31. Dimensionless representation of the burning rate c_l for planar systems of parallel droplet streams as a function of the distance s_i from the center. The results are (a) for arrays with two and three droplet streams and (b) for an array with five droplet streams

tion of the distance s_i from the center stream is shown in this figure. It can be seen, that within the same planar array the values of c_l are the same for droplet streams, which are located in the same distance s_i from the center stream. For larger distances from the center of the array the burning rate is higher and tends with increasing s_i to the result obtained for a single droplet stream.

A higher approximation to technical sprays are droplet clouds or droplet groups. Theoretical investigations on the combustion of droplet groups have been performed by different authors [396-399]. Different modes of combustion are distinguished as reported by Chiu et al. [400]. Group combustion has been studied for instance experimentally by Akamatsu et al. [401].

6.4.2 Droplet Temperature

Droplet temperature is an important parameter in droplet combustion. Especially in the initial phase of the combustion process an increase of droplet temperature is observed, as can be seen from the theoretical results depicted above. After the initial phase the surface temperature of the droplet is close to the boiling temperature. Boiling temperatures of different n-alcanes are shown in Fig. 6.32. The data is from Ref. [3]. The boiling temperature increases with increasing chain length of the molecules. The temperature of

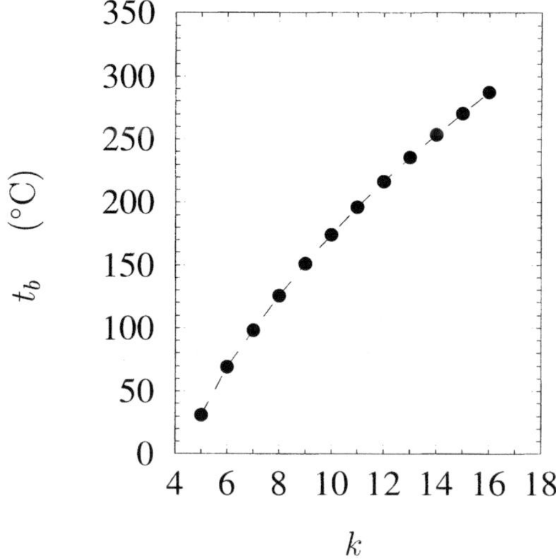

Fig. 6.32. Boiling points of hydrocarbons or n-alcanes as a function of number k of carbon atoms at normal pressure

evaporating suspended water droplets has been measured by Richards and Richards [255] using microencapsulated beads of thermochromic liquid crystals (TLC, see Sect. 4.8.4). A dilute suspension of microencapsulated TLC in distilled water was used.

Optical measurements of the droplet temperature in flames can be performed for instance with the rainbow technique described in Sect. 4.7.4. Measurements with this technique have been reported in Sect. 6.3.2 for non-burning and in Ref. [402] for burning droplets. Measurements of freely moving droplets can be performed with an instrument, which combines the phase Doppler and rainbow technique [403].

It should be mentioned, that measurements with the rainbow refractometry are influenced by temperature gradients. When the droplet surface temperature is known, the temperature gradient at the droplet surface may be determined [240]. Temperature measurements with coherent anti-Stokes Raman spectroscopy in the gas phase around a burning droplet stream have been performed by Zhu and Dunn-Rankin [404].

6.4.3 Flame Propagation

An important topic for technical applications is, the propagation of a flame front in a spray. This problem has been studied theoretically for example in Refs. [405-406]. Experimental investigations have been performed in drop towers under nearly zero gravity conditions for instance by Brzustowski et al. [407]. These authors studied the flame propagation along linear droplet arrays. Sangiovanni and Dodge studied the flame structure of monodisperse droplet streams burning in the post-combustion zone of a flat flame burner.

Depending on the inter-droplet spacing single droplet combustion or group combustion may occur [408].

More complicated droplet systems are the monodisperse planar droplet arrays described in Sect. 3.3.2. An interesting question is, how a flame propagates in a planar droplet array from a droplet stream, which has been ignited, to the neighboring droplet streams [149]. In Fig. 6.33 the quasi-stationary flame in a planar monodisperse droplet array consisting of three parallel droplet streams is shown. The droplet stream at the outer left hand side is preheated and ignited by a heating coil. This stream is called first stream or stream number one; the next stream to the right is called the second stream or stream number two, and so on. After ignition each stream is surrounded by a separate flame front. The flame fronts merge further upstream to form a common flame front surrounding all streams. The flame spreads from the first stream to the second one, and so on.

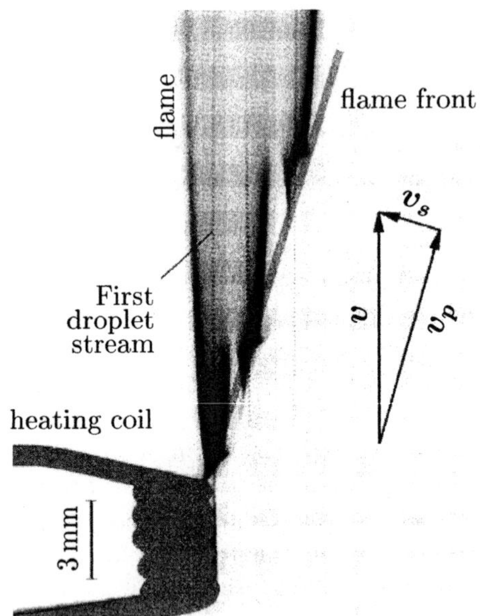

Fig. 6.33. Flame propagation across planar droplet array, which consists of three parallel monodisperse droplet streams burning in the ambient air at room temperature. The velocity components v_p parallel and v_s perpendicular to the flame front are shown

Although the flame between the droplet streams may be a premixed flame, a diffusion flame, or a mixture of both, a virtual flame front can be defined, similarly as in the theoretical description in Ref. [405]. This virtual flame front is the line connecting the points of ignition on the droplet streams, as shown in Fig. 6.33, where the components of the droplet velocity parallel and perpendicular to this virtual flame front are depicted.

The velocity component v_s perpendicular to the flame front characterizes flame propagation. Measurements of the velocity of flame propagation v_s have been performed for different droplet arrays with different spacings s_a

between neighboring streams. Ethanol has been used as fuel. The influence of the initial droplet temperature, size, and volume flux will be shown. The flame has been imaged by a CCD camera. From about hundred frames taken from the same configuration a mean value for the velocity of flame propagation v_s has been determined. For this purpose the droplet velocity and the inclination of the apparent flame front have been measured.

From measurements with different droplet velocities the same velocity v_s of flame propagation has been obtained, when all other parameters were held constant. In Fig. 6.34 results for a droplet array with five droplet streams are shown. A virtual flame front has been defined between each pair of neighboring droplet streams and the corresponding velocity v_s of flame propagation has been determined.

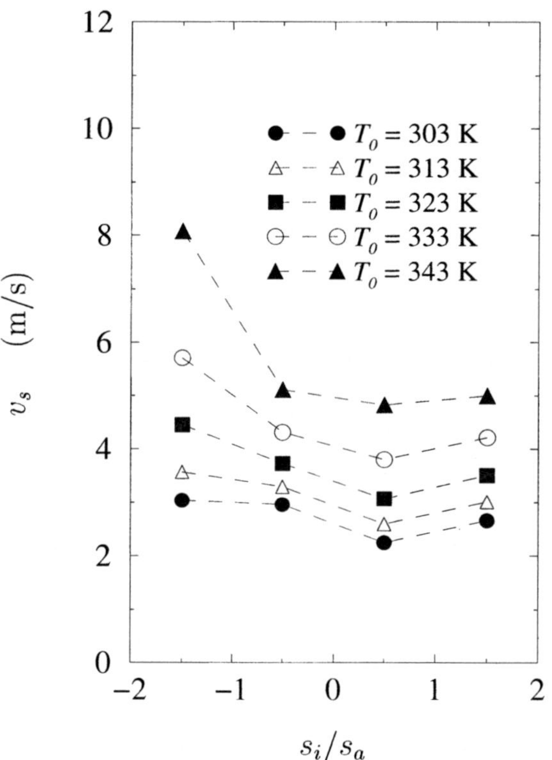

Fig. 6.34. Velocity of flame propagation v_s between neighboring droplet streams of an array consisting of five parallel droplet streams with spacing $s_a = 1200\,\mu$m. Shown are results for five different initial droplet temperatures T_0. The distance s_i is the distance of a droplet stream from the center, whereas s_a is the spacing between neighboring droplet streams. The initial droplet diameter is $d_0 = 53\,\mu$m

In the representation of Fig. 6.34 the velocity of flame propagation between the first and second droplet stream is plotted at $s_i/s_a = -1.5$, the velocity of flame propagation between the second and third stream at $s_i/s_a = -0.5$, and so on. Here s_i is again the distance of a droplet stream from the center of the droplet array. Shown are results for two initial droplet

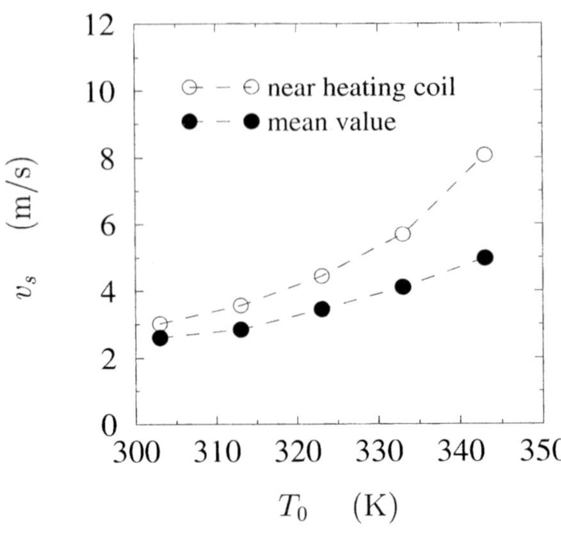

Fig. 6.35. Velocity v_s as a function of initial droplet temperature T_0. The data is based on the result of Fig. 6.34. Shown are the velocity of flame propagation between the first and second droplet stream, which are near the heating coil and the mean value of the three remaining velocities v_s between neighboring droplet streams a five stream droplet array

temperatures. It can clearly be seen, that the flame propagation is faster for higher initial temperatures.

The highest values of the velocity of flame propagation are found between the first and second droplet stream. This may be caused by the influence of the heating coil or by cooling effects before ignition in the streams with larger distances from the ignition source. The increase of the velocity v_s with increasing initial droplet temperature T_0 appears more significant in Fig. 6.35. Two sets of data are shown, namely the velocity of flame propagation between the first and second droplet stream, which shows higher values and the mean values of the velocities v_s between the other pairs of neighboring droplet streams. For both data sets the increase of v_s with initial temperature is clearly recognizable.

Results for a planar system with three droplet streams are presented in Fig. 6.36. Shown is the velocity of flame propagation between the first and second droplet stream as a function of the spacing s_a between neighboring streams for different initial droplet temperatures. It can be seen again, that the flame propagation is faster for higher initial temperatures. With increasing spacing s_a the velocity of flame propagation increases first and decays for larger spacings. There seems to exist a weak maximum in the range $1000\,\mu\text{m} < s_a < 1500\,\mu\text{m}$. This observation may be explained by stronger mutual hindering of evaporation for small spacings s_a and by the difficulty of flame spreading over larger spacings s_a. Mutual interaction between neighboring streams can be recognized in Fig. 6.29.

The influence of the droplet size on flame propagation is shown in Fig. 6.37 for two different initial droplet temperatures. Practically no influence of droplet size has been found, when the volume flux of the droplet streams was

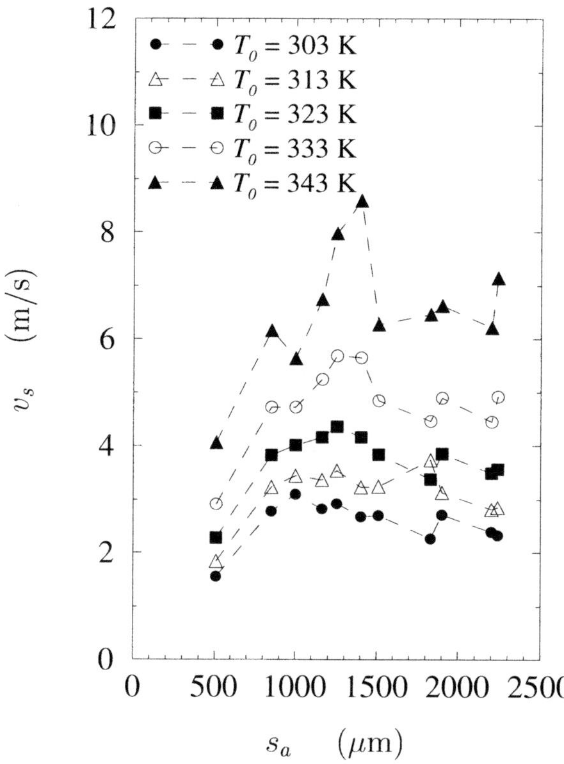

Fig. 6.36. Velocity of flame propagation v_s as a function of spacing s_a between neighboring droplet streams for different initial droplet temperatures T_0. Shown are results between the first and the second droplet stream of an array with three streams with the initial droplet diameter $d_0 = 54\,\mu m$

kept constant. This condition can easily be satisfied by varying the excitation frequency of the droplet stream generator, as described in Sect. 2.3. The results of Fig. 6.37 have been obtained for constant initial droplet velocity. Therefore large droplets with a large spacing lead to the same velocity of flame propagation as small droplets with small spacing. This means, that rather the amount of liquid per unit length V_l within a droplet stream than the droplet size is the parameter, which influences the velocity of flame propagation v_s. The velocity of flame propagation as a function of volume flux $\dot{V}$ for different initial droplet temperatures is shown in Fig. 6.38. As the initial droplet velocity was held constant, an increase in volume flux resulted directly in an increase of the amount of liquid per unit length V_l. Under the present conditions an increase of V_l leads first to an increase of the velocity of flame propagation and then, for higher values, to a leveling off or even to a very weak maximum of v_s.

Experiments on flame propagation in planar sprays have been described in Ref. [409]. There the distance and size of the droplets are described with distribution functions. Flame propagation in droplet clouds and sprays has been studied by different authors with different experimental setups [410-413]. It should be emphasized, that he flame speed or velocity of flame propagation

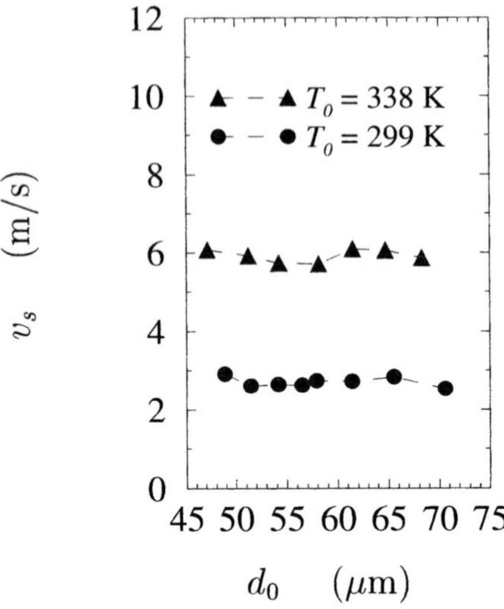

Fig. 6.37. Velocity of flame propagation v_s between the first and the second droplet stream of an array consisting of three streams as a function of initial droplet diameter d_0. For a specified initial droplet velocity the volume flux $\dot{V}$ has been held constant by regulating the initial droplet spacing s_0 appropriately. The spacing between neighboring droplet streams was $s_a = 1200\,\mu m$. Shown are results for two different initial droplet temperatures

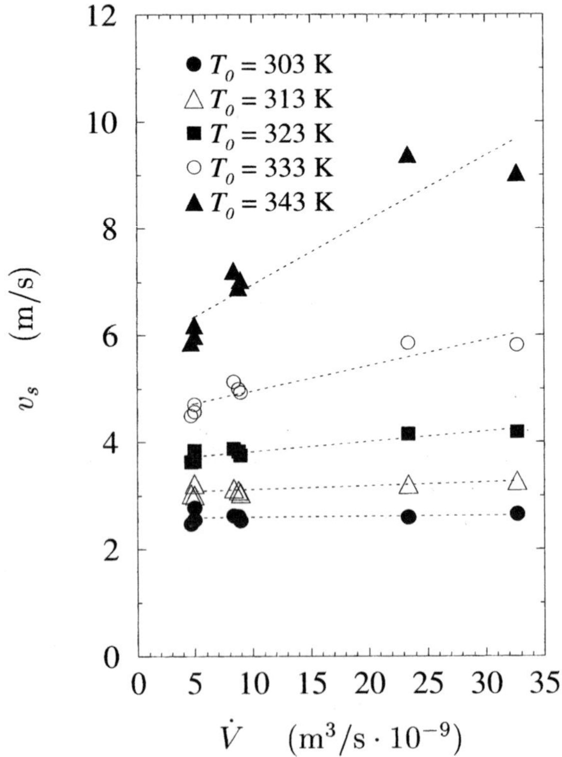

Fig. 6.38. Velocity of flame propagation v_s between first and second droplet stream of an array consisting of three streams versus volume flux $\dot{V}$ for different initial droplet temperatures. The initial droplet velocity was held constant. The volume flux is the value for one droplet stream of the array. The spacing between neighboring droplet streams was $s_a = 1600\,\mu m$

has been defined in different ways and for different experimental configurations. A comparison of results is therefore difficult.

6.4.4 Microexplosions

Disruption or microexplosions of droplets are observed during droplet heating processes, for instance in spray flames. The occurrence of microexplosions depends essentially on the droplet substance. The decomposition of organic azides when heated has been studied by Law et al. [414]. Usually microexplosions are observed with emulsified or multicomponent droplets, as described by many authors [415-420]. The disruption or microexplosion is caused in these cases by superheating of at least one component of the droplet liquid, followed by a sudden vaporization of the superheated component.

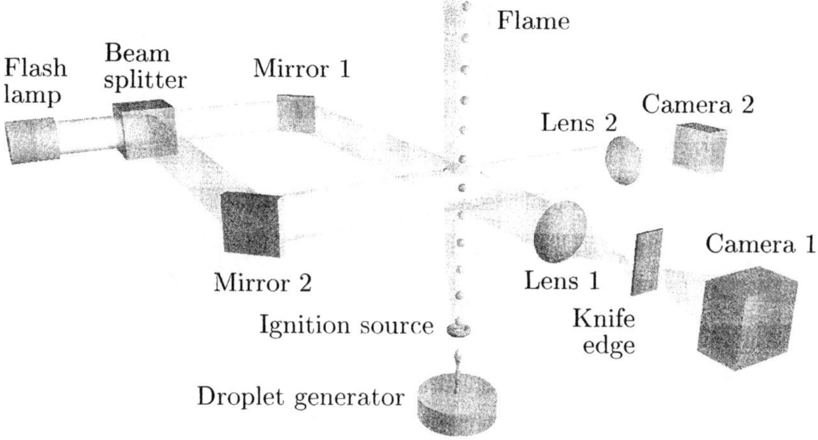

Fig. 6.39. Experimental setup for the observation of microexplosions from two directions. One direction of observation allows to take schlieren images due to the knife edge. The excitation frequency of the droplet generator was modulated in order to obtain larger distances between the droplets, as described in Sect. 2.3

Microexplosion phenomena can be observed with burning monodisperse droplet streams. In the example shown in Figs. 6.40 and 6.41 the droplets consist initially, when the liquid jet emerges from the droplet stream generator, of a homogeneous mixture of 33.3 %vol n-pentane and 66.7 %vol n-hexadecane. The experimental setup, shown schematically in Fig. 6.39, allows observations from two directions perpendicular to each other. The video technique described in Sect. 4.4 has been used in combination with electronic mixing

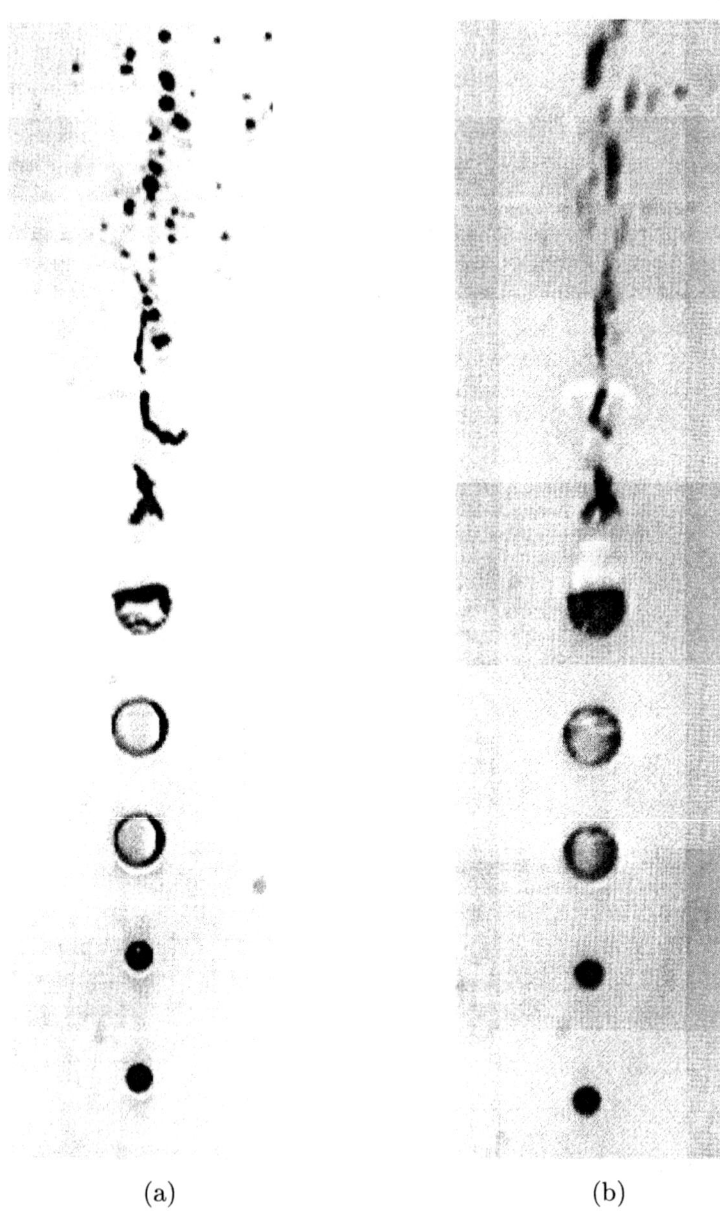

(a) (b)

Fig. 6.40. Photographs of burning monodisperse droplet stream from two directions perpendicular to each other (a) and (b). Details of microexplosion can be observed. The droplets move from bottom to top

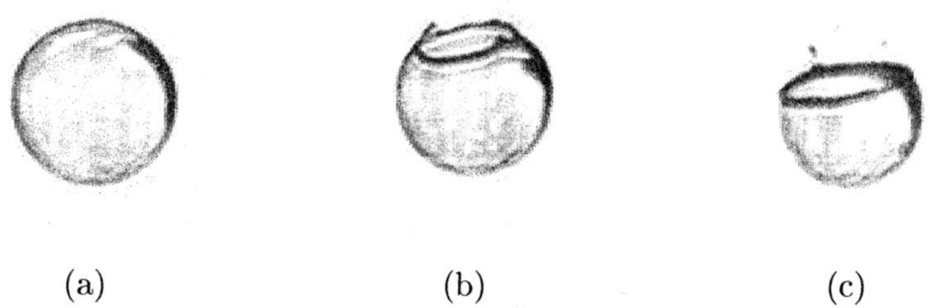

Fig. 6.41. Three photographs of different droplets at a different state of microexplosion process. In (a) a bubble has been formed due to rapid vaporization of the droplet kernel. In (b) and (c) the process of the collapse of such bubbles is shown

of the video signals in order to record the images of both cameras simultaneously on the same video frame. With the knife edge, positioned in the focal plane of lens 1, schlieren images of the phenomena have been obtained. Above the heated coil the droplet stream is surrounded by a laminar diffusion flame, as shown in Fig. 6.27. The occurrence of microexplosions of the droplets depended on the excitation of the droplet generator and the distance of the heating coil from the droplet generator. Microexplosions result in broadening of the flame and explosion or disruption of droplets. Photos of the droplet stream with microexploding droplets are shown in Fig. 6.40.

More details of the microexplosion process can be recognized in the frames shown in Figs. 6.41 and 6.42, which have been taken with a higher magnification. When a droplet enters the flame, it is heated and the more volatile

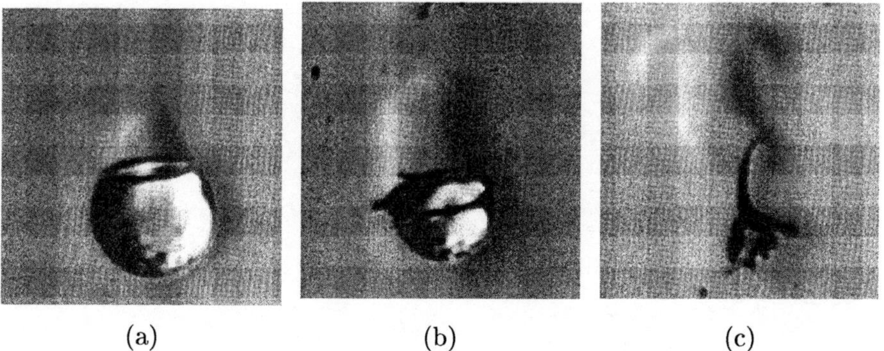

Fig. 6.42. Three photographs of different droplets in different states of microexplosion processes. Due to the schlieren system the vapor emerging from the collapsing bubbles can be recognized. In (a) and (b) the bubbles have just started to collapse. In (c) only fragments of a collapsed bubbles and the emerged vapor cloud are left

component evaporates from the outer layer of the droplet, which results in an enrichment of the less volatile component in this layer. The surface temperature reaches values, which are close to the higher boiling temperature in the outer layer. Therefore the kernel of the droplet, with a higher concentration of the more volatile component and therefore with a lower boiling temperature, is superheated. The superheated kernel may vaporize explosively, which will result in an increase of the droplet due to internal vapor production, as can be seen from Fig. 6.41(a).

Mostly the enlarged droplets or bubbles burst on the upstream side, probably due to shear forces, as can be seen in Fig. 6.41(b) and (c), which show two different exploding droplets at a later state of the process. In Fig. 6.42 three other droplets are presented at different states of microexplosion processes. These photographs have been taken with camera 1 with the knife edge positioned in the focal plane of lens 1. This schlieren optic allows to observe, due to the different refractive indices, the fuel vapor emerging from the bursting bubbles. In Fig. 6.42(a) and (b) the bubble has just started to collapse; state (b) is slightly later than (a). In (c) only fragments of a collapsed bubble are left. The experimental techniques used here allow to observe microexplosion processes in detail. It depends on the droplet system and on the ambient conditions if microexplosions of multicomponent or emulsified droplets are possible. The ambient pressure is an essential parameter.

6.5 Freezing

Freezing of liquid droplets, sublimation of frozen droplets, and crystal growth

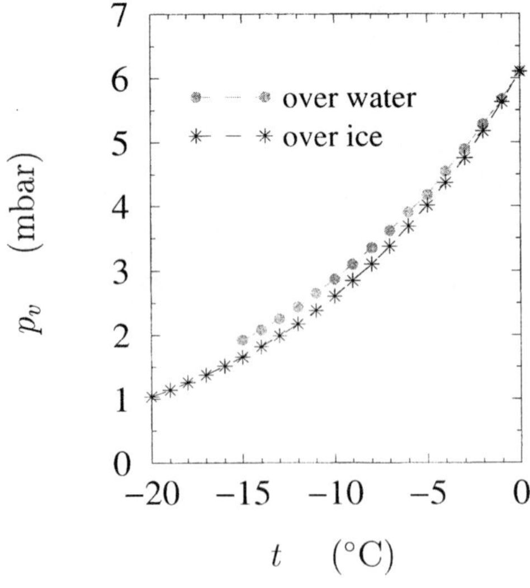

Fig. 6.43. Vapor pressure over liquid water and ice as a function of temperature

on frozen droplets are important processes in clouds of the upper atmosphere of the Earth. A detailed description of these processes may be found for instance in Ref. [20]. Ice crystals occur in natural ice clouds and in contrails of jet aircraft flying at high altitude [421-423]. For the understanding of phase transition processes with ice crystals it is important, that the vapor pressure over water is higher than over ice as can be seen in Fig. 6.43. The data are from Ref. [3]. In clouds, where liquid droplets as well as frozen droplets or ice crystals exist, the liquid droplets will evaporate while ice crystals will grow due to their lower vapor pressure.

Field experiments [424, 425] and laboratory experiments [426, 427] have been performed to study frozen droplets or ice crystals in the atmosphere and to simulate relevant processes. Optical properties of frozen droplets or ice crystals are important for the radiation balance of the Earth. Scattering patterns allow to characterize single particles or clouds.

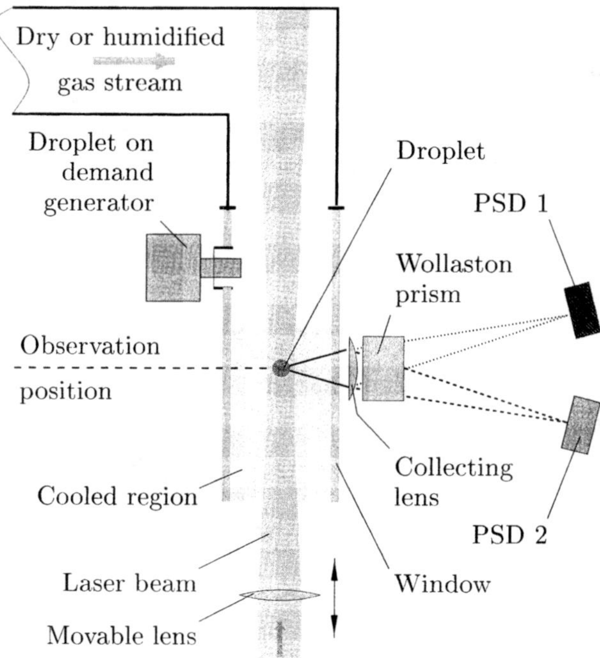

Fig. 6.44. Schematic view of experimental setup used at ITLR to study freezing of liquid droplets and to observe the polarization behavior of the scattered light. The droplet is stabilized near the focus of the movable lens. Moving the lens in the indicated direction allows to manipulate the vertical droplet position

Figure 6.44 represents a schematic view of an experimental arrangement for studying the freezing of liquid droplets and optical properties of frozen droplets under laboratory conditions. This setup for optical levitation of liquid droplets and solid particles is similar to the experimental arrangement shown in Fig. 6.2. However, several devices have been added to allow cooling and freezing of levitated droplets. An essential feature of this setup is, that the lower section of the observation chamber can be cooled. Minimum temperatures of −60°C can be achieved. With the droplet on demand generator single droplets are injected into the chamber. Then a liquid droplet is stabilized and levitated above the cooled section of the chamber. Moving the lens downwards the droplet is brought in the observation position in the cooled section of the chamber.

Experiments have been performed with temperatures below 0°C. The levitated droplet becomes supercooled and evaporates, when it is bathed in a dry air stream [374]. The air stream is applied, in order to avoid vapor accumulation in the vicinity of the droplet. A theoretical study of freezing of supercooled droplets may be found for instance in Ref. [428].

One has to distinguish between homogeneous freezing of nearly pure water and freezing initialized by so-called ice forming nuclei [20]. For pure water the mean freezing temperature decreases with droplet size [20, 425]. In the Earth's atmosphere no liquid droplets have been observed with temperatures below −40°C.

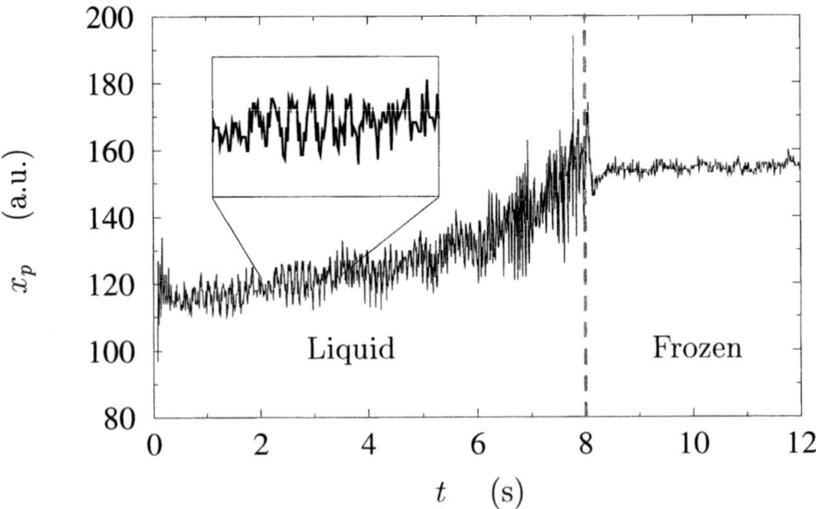

Fig. 6.45. Position x_p of optically levitated droplet as a function of time. During the experiment the temperature has been decreased, in order to obtain droplet freezing. The vertical dashed line indicates the phase change liquid-solid

Monitoring the vertical droplet position by means of a PSD sensor, as described in Sect. 6.2.1, allows to detect the freezing of a droplet, which is cooled by decreasing the temperature in the observation chamber. The output signal of the PSD sensor obtained from a freezing droplet is shown in Fig. 6.45. As long as the droplet is liquid regular vertical oscillations in the position of the evaporating droplet are observed, which are caused by partial wave resonances. As soon as the droplet is frozen the regular oscillations disappear and only random oscillations are observed. In addition the pattern of the scattered light changes. The regular fringes observed in the forward hemisphere of a liquid droplet disappear or are disturbed [380, 427].

6.5.1 Droplet Size and Shape

The evolution of size and shape of a droplet during and after freezing has been studied using a video technique in combination with a flash lamp for illumination, similar to the description in Sect. 4.4. A flash lamp has to be used as the frozen droplets or ice crystals are inhomogeneous or of irregular shape and move therefore up and down and rotate fast. The flash duration is very

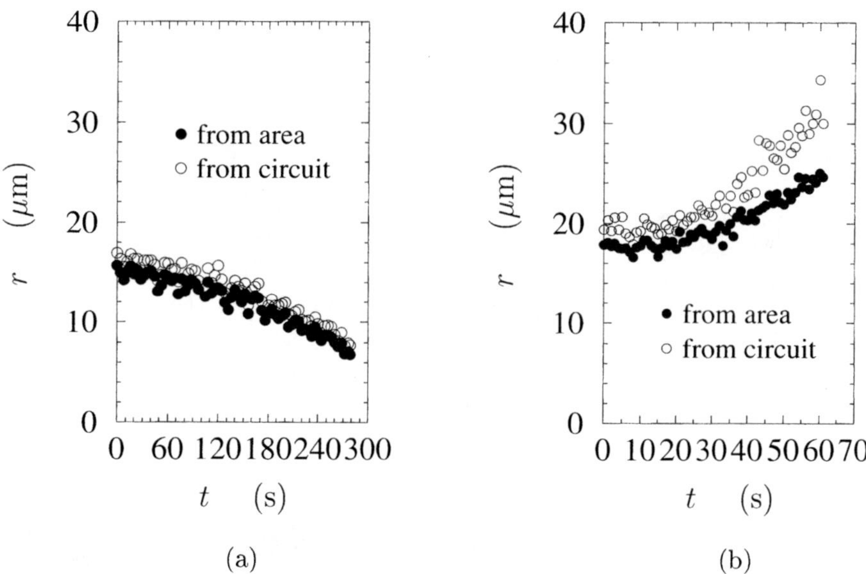

Fig. 6.46. Development of droplet size (a) for a sublimating droplet and (b) for a droplet with crystal growth on it. The size has been determined from a video tape using image processing. In (a) as well as in (b) both, the equivalent radius r_a obtained from the area and the equivalent radius r_c obtained from the circumference are shown

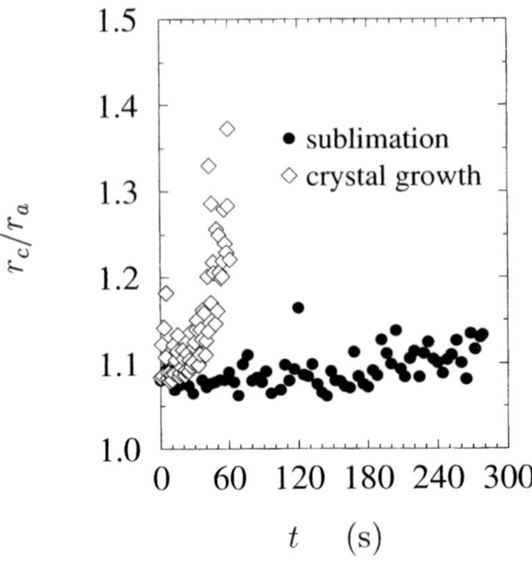

Fig. 6.47. Ratio r_c/r_a as a function of time for sublimating droplet and for droplet with crystal growth on it

short, approximately 200 ns, in order to obtain sharp images. The droplets are moved downwards in the observation chamber from the first position of levitation, where they are liquid, into the observation position, where the temperature is low enough, that freezing occurs. The droplet size has been determined from the video frames by two evaluation methods using image processing, as described in Sect. 4.6.1. In Fig. 6.46 results for a sublimating droplet and for a droplet with growing ice crystals are presented.

Two equivalent droplet radii are obtained from the contour line of the droplets. One is based on the length of the circumference of the droplet, the other one on the area enclosed by the contour line. The temperature of the gas stream through the chamber decreases in the cold region. Sublimation of the droplet is obtained with a dry gas stream. When the gas stream is humidified, supersaturation is obtained at the position of the droplet, which results in crystal growth on the levitated frozen droplet. It can be seen in Fig. 6.46(a), that in the case of a sublimating frozen droplet both equivalent

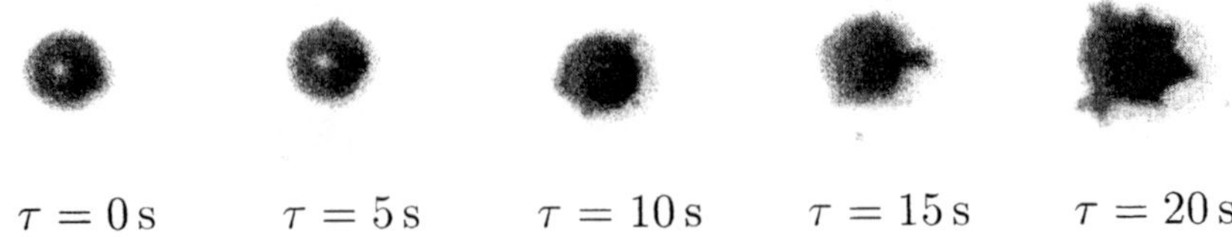

$\tau = 0\,\text{s}$ $\tau = 5\,\text{s}$ $\tau = 10\,\text{s}$ $\tau = 15\,\text{s}$ $\tau = 20\,\text{s}$

Fig. 6.48. Photographs of frozen droplet with growing ice crystals at different times. The orientation of the growing crystal changes, since the droplet rotates

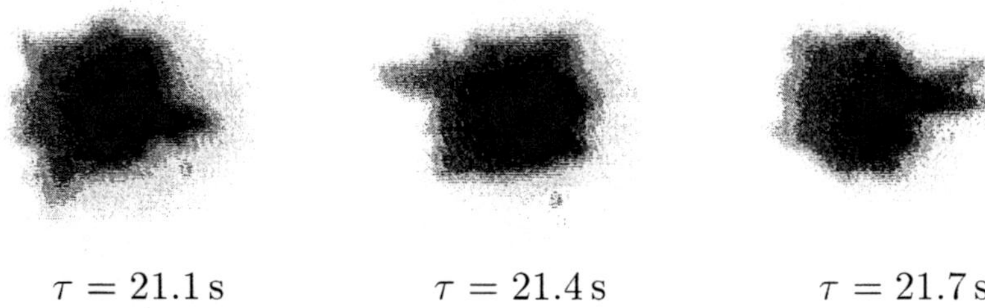

$\tau = 21.1\,\mathrm{s}$ $\tau = 21.4\,\mathrm{s}$ $\tau = 21.7\,\mathrm{s}$

Fig. 6.49. Different projections of frozen droplet with ice crystals. The photographs have been taken at approximately the same time. An impression of the three-dimensional structure is obtained due to the rotation of the frozen droplet

radii decrease; the ratio r_c/r_a remains approximately constant, as can be seen from Fig. 6.47. In the case of crystal growth both radii increase. However, the radius r_c, which is based on the length of the contour, increases faster since the droplet with crystals is not spherical. The ratio r_c/r_a increases therefore with time.

The shape of the droplets can be obtained from video frames. If the ambiance is not saturated sublimation of the frozen droplets occurs and it can be observed that the shape of the droplets remains approximately spherical during the time of observation. Shadows of frozen droplets in a supersaturated ambiance are shown in Figs. 6.48 and 6.49. The ice crystals, which grow in this case on the frozen droplets can be recognized in Fig. 6.48, where a series of video frames of the same droplet is shown. Different projections at approximately the same time can be considered, in order to provide an impression of the three-dimensional structure of the ice crystals. Due to the fast rotation of the crystal, these different projections can be obtained from a sequence of video frames, as shown in Fig. 6.49.

6.5.2 Depolarization

Properties of the light scattered by frozen droplets or ice crystals are important for the characterization of particles with remote sensing methods. Experimental as well as theoretical investigations on light scattering by ice crystals may be found for instance in Refs. [429, 430]. The depolarization of laser light caused by ice crystals is one of the effects used for characterization of the particles. With a ground-based scanning lidar the depolarization of clouds or contrails has been measured [431]. In a laboratory experiment the depolarization of an artificial cloud has been studied by Sassen [432]. The polarization behavior of single optically levitated ice crystals has been studied in Ref. [433].

Frozen droplets have been levitated in the observation chamber shown in Fig. 6.44. The light scattered at $\theta = 90°$ has been observed with the sketched

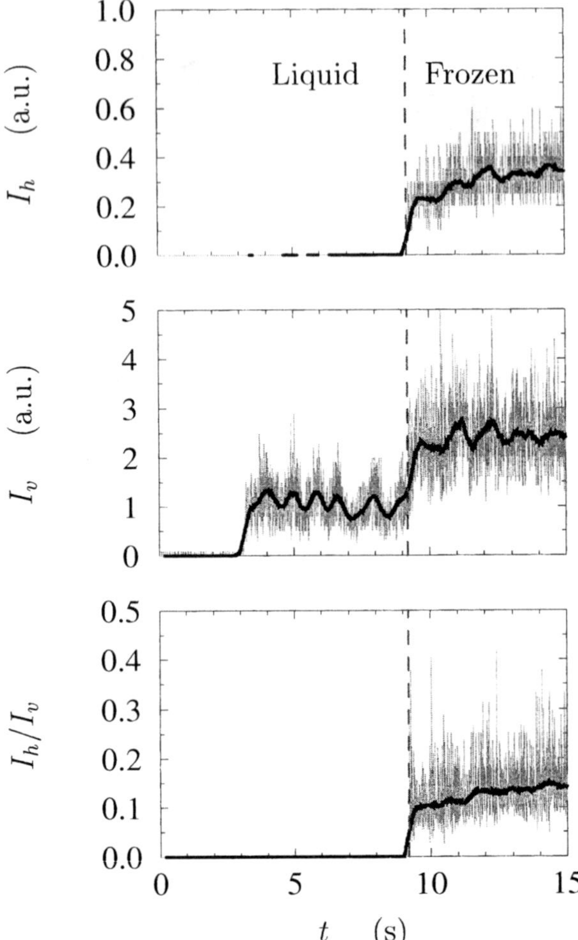

Fig. 6.50. Intensity of light scattered by freezing water droplet. The top diagram shows the intensity I_h of scattered light with polarization perpendicular to the polarization of the incident light. The center diagram shows the intensity I_v of scattered light with polarization parallel to the polarization of the incident light. The bottom diagram shows the ratio I_h/I_v. The sample rate was 4 ms. The solid lines represents running averages over hundred neighboring measured values. The vertical dashed line indicates the moment of freezing

optical arrangement, which consists of a Wollaston prism, a lens, and two PSD sensors. Here the PSD sensors are used to detect the intensity of the scattered light. The incident laser light has linear polarization. The scattered light collected by the lens is split by the Wollaston prism into two components, which are polarized perpendicular to each other. By adjustment of the prism it can be achieved, that only sensor 1 receives a nonzero light intensity, when the scattered light is polarized in the same direction as the incident light. The output signals of both PSD sensors are recorded and stored with a computer system.

With the described technique the polarization behavior of freezing water droplets has been studied. The temperature at the observation position of the levitated droplets was approximately $-35°$C. The droplets remained liquid at this temperature since the droplets consisted of practically pure water.

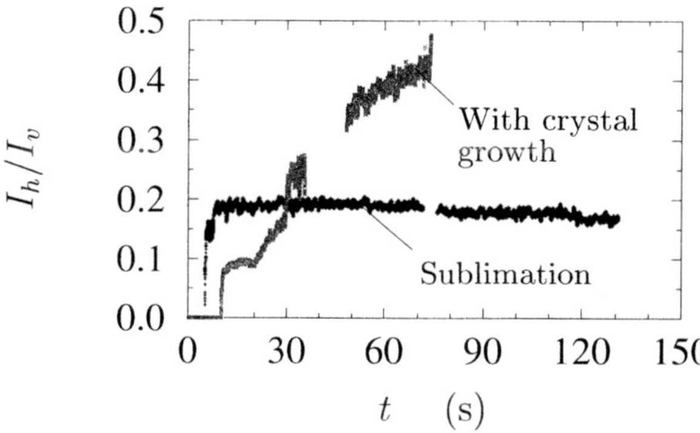

Fig. 6.51. Mean values of the ratio I_h/I_v as a function of time for a sublimating droplet and for a droplet with crystal growth on its surface

Freezing could be initialized by lowering the velocity of the gas stream, which resulted in a decrease of the temperature by approximately 6°C at the droplet position. The freezing process, however, is very fast and cannot be resolved with the measurement techniques of the present experiment, whereas the polarization behavior can be deduced from the output signals of the PSD sensors.

An example for the light scattered by a freezing droplet is shown in Fig. 6.50. The center diagram shows the intensity I_v of the scattered light polarized in the same direction as the incident light. At the beginning, during the first three seconds, the measured intensity is zero, since no droplet is levitated; as soon as a droplet is in the observation position, the output signal increases. The values, which have been sampled every 4 ms, are indicated by symbols. The measured values show irregular oscillations, due to intense rotation of the droplet. The solid line represents running averages. Each average value has been determined from hundred single measurements sampled in sequence.

When the droplet freezes the intensity I_h of the scattered light polarized in the other direction increases, as shown in the top diagram of Fig. 6.50. The appearance of light with the other polarization direction indicates clearly the instant of droplet freezing. The ratio of both intensities I_h/I_v is shown in the diagram at the bottom. It seems, that the polarization ratio is suitable for the characterization of freezing behavior.

Examples for the mean values of the polarization ratio I_h/I_v for single sublimating droplets and for single droplets with growing ice crystals are shown in Fig. 6.51. The interruptions of the measurements are caused by a mirror, which is introduced into the light path to take video frames of the levitated particle for the detection of size and shape. The polarization ratio of a sublimating frozen droplet remains approximately constant, whereas the polarization ratio increases, when ice crystals are growing on frozen droplets which are initially spherical. This increase becomes weaker with time and there seems to be a level off for values between $I_h/I_v = 0.4$ and $I_h/I_v =$

0.5. Results can be obtained until the particle is too large to be levitated by the radiation pressure forces. Bulk measurements on artificial clouds at higher temperatures are presented in Ref. [432]. Lidar measurements of the polarization ratio in contrails show low values for the polarization ratio of approximately 0.2 for very young contrails. For older contrails, in which ice crystals are expected, values around 0.5 have been reported [431].

6.6 Droplet Streams in High Pressure Environments

The most characteristic feature of a droplet is the interface which separates the droplet liquid from the surrounding medium. The existence of the interface depends on the surface tension, which can be considered as a manifestation of the atomic structure of matter. The equilibrium states of single component droplets exist in the vapor-liquid region of the p, v, T-surface depicted in Fig. 1.14. Above the critical temperature T_{crit} (or the critical pressure p_{crit}) thermodynamic equilibrium between two distinct phases does not exist. In the critical point the density difference between the droplet liquid and ambient vapor becomes zero. Inspection of the temperature-entropy diagram reveals that the enthalpy of evaporation tends to zero when the critical point is approached. From Eq. 1.8 it follows that the surface tension vanishes in the vicinity of the critical point. A fascinating task is the investigation of these phenomena under transient conditions. It should be mentioned that the phase equilibrium for very small droplets according to Eq. 1.80 depends on the curvature of the interface between droplet and vapor. Many technical applications deal with droplets consisting of mixtures. In this case the thermodynamic description is of course more complicated, since the mixture-critical point must be considered [434].

From Table 1.4 it can be seen that the critical pressure is rather high for the listed substances. It is not easy to produce and observe droplets under such conditions in a laboratory experiment. An experimental apparatus for studying monodisperse droplet streams at elevated pressures is depicted in Fig. 6.52. The droplets are produced with a droplet stream generator. The driving pressure for the droplet generator, which is determined by the pressure difference between liquid reservoir and test chamber, ranges typically from a fraction of one bar up to approximately ten bar. This means, that the liquid reservoir must be pressurized [435]. Four windows were provided to observe the droplets in the test chamber (8). The liquid reservoir (4) is pressurized with the piston (3) in order to obtain the driving pressure for the droplet stream generator. The temperature can be measured with a thermocouple within the piston.

Some photos of an ethane droplet stream, taken with this apparatus under nearly critical conditions are shown in Fig. 6.53. In order to avoid condensation of ethane on the windows the test chamber has been flushed with nitrogen. The temperature and pressure were (a) $\Delta T = T - T_{crit} = -0.1°\text{C}$,

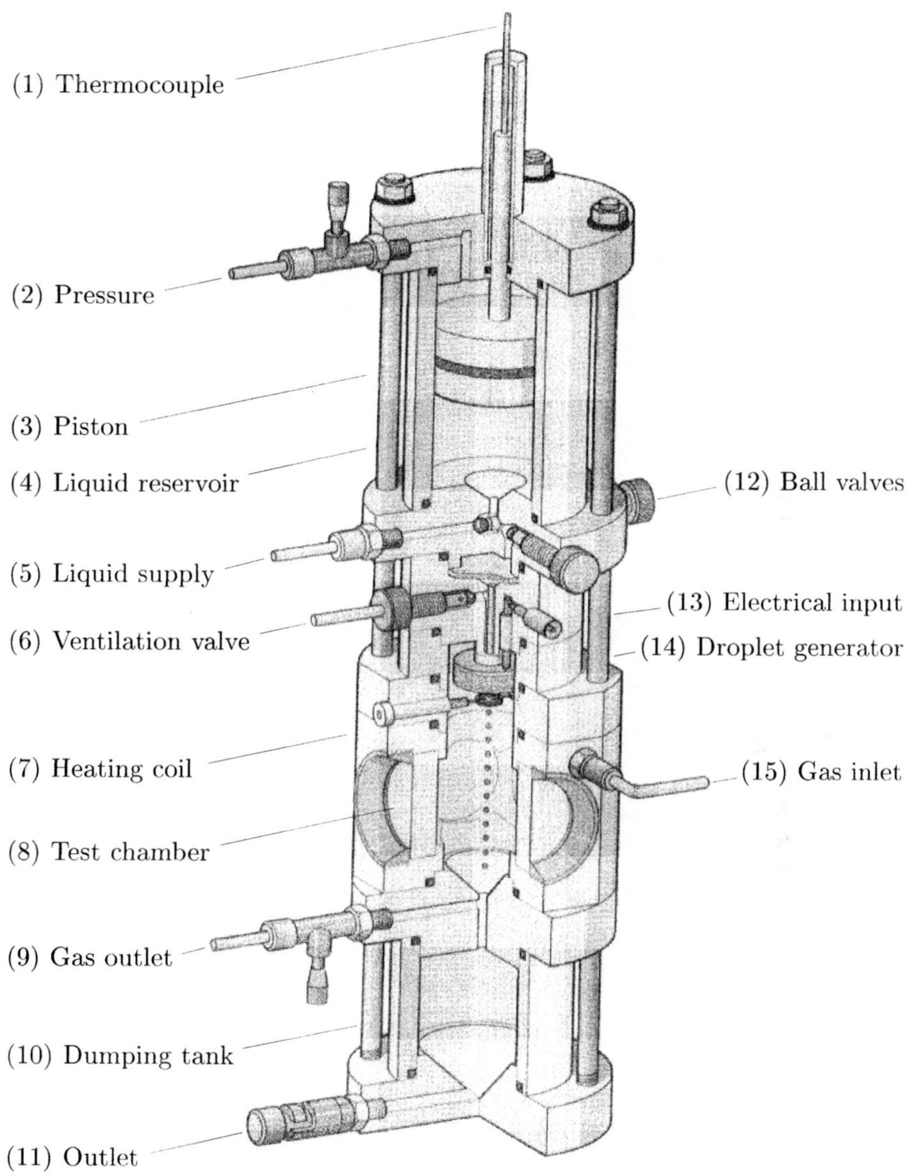

Fig. 6.52. Apparatus used at ITLR for studying droplet streams in gaseous environments at pressures up to 100 bar (from G. Bauer, ITLR)

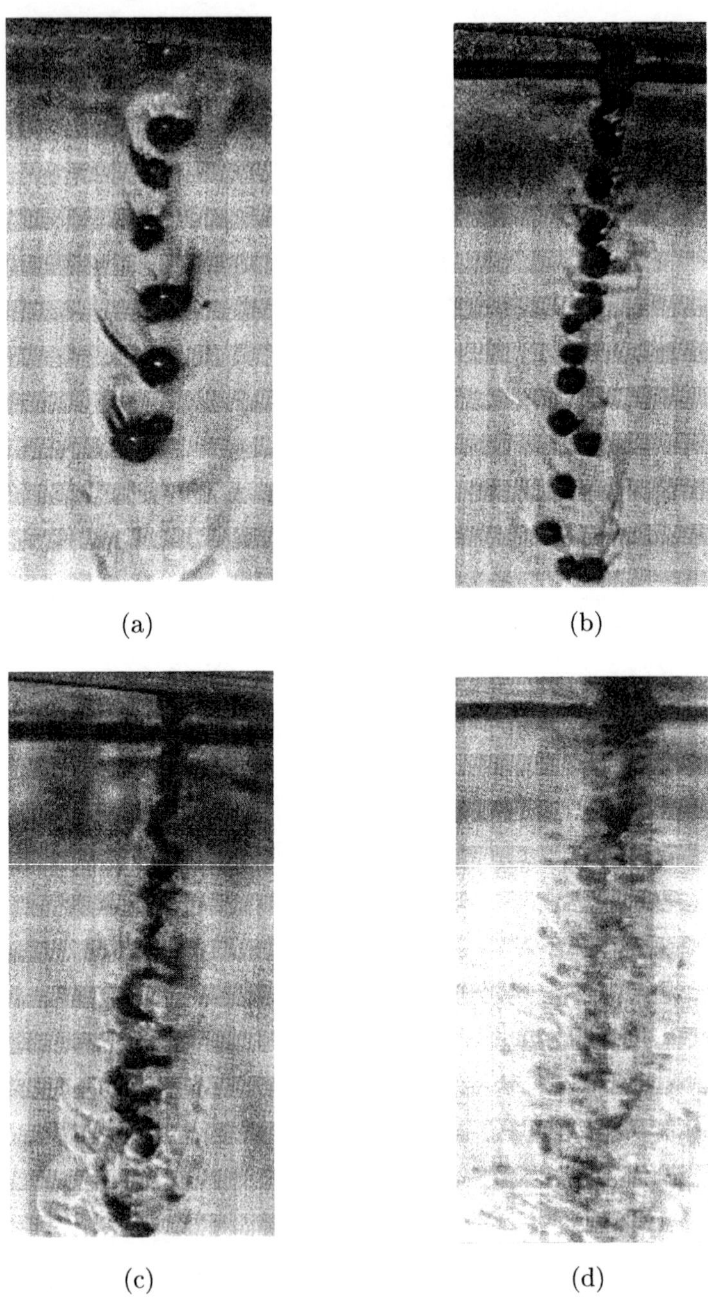

Fig. 6.53. Ethane droplet streams near the critical point $p_{crit} = 48.8\,\mathrm{bar}$, $T_{crit} = 32.3°\mathrm{C}$. The photos are from droplets in the apparatus of Fig. 6.52 (from G. Bauer, ITLR)

$\Delta p = p - p_{crit} = -0.6\,\text{bar}$, (b) $\Delta T = 0.3°\text{C}$, $\Delta p = -0.1\,\text{bar}$, (c) $\Delta T = 0.4°\text{C}$, $\Delta p = -0.1\,\text{bar}$, (d) $\Delta T = 0.4°\text{C}$, $\Delta p = 0.1\,\text{bar}$. The diameter of the droplets is approximately $140\,\mu\text{m}$ and the droplet velocity $3\,\text{m/s}$.

Practical applications of droplets at elevated pressures and temperatures include soot formation in Diesel engines and gas turbines [436], the determination of thermophysical properties in autoclaves [437], the operation of liquid-propellant rocket engines [438-441]. Recent publications combine experimental work with numerical studies [442] or applied Raman scattering to study mixing under supercritical conditions [443].

7. Miscellaneous Applications

7.1 Introduction

Knowledge about the dynamic behavior of single droplets and droplet systems is important in numerous natural and technical processes. Some of the applications have been investigated since many years, others are basic for new technologies, such as ink-jet printing or droplet-based manufacturing with all its applications.

7.2 Internal Combustion Engines

In Diesel engines or oil burners fuel is delivered to the combustion region as liquid particles. The control of the droplet size and the dispersal of the droplets in the combustion chamber are important questions. Mixture formation in carburetor engines and in fuel injection engines has been discussed in the literature [444]. Thermal and mechanical relaxation processes, evaporation of liquid mixtures, diffusion and interaction processes with solid walls are important for modeling the combustion. Particulate matter such as soot or fly-ash may form in the combustion process. Homogeneous nucleation of carbon and unburned fuel may be the formation process for soot, whereas fly-ash results from non-combustible fuel constituents. Soot and other particulate matter carry a part of the combustion energy.

7.3 Fire Suppression

Kim et al. investigated how the fire intensity and burning rate of fuel are affected by water sprays [445]. These authors report, that water sprays are able to enhance an oil fire. Under the conditions of the performed experiments cooling of the fuel surface was responsible for fire extinction. A special topic is aircraft fire safety. Over the last decades the primary fire extinguishing agent has been Halon. Alternative agents are being tested, since the production of Halon has been prohibited by international agreement. Marker et al. tested an onboard cabin and a cargo compartment water spray system [446]. The mean droplet diameter was approximately 100 μm in the cabin water spray. Smaller droplets were not employed in the cabin since they might be respirable.

7.4 Spray Cooling

The evaporation of liquid droplets is of practical interest for spray cooling. Water spray cooling is used for example in steel industry for quenching of metal sheet. Due to the high velocity of the steam layer between droplet and hot surface the heat transfer rate is very high.

The use of fine mists or aerosol sprays has been suggested to carry heat away from electronic chips, power semiconductors, and computers. The objective is to build efficient small and lightweight cooling systems. The benefits of surfactants and dissolved gases in the droplets have been discussed [447].

7.5 Fuel-Coolant Interaction in Nuclear Reactors

Fuel-coolant interactions may occur when during an accident in a light water reactor water comes in contact with molten core material, which may rapidly fragment and transfer heat to the coolant. The produced steam may cause locally high pressures. When the pressure rise is sufficiently fast a steam explosion may occur. Characteristic of this so-called energetic fuel-coolant interaction is the occurrence of shock waves. In non-energetic fuel-coolant interactions the pressure rise is slower and shock waves do not occur [448-450].

7.6 Cleaning with High Pressure Jets

High pressure jets are suitable to loosen dirt and for cleaning objects such as vehicles, machines, facades of buildings or swimming pools. Due to the high impact pressure a very good cleaning effect is achieved even for sticky and heavy dirt. Due to the high pressure such devices save energy and water.

7.7 Medicine and Health

Inhaled particles may be captured in the respiratory system. Larger particles are deposited in the nose or mouth, the smallest in the alveoli of the lung. In the environment of the respiratory system the composition and the size of the particles may change by evaporation or condensation.

The delivery of aerosolized drugs by nebulization is frequently used in the treatment of respiratory diseases. Inhalation delivery of medicinal agents offers different therapeutic advantages. Pharmaceutical aerosols produced by jet and ultrasonic nebulizers are used for the delivery of drugs via the lung to the respiratory tract. Spacer devices provide a reservoir for the aerosol from which the patient may breathe.

For the treatment of asthma metered dose inhalers have been developed; those portable devices produce very fine sprays of high repeatability in dosage for a range of drugs with different liquid properties. The restrictions of CFC propellants usage have to be taken into account.

The nebulization of suspensions consisting of fine insoluble particles dispersed in a liquid has been discussed in the literature [451]. Attempts have been made to nebulize for example insulin and heparin [452, 453]. Effects of exogenous surfactants delivered as an aerosol have been studied [454].

Medical nebulizers are used extensively in clinical medical practice for the delivery of different drugs. Jet nebulizers [455, 456], ultrasonic nebulizers [457], spinning disk nebulizers, [458] and electrosprayers [126] have been discussed by different researchers. These devices which operate continuously for periods of time of from several minutes to hours produce polydisperse aerosols of therapeutically useful size in the range of $2-5\mu m$. They deliver drug solutions and suspensions. A true solution is a homogeneous one-phase system consisting of molecular dispersion of two or more components. Examples are mixtures of water and alcohols or solutions of sodium chloride in water. A suspension is a dispersed system consisting of a liquid and insoluble solid particles. The liquid is often water or alcohol.

The size of the droplets has a strong effect on their deposition in the human lung. The ambient relative humidity affects the evaporation, the size and the deposition of the droplets as they travel through the lung. Droplets with diameters below $3\mu m$ are deposited mainly in the lower respiratory tract, larger droplets in the mouth and oropharynx.

Detailed aerodynamic simulations for real airway geometries have been performed to study transport [459] and the deposition [460] of particles in the human lung. Other investigations have been devoted to the prediction of the aerosol aspiration efficiency of the human head [461].

7.8 Bioaerosols

Bioaerosols comprise particles of biological origin or activity. These particles which may be solid or liquid carry viruses, bacterial cells, spores, fungi, protozoa, mites, or biologically active molecules [462]. Exposure to bioaerosols may cause respiratory disease and allergic reactions in susceptible individuals.

7.9 Production of Small Solid Particles

In spray drying processes a liquid containing dissolved or suspended solid material is sprayed into a hot drying medium, usually air. The liquid evaporates and solid particles form. Suitable products for spray drying include pigments, kaolin, calcium carbonate, herbicides, fertilizers, food products, pharmaceutics, cosmetics, detergents.

Masters distinguishes four process stages of spray drying: atomization of the feed, mixing of spray with drying medium, evaporation of the moisture, and separation of the dried product from the drying medium [463]. Nozzles and rotary atomizers are used to form the spray. The desired size distribution of the dried product is obtained by controlling the atomization conditions. Small droplets of equal size guarantee short drying times and low droplet temperatures.

The prediction and control of droplet trajectories in the drying chamber are important for the performance of the dryer. Deposits of partially dried product at the wall may be a problem, especially in small chambers.

The evaporation of the droplets liquid involves heat and mass transfer. The evaporation rate of the spray droplets depends on temperature, moisture content, heat and mass transfer properties of the surrounding air. The transfer rates are usually expressed in terms of the dimensionless parameters Reynolds, Prandtl, Schmidt, Nusselt, and Sherwood number. Important questions are distortion, disintegration and coalescence of droplets in turbulent flow, the formation of hollow particles, and the prediction of the distribution function of the product.

When the desired moisture content is reached the particles are separated from the air. The selection of the separating equipment depends on many practical requirements, such as properties of the powder and pollution control, or separation efficiency and value of the product. In a variation of the described process the droplets of the spray are brought into contact with freezing air. The water is removed from the frozen droplets by sublimation under vacuum. This process is called spray freeze drying.

Other technologies for the production of solid particles in the micrometer range use sub- and supercritical fluid extraction or gas saturated solutions [464], thermal processes [465, 466], an electric arc [467], or atomization by ultrasonic waves [468].

7.10 Applications in Agriculture

The efficient and safe application of agricultural sprays for pest control and crop protection depends essentially on the size spectrum of the droplets. The use of sprays with a narrow size distribution should enable a reduction of the application rate and of the volume of spray to be applied and less wastage of the spray. Transport, distribution, on-target deposition and drift of spray clouds are affected by the size distribution of the spray. Important basic phenomena are atomization and characterization of agricultural sprays including atomization of non-Newtonian liquids, changes of the size distribution due to evaporation, droplet spreading upon impact on the target.

Rotary atomizers may be used as hand-held devices, as tractor-mounted boom sprayer or as aircraft-mounted applicators. The range of droplet size produced with rotary atomizers has been discussed for example by Frost [469]. Several models for predicting the dispersion of agricultural sprays have been proposed by different researchers. A review article on this subject has been published by Hewitt [470]. It has been pointed out in the literature that non-Newtonian effects are important for understanding the behavior of agricultural sprays [471, 472].

7.11 Acid Rain

The observation of widespread environmental effects of acid rain has focused attention on the subject of atmospheric chemistry [473]. It is assumed that acid rain forms in the clouds where nitrogen oxides and sulfur dioxide are transformed by different physical and chemical processes into a dilute solution of nitric acid and sulfuric acid. Sunlight increases the rate of most of these processes. The oxidation step required for the formation of the highly soluble acids may occur as gas-phase oxidation or as aqueous-phase oxidation. The cloud water is highly dispersed and has a high surface-to-volume ratio. This means that the uptake of gaseous species may be fast. Rain, fog, snow and other forms of precipitation contain mild solutions of sulfuric and nitric acids.

The oxides of nitrogen and sulfur that cause the increased acidity of precipitation are due to increased emissions from coal and oil-fired plants, from automobile exhaust and from other industrial sources. Natural sources of acid-forming sulfur compounds are hot springs, bacterial decomposition of organic matter.

It has become apparent that acidic precipitation has caused severe damage to numerous ecosystems and also affected the integrity of structures and monuments of great historical value.

7.12 Cloud Physics

The physical understanding and description of cloud formation and dynamics requires theoretical knowledge from thermodynamics, fluid dynamics, kinetic theory of gases, aerosol science, interaction forces and electrical processes in droplet systems. Clouds may scavenge the atmosphere through capture of particles by droplets. Cloud droplets and particles in the atmosphere are responsible for the reflection of radiation back to space and determine the albedo [20, 474]. The Earth's radiation balance is influenced by the presence of aerosols. Aerosols have in general a cooling effect by reflecting radiation from the sun back to space.

The microphysical phase transition processes which control the conversion of water vapor to cloud droplets, ice particles and precipitation are responsible for cloud radiative properties and affect the longwave and the shortwave radiation balance of the Earth. They influence the hydrological cycle. The liquid water contents of clouds and their variation with temperature depends on the particle size. The droplet radii depend on continental and maritime differences [475].

7.13 Air Traffic and Condensation Trails

The impact of contrails from engine exhaust of high-flying aircraft on the Earth´s atmosphere has been often discussed in the last years [476]. Water vapor in the exhaust jets of aircraft causes a strong local increase of relative humidity of the ambient air and leads to the tendency to form additional clouds. White condensation trails, often called contrails, form immediately behind aircraft cruising at high altitudes in cold ambient air. It is assumed that contrails form by isobaric mixing of hot exhaust gases and cold ambient air [477]. The resulting mixture may become supersaturated with respect to liquid water. Condensation nuclei, mainly from the exhaust jet, lead to the formation of liquid water droplets which may freeze and form ice particles [478]. In a dry atmosphere these ice particles will evaporate, which means that the contrails will disappear soon. When the air is supersaturated with respect to the ice phase the ice particles can grow and the contrails may have a long life time.

Additional clouds cause an increase of the Earth´s albedo, as sunlight is reflected by the clouds. The water in the clouds absorbs infrared radiation from the surface of the Earth. The combination of these effects may cause changes of the mean temperature of the Earth. Due to the increasing worldwide air traffic the investigation of the impact of contrail formation on the Earth´s atmosphere and on the climate has become an important issue.

Atmospheric visibility may be influenced by dispersed liquid and solid particles from industrial or natural sources. The reduction of visibility depends essentially on the particle size distribution.

7.14 Effects of Heavy Rain and Ice Accretion on Aircraft Wings

Rain and ice can create aerodynamic roughness of the surface of aircraft wings [479, 480]. Surface roughness tends to encourage transition from laminar to turbulent flow. The turbulence leads to a decrease of the lift coefficient and to an increase of the drag coefficient. The presence of extended regions of roughness can seriously degrade the aerodynamic performance of an aircraft. The investigation of icing situations is very important for aircraft safety and certification.

The effects of ice accretion on aircraft and aircraft components has been studied by different researchers. It has been found that ice accretion on airfoils depends on atmospheric conditions, such as temperature and water content of the atmosphere and droplet size. Aircraft ice formed from large droplets accretes further downstream on aircraft surfaces than that formed from smaller droplets.

Flight testing solely under suitable natural icing conditions would be extremely expensive and time consuming. The simulation of icing situations in laboratory experiments, in flight and by computation is therefore very important for aircraft safety and certification [285, 481, 482]. Erosion damage caused by rain drops has been studied for example by Jenkins [483].

7.15 Applications in Space

The injection of liquids into a vacuum is of both basic scientific and technological interest. The exposure of liquids to high vacuum has been studied in different laboratories [484-486]. Many researchers have used the kinetic theory of gases for describing the mass transport from liquids evaporating in a surrounding vacuum. Some physical aspects of liquids in vacuum have been summarized by Hickman [487].

Technological applications include the rejection of waste water from spacecraft, fuel venting in space or atomization of metals in vacuum. During long duration space missions it becomes necessary to eject liquid wastes from spacecraft into the space environment. These ejected fluids form ice clouds in the vicinity of the spacecraft. The behavior of water in a space environment is important because ice particles formed from water may interfere with optical instrumentation. Important questions are the cone angle of the spray, the velocities of individual particles within the spray, the size distribution of the particles, the thermodynamic state of the particles.

Legge has investigated the behavior of liquid jets in a hypersonic low-density wind tunnel. Other investigators studied the evaporation of liquid droplets in a vacuum as a function of Knudsen number and compared the results with solutions of the Boltzmann equation [212, 488].

One of the important issues in the design of spacecraft or space based facilities is the thermal control system used to remove low temperature heat from the base. Thermal control is mainly a cooling problem wherein waste heat must be rejected to space, in order to maintain the components of the base within the operating temperature range. It has been proposed to use free-flying streams of monodisperse droplets as lightweight heat rejection system in space [138, 486]. Hertzberg suggested a lightweight radiator for heat rejection in space. Fluid mechanical, thermodynamic and physical effects, cavitation and flashing phenomena as well as directional and speed stability of droplet streams in a high vacuum have been studied [489]. A lightweight thermal heat rejection system with recoverable free-flying droplet streams as radiating elements or as material transport system have been discussed by Muntz as potential use of controlled droplet streams in space [138].

7.16 Droplets in a Microgravity Environment

Droplet oscillations, combustion phenomena, or thermodynamic processes near the critical point may be influenced by the Earth´s gravity. In some cases it is desirable to conduct experiments under conditions in which the effects of gravity are greatly reduced. Physical phenomena, which are normally masked by effects due to the Earth´s gravity can be explored in a microgravity environment. Microgravity research of droplets has been performed in various ground- and space-based facilities such as drop towers, reduced gravity aircraft, sounding rockets, the space shuttle or space stations.

The study of combustion in weightlessness allows to unmask characteristics of physical combustion phenomena which are hidden on Earth by gravity effects. Droplets floating in zero-g environment burn in a spherical shape. This means that combustion processes can be studied in their purest state without the influence of gravity [193]. Some experiments have been mentioned in Sect. 3.4.11.

7.17 Droplets as Chemical Reactor

Simpson and al. used colliding droplet streams for a new rapid mixing technique to study the kinetics of fast chemical reactions in the liquid phase. Droplet streams of two reagents were mixed by binary droplet collisions. The resulting stream of coalesced droplets was investigated by Raman spectrometric detection. A time resolution of $10\,\mu s$ is estimated [490]. Chemical reactions in optically levitated droplets have been studied in Refs. [491-493] using Raman spectroscopy

7.18 Atomization of Alloy Powders

Powder metallurgy has become attractive for both commercial and military applications in terms of material and energy conservation, and homogeneity of composition and microstructure. Various atomization techniques are known such as inert gas and soluble gas atomization, water and steam atomization, rotating electrode processes, centrifugal atomization, ultrasonic atomization, roller atomization and vibrating electrode atomization have been discussed in the literature [494]. Applications to aluminum alloys have been discussed in a review article by Lavernia et al. [495].

A large range of different powders are made by atomization. Quality demands of the products include size, shape, microstructure, surface smoothness. Swirl jets have been used for example for lead shot production. But it is difficult to pressurize liquid metals with high melting points. Droplet stream generators are able to produce monosized droplets of liquid metal, but it is problematic to keep hundreds of orifices open, which are needed for a significant volume rate. Centrifugal atomization has been used for atomizing different materials. Ultrasonic atomization has been used for producing electronic solder powders. In the technique of water atomization the molten metal is poured into water sprays. The molten metal disintegrates and is quenched. The resulting physico-chemical processes are extremely complex and obviously only poorly understood yet.

7.19 Ink-Jet Printing

Ink-jet technology has been developed probably since the seventies. The market for printers for personal use has grown explosively in the last decade. These personal printers are often attached to a single computer and reside on the desktop. The trend has been to quieter nonimpact printers at low cost. Different ink-jet implementations have been tried, especially droplet-on-demand devices based on piezoceramic and thermal ink-jet technology. Thermal ink-jet printing has been most successful in the last years.

Droplet formation and bubble growth with respect to droplet-on-demand generators has been studied theoretically and experimentally in a number of papers [496-503]. Other studies of ink-jet printing have been devoted to the initial stages of droplet impact [504] or to technical aspects [505, 506].

7.20 Solder Jet Technology

Very fine solder droplets can be placed on electronic circuit boards by using ink-jet printing technology based on piezoelectric droplet on demand generators. Solder droplets in the diameter range $25 - 125\,\mu$m can be deposited at rates up to $2000\,\mathrm{s}^{-1}$. Typical droplet diameters are approximately $100\,\mu$m. With increasing circuit density smaller droplets will be needed. In this non-contact solder technology each droplet is dispensed under digital control. By modulating the excitation signal of the droplet generator it becomes possible to vary the droplet size in a certain range. The solder droplets travel typically with a velocity of 1 m/s. The maintenance of a locally low oxygen environment near the dispensing device is essential.

Another solder jet technology uses the principle of the vibrating orifice generator. A continuous metal jet is produced by applying a backpressure to a reservoir of molten solder. A stream of uniform droplets is obtained by applying a mechanical vibration as described in Sect. 2.3. With an electrical field the droplet can be charged and deflected to the substrate.

7.21 Picoliter Fluid Dispenser in Biotechnology

For many applications in biotechnology the dispensing of extremely small liquid volumes of the order of picoliters is essential. Examples are biomedical materials or diagnostic reagents. The use of ink-jet technology has been suggested for dispensing fluids on picoliter scale.

7.22 Micro-Jet Printing of Microlenses

Drop on demand generators have been used to fabricate microlenses by depositing tiny droplets of UV-curing optical material on a substrate. Applications, which may be found in optoelectronics, include collimation of the outputs of LED and diode lasers. The printing of lens arrays with spacings in the micrometer range has been achieved. For certain applications a small ellipticity in lenslet shape is required. Hemi-elliptical lenslets can be printed by depositing droplets along a line, which join before curing [507].

7.23 Applications in Medicine

For laser surgical and dental procedures, that involve tissues not suitable for laser methods because of low absorption of optical energy, the use of ink-jet fluid microdispensers in combination with laser optics has been proposed. In this technique a small amount of dye as an absorber of optical energy is deposited for each laser pulse. Thus the ablation process of tissue or bone becomes independent of the absorption properties.

7.24 Droplet-Based Manufacturing

Droplet-based manufacturing of small objects has been described recently by different authors [508-512]. Using various microdroplet ejection systems the possibility of making structural components by depositing molten drop by molten drop has been explored in some laboratories. In comparison with other existing freeform fabrication technologies, such as stereolithography, laser sintering or rapid solidification spray forming [495], this new technique is different because the deposition of the droplets can be controlled with high precision for fabrication of three-dimensional structures with low tolerances.

The final part can be made from the raw material in one integrated operation without the need for polishing, cutting or other finishing measures. It may even become possible in the future to design a structural element on a computer screen and print it out afterwards as material object. This means that fewer machining steps are needed than with conventional techniques. Objects built up by precision deposition of molten microdroplets under controlled thermal conditions have finer grain structure and therefore superior properties than those that are cast.

Controlled droplet streams of high angular stability and low speed dispersion have been used by Orme for a droplet based net-form manufacturing technique for small objects. The droplet streams can be precisely directed to the target for fabricating three-dimensional structures [512]. Different droplet configurations can be generated by applying amplitude-modulated disturbances to the droplet generator. This technique has been described in Sect. 2.3. The droplets, which are ejected into either a vacuum or an inert gas environment, carry electrical charges and can be deflected by applying electrical fields for controlling the deposition of the droplets on the substrate.

Chun´s precision droplet-based manufacturing of objects is based on ink-jet printer technology. The droplet generators used by Chun produce small droplets of low temperature metals. Uniformity of the droplets allows better control of the resulting microstructure and the investigation of basic processes [513].

Other investigators have studied basic modes of droplet deposition and solidification, such as columnar deposition, sweep deposition, or repeated sweep deposition and described the required operation conditions and methods for predicting the shape of the resulting solid structures [509].

Applications of droplet-based manufacturing include the production of small objects, precision made powders, printing of customized electronic circuits, the production of glass beads for reflectors on road signs and solder droplets of electronic packaging. It might also become possible to produce alloys, that are impossible to make with conventional techniques, by mixing droplet streams from independent generators with different materials and thermodynamic states.

All these technologies which can be described as three-dimensional printing are in their infancy at present and a great deal of fundamental work has to be done. Future work should include aerodynamic drag and heat transfer along the droplet trajectories for determining the droplet temperature upon arrival at the substrate, heat conduction after impact, droplet undercooling, remelting of deposited material, and microstructural evolution of the sprayed deposits. Blazdell et al. describe the computer aided manufacturing of ceramics using multilayer jet printing [514].

7.25 Thermal Spraying

The term thermal spraying describes a wide class of processes in which droplets of molten metals, glasses or polymers are sprayed onto a surface in order to produce coatings, net shapes or to create composite materials with new properties. The heat source for melting the material feedstock can be provided by a combustion flame or by an electric arc. The molten particles are accelerated by the flame and transported to the surface. Plasma spraying can be used to deposit metal coatings and to build near net-shape parts by applying different coatings. The quality of the coatings depends critically on the droplet impact. Rebounding of the droplet and ejection of liquid as well should be avoided.

7.26 Compound droplets

Compound droplets consist of two or more phases. These liquid particles, which are suspended in yet another phase, may occur as fluid particles within a drop, as fluid particles attached to a drop or as two separate fluid particles [268]. It follows that phenomena observed with compound droplets are in general complicated [515, 516]. There are many practical situations in which compound droplets are important. A short description of this topic may be found in Chap. 8 of Ref. [268].

Applications of compound droplets include liquid-liquid extraction and direct-contact heat exchange. For direct-contact heat exchange two immiscible fluids are in contact without a separating wall. As a result the needed temperature difference is small. This technique is used in ocean water desalination, in geothermal heat recovery and in energy storage. Compound encapsulated droplets consisting of a spherical gas bubble coated with a thin liquid film could be used for direct heat and mass transfer or as liquid membranes. When the liquid film is solidified a solid shell forms which could be used as lightweight structural material [517].

7.27 Microencapsulation

Microcapsules consist of certain small amounts of substances, such as chemical agents, dyes or inks wrapped in protective coatings for the purpose of shielding from the surrounding environment. The size of these capsules ranges from one micrometer to some millimeters. The contents of the capsule are released at a later time, which is appropriate to the application. Encapsulation techniques are applied for example in food, pharmaceutical and detergent industries [518]. Theoretical and experimental investigation have been devoted to the fluid-dynamic behavior of encapsulated droplets [517].

7.28 Submicron Particles and Microspheres

Submicron spherical particles are of great practical interest for powder coatings, microcapsules, drug delivery, and the production of ceramics and other powder-based material [466]. Centrifugal disc atomization has been applied for manufacturing particles of narrow size distribution. The nearly molecular dimensions of nanoparticles provide a potential for the development of new technologies and applications. Improved electronic, optical, and mechanical properties have been associated with ceramic materials consolidated from nanoscale powders. These ceramic powders should possess a narrow size distribution and uniform chemical composition.

Known processes for producing nanoscale materials comprise solvent evaporation processes, plasma processes, or combustion flames, electrical heating, or lasers. These processes are also classified as spray pyrolysis. Another method uses vapor-phase nucleation in hypersonic free jet expansion [519]. The production of protein nanoparticles has been discussed by Gomez et al. [520]. Physical, chemical, and mechanical properties of nanophase materials have been described by Siegel [521].

7.29 Slurry Droplets

Slurry fuel droplets contain liquid fuels and solid fuel particles [522]. Pulverized coal mixed with liquid fuels or water form slurries which can be pumped, transported and handled like premium fuels [268]. Coal water slurries have been investigated over the last years as an alternative fuel for industry and power station boilers. These slurry fuels have attracted special interest and have been investigated as alternate fuels because their cost, pipeline transportation and handling are convenient and economical.

Investigations carried out in different countries include production techniques, rheological properties and NO_x-emissions. Different authors have investigated burning rates of single slurry droplets [523]. Maloney and Spann suspended single slurry droplets electrodynamically and heated them with pulsed laser beams in order to study evaporation and explosive boiling characteristics [524]. Law et al. and Miyaska and Law studied droplet combustion of coal/oil and of coal/oil/water mixtures [525, 526]. Walsh et al. investigated ignition and combustion of coal-water slurries in a turbulent diffusion flame [527].

Sometimes a small amount of water is added to the fuel in order to achieve shattering of the droplet when heated. Important issues are gas eruption from the interior of the droplet, mass ejection or total droplet disruption. Practical aspects of slurry droplet combustion have been studied in the last years [528-531].

7.30 Emulsions

Emulsions consist of fine droplets of one liquid dispersed in another liquid. The droplets are called the disperse or internal phase. The liquid in which the droplets are suspended is called the continuous or external phase. The liquids which form the two phases are immiscible. The droplet size is in general larger than $0.1\mu m$. Emulsifying agents or emulsifiers are often added in order to increase the stability of the emulsion.

The classic emulsion is a mixture of oil and water. Emulsions in which oil droplets are the internal phase are called oil-in-water emulsions. Water-in-oil emulsions consist of water droplets dispersed in oil. These two types of emulsions are usually labeled o/w or w/o.

Emulsions have numerous applications in technology and everyday life. Examples are emulsion paints, polishes, agricultural sprays, medical and cosmetic emulsions, food emulsions such as butter, margarines, or salad dressings [532]. It should be mentioned, that emulsions are nonequilibrium systems.

Microemulsions may be considered as dispersions of supramolecular species with a characteristic length scale of $0.01\mu m$. The structure of these systems is determined by a minimum of their free energy [532, 533].

7.31 Aerosols

Aerosols consist of tiny solid or liquid particles dispersed in a gaseous medium. Practical examples of aerosol systems are fumes, dusts, mists or smokes. These systems represent obviously conditions of highly dispersed matter. It is obviously problematic to give a generally accepted definition of an aerosol system [534].

Due to the small dimensions the particle Reynolds number is in general small and the Knudsen number is not negligible. Effects which are influenced by the Knudsen number become important. Correction factors must be used to describe the influence of the Knudsen number on the drag, the heat transfer or condensation and evaporation processes. The vapor pressure becomes dependent on the particle size.

Thermophoretic, photophoretic and diffusiophoretic forces must be taken into account when describing the motion of aerosol particles. Aerosol science is a wide field with its own vast literature, journals and books [535, 536].

Important issues in aerosol science are for example medical aerosols, stratospheric aerosols, aerosols and climate, interaction between light and aerosol particles, optical and electrical properties of aerosols, radioactivity of aerosols, deposition and removal of aerosol particles, filtration mechanisms, interaction forces between aerosol particles, toxic fibers and health hazards due to aerosols [534]. For better understanding of stratospheric aerosols research on nonequilibrium composition, freezing of droplets, heterogeneous reactions and processes on ice surfaces is important.

7.32 Contact Angles and Wettability

It has been pointed out in Chap. 1.3 that contact angles between liquids and solids play an important role in numerous natural and technical processes [15]. In many situations spreading of liquids on surfaces of solids is requested. Examples are insecticides on leaves, oils on metallic surfaces, paints and inks, soldering and brazing, blending of polymers, reinforcement of polymers with fiber, biocompatibility of polymers, powder coating, cleaning and dyeing of fabrics, detergents. Detergency has been defined as removal of foreign material from solid surfaces by surface chemical means [7]. The washing of clothes is a common example of detergent action.

Large contact angles are of interest for waterproofing fabrics. A natural example for the application of contact angles are feathers of birds, a technical application of contact angles is the separation of different types of solid particles by flotation. As another application of water repellency Adamson mentions the prevention of the deterioration of blacktop roads, which consist of crushed rock coated with bituminous material. Here it has to be avoided that water tends to spread into the interface between bitumen and stone and to detach the aggregate from the binder. It has been pointed out, that wetting plays an important role for the adhesive qualities of binders on road aggregates, namely the affinity for coating wet aggregates and the stripping of coated material by water. Effects of additives, so-called dopes, for modifying the adhesion properties of water and binders, have been studied [537].

For many practical applications the spreading of lubricating oil is as important as the oils viscosity. For some applications such as, mechanical clocks, or fine mechanical meters nonspreading properties of oil are desired.

In many technological applications the condensation of vapors plays a major role. Condensation can be observed when the walls of a container are cooled below the saturation temperature. The vapor may condense as a continuous film or as small droplets increasing in size until they run down the wall due to gravity. This dropwise condensation will enhance the heat transfer [327]. For the prediction of bubble formation and boiling heat transfer in liquids above a heated surface it is important to know if the surface is wetted. Bubble formation depends on the condition of wetting.

Measurements of contact angles may give information on physical and chemical properties of solid surfaces, for example monolayers [7].

7.33 Droplet Separation Technology

Due to the increasing public interest in environmental protection droplet separation technology has become an important issue in industry. Droplet separators are used to remove droplets from process gases, especially from waste gas streams. The droplets to be separated from the gas flow cover a very wide size range usually between 0.2 and 20 μm [319]. In impingement separators the droplets are transported to a solid surface and deposited there. They coalesce with liquid drops, films or sheets. A large variety of separators like cyclones, wave-plate separators, packed bed separators, filters, wet scrubbers, and swirl tubes are in use. The separation of large droplets can be achieved with lower gas velocities. Too large velocities may lead to droplet break-up and to the formation of fine droplets, which are difficult to remove.

References

1. H.S. Green. The structure of liquids. In S. Flügge, editor, *Structure of Liquids*, Vol. 10 of *Encyclopedia of Physics*. Springer-Verlag, Berlin, 1960.
2. S. Ono and S. Kondo. Molecular theory of surface tension in liquids. In S. Flügge, editor, *Structure of Liquids*, Vol. 10 of *Encyclopedia of Physics*. Springer-Verlag, Berlin, 1960.
3. R.C. Weast, editor. *CRC Handbook of Chemistry and Physics*. CRC press, Boca Raton, 66th edition, 1985.
4. P.L. Altman and D.S. Dittmer. *Blood and Other Body Fluids*. Federation of American Societies for Experimental Biology, Bethesda, 3rd edition, 1971. Prepared under the auspices of the commitee on biological handbooks.
5. P.-O. Glantz. The surface tension of salvia. *Ondontologisk Revy*, Vol. 21, No. 2, pp. 119–127, 1970.
6. W. Blanke, editor. *Thermophysikalische Stoffgrößen*. Wärme- und Stoffübertragung, editor U. Grigull. Springer-Verlag, Berlin, 1989.
7. A.W. Adamson. *Physical Chemistry of Surfaces*. John Wiley & Sons Inc., New York, 4th edition, 1982.
8. J.J. Bikerman. *Physical Surfaces*. Academic Press, New York, 1970.
9. T. Iida and R.I.L. Guthrie. *The Physical Properties of Liquid Metals*. Clarendon Press, Oxford, 1988.
10. N.N. Kochurova and A.I. Rusanov. Dynamic surface properties of water: Surface tension and surface potential. *J. Colloid and Interface Sci.*, Vol. 81, No. 2, pp. 297–303, 1981.
11. D.J. Woodland and E. Mack Jr. The effect of curvature of surfaces on surface energy. Rate of evaporation of liquid droplets. Thickness of saturated vapor films. *J. Am. Chem. Soc.*, Vol. 55, pp. 3148–3161, 1933.
12. J.L. Shereshefsky and S. Steckler. A study of the evaporation of small drops and of the relationship between surface tension and curvature. *J. Chem. Phys.*, Vol. 4, No. 2, pp. 108–115, 1936.
13. N.A. Fuchs and A.G. Sutugin. *Highly Dispersed Aerosols*. Ann Arbor Science Publishers, Ann Arbor, 1970.
14. J.L. Moillet. *Waterproofing and Water-repellency*. Elsevier, Amsterdam, 1963.
15. J.C. Berg. *Wettability*, Vol. 49 of *Surfactant Science Series*. Marcel Dekker Inc., New York, 1993.
16. J.C. Slattery. *Interfacial Transport Phenomena*. Springer-Verlag, New York, 1990.
17. A.U. Bailey. *Electrostatic Spraying*. John Wiley & Sons Inc., New York, 1988.
18. J. Bellan and K. Harstad. Electrostatic dispersion and evaporation of clusters of drops of high-energy fuel for soot control. In *Proc. 26th Symp. (Int.) on Combustion*, pp. 1713–1722. The Combustion Institute, 1996.

19. J.R. Adam, N.R. Lindblad, and C.D. Hendricks. The collision parameter coalescence and disruption of water droplets. *J. Appl. Phys.*, Vol. 39, No. 11, pp. 5173–5180, 1968.
20. H.R. Pruppacher and J.D. Klett. *Microphysics of Clouds and Precipitation*. D. Reidel Publishing Company, Dordrecht, 1978.
21. C. Pounder. Charge-carrying particles from Leidenfrost boiling, part I. Investigation of the Leidenfrost phenomenon and the discovery and examination of particles. *J. Electrostatics*, Vol. 9, pp. 159–175, 1980.
22. C. Pounder. Charge-carrying particles from Leidenfrost boiling (an aspect of saline contact charging), part II. The electrical properties of Leidenfrost drops and associated particles. *J. Electrostatics*, Vol. 9, pp. 177–182, 1980.
23. L. Bergmann and C. Schaefer. *Lehrbuch der Experimentalphysik, Band II, Elektrizitätslehre*. Walter de Gruyter & Co., Berlin, 3rd edition, 1958.
24. Sir W. Thomson. On a self-acting apparatus for multiplying and maintaining electric charges with applications to illustrate the Voltaic theory. *Proc. R. Soc. London*, Vol. 16, pp. 67–72, 1867.
25. S. Chandrasekhar. *Hydrodynamic and Hydromagnetic Stability*. The International Series of Monographs on Physics. Clarendon Press, Oxford, 1961.
26. H. Lamb. *Hydrodynamics*. Cambridge University Press, Cambridge, 6th edition, 1957.
27. L.D. Landau and E.M. Lifschitz. *Fluid Mechanics, Vol.6 of Course of Theoretical Physics*. Pergamon Press, New York, 1959.
28. L.D. Landau and E.M. Lifschitz. *Lehrbuch der Theoretischen Physik VI, Hydrodynamik*. Akademie-Verlag, Berlin, 1966.
29. E. Trinh, A. Zwern, and T.G. Wang. An experimental study of small-amplitude drop oscillations in immiscible liquid systems. *J. Fluid Mech.*, Vol. 115, pp. 453–474, 1982.
30. G. Brenn. *Untersuchungen zu Schwingungsbewegungen von Flüssigkeitstropfen*. PhD dissertation, Universität Stuttgart, 1990.
31. C.A. Miller and L.E. Scriven. The oscillations of a fluid droplet immersed in another fluid. *J. Fluid Mech.*, Vol. 32, No. 3, pp. 417–435, 1968.
32. G.B. Foote. A numerical method for studying liquid drop behaviour: Simple oscillations. *J. Computational Physics*, Vol. 11, pp. 507–530, 1973.
33. J.A. Tsamopoulos and R.A. Brown. Nonlinear oscillations of inviscid drops and bubbles. *J. Fluid Mech.*, Vol. 127, pp. 519–537, 1983.
34. E. Trinh and T.G. Wang. Large amplitude drop shape oscillations. In *Proc. 2nd Int. Colloq. on Drops and Bubbles*, pp. 82–87, 1982.
35. A. Prosperetti. Normal-mode analysis for the oscillations of a viscous liquid drop in an immiscible liquid. *J. Méc.*, Vol. 19, pp. 149–182, 1980.
36. W.A. Reid. The oscillations of a viscous liquid drop. *Quart. Appl. Math.*, Vol. 18, pp. 86–89, 1960.
37. G. Brenn and A. Frohn. An experimental method for the investigation of droplet oscillations in a gaseous medium. *Experiments in Fluids*, Vol. 15, pp. 85–90, 1993.
38. Lord J.S.W. Rayleigh F.R.S. On the equilibrium of liquid conducting masses charged with electricity. *Philos. Mag.*, Vol. 14, pp. 184–186, 1882.
39. A.T. Shih and C.M. Megaridis. Thermocapillary flow effects on convective droplet evaporation. *Int. J. Heat Mass Transfer*, Vol. 39, No. 2, pp. 247–257, 1996.
40. F.L.A. Ganzelvles and C.W.M. van der Geld. Marangoni convection in binary drops in air cooled from below. *Int. J. Heat Mass Transfer*, Vol. 41, No. 10, pp. 1293–1301, 1998.

41. C.K. Law. Multicomponent droplet combustion with rapid internal mixing. *Combustion and Flame*, Vol. 26, pp. 219–233, 1976.
42. M. Renksizbulut and R.J. Haywood. Transient droplet evaporation with variable properties and internal circulation at intermediate Reynolds numbers. *Int. J. Multiphase Flow*, Vol. 14, No. 2, pp. 189–202, 1988.
43. S.M. Ghiaasiaan and Eghbali. Transient mass transfer of a trace species in a spherical droplet with internal circulation. *Int. J. Heat Mass Transfer*, Vol. 37, No. 15, pp. 2287–2295, 1994.
44. J.N. Chung and P.S. Ayyaswami. The effect of internal circulation on the heat transfer of a nuclear reactor containment spray droplet. *Nuclear Technology*, Vol. 35, pp. 603–610, 1977.
45. A.Y. Tong and W.A. Sirignano. Multicomponent transient droplet vaporization with internal circulation: Integral equation formulation and approximate solution. *Numerical Heat Transfer*, Vol. 10, pp. 253–278, 1986.
46. S.R. Harris, W.R. Lempert, L. Hersh, C.L. Burcham, D.A. Saville, and R.B. Miles. Quantitative measurements of internal circulation in droplets using flow tagging velocimetry. *AIAA Journal*, Vol. 34, No. 3, pp. 449–454, 1996.
47. H.R. Pruppacher and K.V. Beard. A wind tunnel investigation of the internal circulation and shape of water drops falling at terminal velocity in air. *Quart. J. R. Meteor. Soc.*, Vol. 96, pp. 247–256, 1970.
48. A.H. Lefebvre. *Atomization and Sprays*. Combustion: An International Series. Hemisphere Publishing Corporation, New York, 1989.
49. J.A.F. Plateau. *Statique Expérimentale et Théorique des Liquides Soumis aux Seules Forces Moléculaires.* Gauthier & Villars, Paris, 1873.
50. Lord J.S.W. Rayleigh F.R.S. On the instability of jets. *Proc. London Math. Soc.*, Vol. 10, pp. 4–13, 1878.
51. Lord J.S.W. Rayleigh F.R.S. On the capillary phenomena of jets. *Proc. R. Soc. London*, Vol. 29, pp. 71–97, 1879.
52. Lord J.S.W. Rayleigh F.R.S. Further observations upon liquid jets, in continuation of those recorded in the Royal Society's Proceedings for March and May 1879. *Proc. R. Soc. London*, Vol. 34, pp. 130–145, 1882.
53. Lord J.S.W. Rayleigh F.R.S. On the instability of a cylinder of viscous liquid under capillary force. *Philos. Mag.*, Vol. 34, pp. 145–154, 1892.
54. Lord J.S.W. Rayleigh F.R.S. On the theory of the capillary tube. *Proc. R. Soc. London, Ser. A*, Vol. 92, pp. 184–195, 1915.
55. S. Middleman. *Modeling Axisymmetric Flows.* Academic Press, San Diego, 1995.
56. L.S. Goren, S. Middleman, , and G. Gavis. On expansion of capillary jets of viscoelastic fluids. *J. Appl. Polymer Sci.*, Vol. 3, No. 9, pp. 367–368, 1960.
57. A.C. Merrington. Flow of viscoelastic materials in capillaries. *Nature*, Vol. 152, pp. 663, 1943.
58. P.L. Clegg. *The Rheology of Elastomers.* Pergamon Press, New York, 1958.
59. M. Mooney. *Rheology, Theory and Applications.* Academic Press, New York, 1958.
60. C. Weber. Zum Zerfall eines Flüssigkeitsstrahles. *Z. angew. Math. Mech.*, Vol. 11, pp. 136–154, 1931.
61. S. Temkin and S.S. Kim. Droplet motion induced by weak shock waves. *J. Fluid Mech.*, Vol. 96, pp. 133–157, 1980.
62. S. Temkin and H.K. Mehta. Droplet drag in a accelerated flow. *J. Fluid Mech.*, Vol. 116, pp. 297–313, 1982.
63. H.N. Oğuz and S.S. Sadhal. Effects of soluble and insoluble surfactants on the motion of drops. *J. Fluid Mech.*, Vol. 194, pp. 563–579, 1988.

64. R.M. Griffith. The effect of surfactants on the terminal velocity of drops and bubbles. *Chem. Engng. Sci.*, Vol. 17, pp. 1057–1070, 1962.
65. H. Czyż. On the concentration of aerosol particles by means of drift forces in a standing wave field. *Acoustica*, Vol. 70, pp. 23–28, 1990.
66. G. Rudinger. Relaxation in gas-particle flow. In P.P. Wegener, editor, *Nonequilibrium Flows, Part I*, Vol. 1 of *Gasdynamics Series*. Marcel Dekker, New York, 1969.
67. V.A. Kuz. A vapor pressure equation for droplets. *Langmuir*, Vol. 9, pp. 3722–3723, 1993.
68. D.K. Chattoraj and K.S. Birdi. *Adsorption and the Gibbs Surface Excess*. Plenum Press, New York, 1984.
69. G.S.H. Lock. *Latent Heat Transfer An Introduction to Fundamentals*. Oxford Engineering Science Series. Oxford University Press, Oxford, 1994.
70. N. Dombrowski and G. Munday. Spray drying. In N. Blakebrough, editor, *Biochemical and Biological Engineering Science*. Academic Press, London, 1968.
71. K.K. Kuo. *Principles of Combustion*. John Wiley & Sons Inc., New York, 1986.
72. F.A. Williams. *Combustion Theory*. Combustion Science and Engineering Series. Addison & Wesley, Redwood City, 2nd edition, 1988.
73. G.M. Faeth. Current status of droplet and liquid combustion. *Prog. Energy Combustion Sci.*, Vol. 3, pp. 191–224, 1977.
74. W.D. Niven, editor. *The Scientific Papers of James Clerc Maxwell*, Vol. 2. Cambridge University Press, Cambridge, 1980.
75. N.A. Fuchs. *Evaporation and Droplet Growth in Gaseous Media*. Pergamon Press, London, 1959.
76. L. Monchick and H. Reiss. Studies of evaporation of small droplets. *J. Chem. Phys.*, Vol. 22, No. 5, pp. 831–836, 1954.
77. P.N. Shankar. A kinetic theory of steady condensation. *J. Fluid Mech.*, Vol. 40, part 2, pp. 385–400, 1970.
78. S.K. Loyalka. Condensation of a spherical droplet. *J. Chem. Phys.*, Vol. 58, No. 1, pp. 354–356, 1973.
79. T. Matsushita. Kinetic analysis of the problem of evaporation and condensation. *Phys. Fluids*, Vol. 19, No. 11, pp. 1712–1715, 1976.
80. K. Yamamoto and T. Nishitani. Kinetic theory of condensation on and evaporation from a droplet. In *Proc. 14th Int. Symp. on Rarefied Gas Dynamics*, pp. 893–900, 1984.
81. M. Sitarski and B. Nowakowski. Condensation rate of trace vapor on Knudsen aerosols from the solution of the Boltzmann equation. *J. Colloid and Interface Sci.*, Vol. 72, No. 1, pp. 113–122, 1979.
82. M. Birouk, C. Chauveau, B. Sarh, A. Quilgars, and I. Gökalb. Turbulence effects on the vaporization of monocomponent single droplets. *Combustion Sci. Tech.*, Vol. 113-114, pp. 413–428, 1996.
83. P. Ravindran and E.J. Davis. Multicomponent evaporation of single droplets. *J. Colloid and Interface Sci.*, Vol. 85, No. 1, pp. 278–288, 1982.
84. J. Kalkkinen, T. Vesala, and M. Kulmala. Binary droplet evaporation in the presence of an inert gas: An exact solution of the Maxwell-Stefan equations. *Int. Comm. Heat Mass Transfer*, Vol. 18, pp. 117–126, 1991.
85. C.K. Law, T.Y. Xiong, and C.H. Wang. Alcohol droplet vaporization in humid air. *Int. J. Heat Mass Transfer*, Vol. 30, No. 7, pp. 1435–1443, 1987.
86. K.P. Lee, S.H. Wang, and S.C. Wong. Spark ignition characteristics of monodisperse multicomponent fuel sprays. *Combustion Sci. Tech.*, Vol. 113-114, pp. 493–502, 1996.
87. V.K. La Mer. *Retardation of Evaporation by Monolayers: Transport Processes*. Academic Press, New York, 1962.

88. I.S. Costin and G.T. Barnes. Two-component monolayers, I. experimental procedures and selection of a system, II. surface pressure-area relation for the octadecanol-docosyl sulphate system, III. evaporation resistances of the octadecanol-docosyl sulphate system. *J. Colloid and Interface Sci.*, Vol. 51, pp. 94–132, 1975.
89. T.A. Brzustowski, E.M. Twardus, and A. Wojcicki, S. Sobiesiak. Interaction of two burning fuel droplets of arbitrary size. *AIAA Journal*, Vol. 17, No. 11, pp. 1234–1242, 1979.
90. M. Labowsky. Calculation of the burning rates of interacting fuel droplets. *Combustion Sci. Tech.*, Vol. 22, pp. 217–226, 1980.
91. T.Y. Xiong, C.K. Law, and K. Miyasaka. Interactive vaporization and combustion of binary droplet systems. In *Proc. 20th Symp. (Int.) on Combustion*, pp. 1781–1787. The Combustion Institute, 1984.
92. J. Xin and C.M. Megaridis. Modeling of multiple droplet streams in close spacing configurations and sprays. *Atomization and Sprays*, Vol. 7, pp. 267–294, 1997.
93. C.F. Bohren and D.R. Huffman. *Absorption and Scattering of Light by Small Particles.* J. Wiley & Sons, New York, 1983.
94. H.C. van de Hulst. *Light Scattering by Small Particles.* Dover, New York, 1981.
95. J. Popp. *Elastische und inelastische Lichtstreuung an einzelnen sphärischen Mikropartikeln.* PhD dissertation, Universität Würzburg, 1994. Verlag Shaker, Aachen.
96. M. Born and E. Wolf. *Principles of Optics.* Pergamon Press, Oxford, 4th edition, 1970.
97. E. Hecht. *Optics.* Addison-Wesley, New York, 1987.
98. K. Anders, N. Roth, and A. Frohn. Light scattering at the rainbow angle: Information on size and refractive index. In *Proc. 3rd Int. Congr. on Optical Particle Sizing*, pp. 237–242, 1993.
99. H.C. van de Hulst and R.T. Wang. Glare points. *Applied Optics*, Vol. 30, No. 33, pp. 4755–4763, 1991.
100. J.P.A.J. van Beeck. *Rainbow Phenomena: Development of a Laser-Based, Non-Intrusive Technique for Measuring Droplet Size, Temperature and Velocity.* PhD thesis, Technische Universiteit Eindhoven, 1997.
101. H.M. Nussenzveig. The theory of the rainbow. *Scientific American*, Vol. 236, No. 4, pp. 116–127, 1977.
102. G.B. Airy. On the intensity of light in the neighbourhood of a caustic. *Transactions of the Cambridge Philosophical Society*, Vol. 6, pp. 379–402, 1838.
103. J.D. Walker. Multiple rainbows from single drops of water and other liquids. *Am. J. Phys.*, Vol. 44, No. 5, pp. 421–433, 1976.
104. F. Herrmann. *Entwicklung eines optischen Meßverfahrens hoher räumlicher und zeitlicher Auflösung zur Bestimmung von Tropfengrößen.* PhD dissertation, Universität Stuttgart, 1994. Verlag Shaker, Aachen.
105. N. Roth. *Bestimmung des Brechungsindex in Einzeltropfen aus dem Streulicht im Bereich des Regenbogens.* PhD dissertation, Stuttgart University, 1998. Verlag Shaker, Aachen.
106. G. Mie. Beiträge zur Optik trüber Medien, speziell kolloidaler Metallösungen. *Annalen der Physik*, Vol. 4, No. 25, pp. 377–455, 1908.
107. *Proc. 2nd Int. Congr. on Optical Particle Sizing.* Arizona State University, Printing Services, 1990.
108. H. Hügel. *Strahlwerkzeug Laser.* Studienbücher Maschinenbau. Teubner, Stuttgart, 1992.

109. G. Gouesbet, B. Maheu, and G. Gréhan. Light scattering from a sphere arbitrarily located in a Gaussian beam, using a Bromwich formulation. *J. Opt. Soc. Am. A*, Vol. 5, No. 9, pp. 1427–1443, 1988.
110. G. Gouesbet, G. Gréhan, and B. Maheu. On the generalized Lorenz-Mie theory: First attempt to design a localized approximation to the computation of the coefficients g m n. *J. Optics*, Vol. 20, No. 1, pp. 31–43, 1989.
111. G. Gouesbet, G. Gréhan, and B. Maheu. Localized interpretation to compute all the coefficients g m n in the generalized Lorenz-Mie theory. *J. Opt. Soc. Am. A*, Vol. 7, No. 6, pp. 998–1007, 1990.
112. T. Wriedt. A review of elastic light scattering theories. *Part. Part. Syst. Charact.*, Vol. 15, pp. 67–74, 1998.
113. A. Ashkin and J.M. Dziedzic. Optical levitation by radiation pressure. *Appl. Phys. Lett.*, Vol. 19, No. 8, pp. 283–285, 1971.
114. P. Debye. Der Lichtdruck auf Kugeln von beliebigem Material. *Annalen d. Physik*, Vol. 4, No. 30, pp. 57–136, 1909.
115. G. Roosen. A theoretical and experimental study of the stable equilibrium positions of spheres levitated by two horizontal laser beams. *Optics Communications*, Vol. 21, No. 1, pp. 189–194, 1977.
116. K.F. Ren, G. Gréhan, and G. Gouesbet. Radiation pressure forces exerted on a particle arbitrarily located in a Gaussian beam by using the generalized Lorenz-Mie theory, and associated resonance effects. *Optics Communications*, Vol. 108, pp. 343–354, 1994.
117. S.C. Hill and R.E. Benner. Morphology dependent resonances. In P.W. Barber and R.K. Chang, editors, *Optical Effects Associated with Small Particles*, Advanced Series in Applied Physics. World Scientific, Singapore, 1988.
118. P. Chylek. Partial-wave resonances and the ripple structure in the Mie normalized extinction cross section. *J. Opt. Soc. Am.*, Vol. 66, No. 3, pp. 285–287, 1976.
119. P. Chylek, V. Ramaswamy, A. Ashkin, and J.M. Dziedzic. Simultaneous determination of refractive index and size of spherical dielectric particles from light scattering data. *Applied Optics*, Vol. 22, No. 15, pp. 2302–2307, 1983.
120. K. Schaschek, J. Popp, and W. Kiefer. Morphology dependent resonances in Raman spectra of optically levitated microparticles: Determination of radius and evaporation rate of single glycerol/water droplets by means of internal mode assignment. *Ber. Bunsenges. Phys. Chem.*, Vol. 97, No. 8, pp. 1007–1011, 1993.
121. K. Schaschek, J. Popp, and W. Kiefer. Observation of morphology-dependent input and output resonances in time-dependent Raman spectra of optically levitated microdroplets. *J. Raman Spectroscopy*, Vol. 24, pp. 69–75, 1993.
122. G. Schweiger. Raman scattering on single aerosol particles and on flowing aerosols: A review. *J. Aerosol Sci.*, Vol. 21, No. 4, pp. 483–509, 1990.
123. G. Schweiger. Raman scattering on microparticles: Size dependence. *J. Opt. Soc. Am. B*, Vol. 8, No. 8, pp. 1770–1778, 1991.
124. I.N. Tang and H.R. Munkelwitz. Simultaneous determination of refractive index and density of an evaporating aqueous solution droplet. *Aerosol Science and Technology*, Vol. 15, pp. 201–207, 1991.
125. K. Masters. *Spray Drying Handbook*. Longman Scientific & Technical, Harlow, 5th edition, 1991.
126. K. Tang and A. Gomez. Generation by electrospray of monodisperse water droplets for targeted drug delivery by inhalation. *J. Aerosol Sci.*, Vol. 25, No. 6, pp. 1237–1249, 1994.
127. A. Jaworek and A. Krupa. Jets and drops formation in electrohydrodynamic spraying of liquids. a systematic approach. *Experiments in Fluids*, Vol. 27, pp. 43–52, 1999.

128. A. Jaworek and A. Krupa. Classification of the modes of ehd spraying. *J. Aerosol Sci.*, Vol. 30, No. 7, pp. 873–894, 1999.
129. A. Gomez and K. Tang. Charge and fission of droplets in electrostatic sprays. *Phys. Fluids*, Vol. 6, No. 1, pp. 404–414, 1994.
130. *Electrosprays: Theory and Applications*, Vol. 25 of *Special Issue, J. Aerosol Sci.*, 1994.
131. G. König, K. Anders, and A. Frohn. A new light-scattering technique to measure the diameter of periodically generated moving droplets. *J. Aerosol Sci.*, Vol. 17, pp. 157–167, 1986.
132. L. Prandtl, K. Oswatitsch, and K. Wieghardt. *Führer durch die Strömungslehre.* Vieweg, Braunschweig, Wiesbaden, 8th edition, 1984.
133. A.S.M. Nuruzzaman, A.B. Hedley, and J.M. Beér. Combustion of monosized droplet streams in stationary self-supported flames. In *Proc. 13th Symp. (Int.) on Combustion*, pp. 787–799. The Combustion Institute, 1970.
134. R.N. Berglund and B.Y.H. Liu. Generation of monodisperse aerosol standards. *Environ. Sci. Tech.*, Vol. 7, pp. 147–153, 1973.
135. M. Orme and E.P. Muntz. New technique for producing highly uniform droplet streams over an extended range of disturbance wavenumbers. *Rev. Sci. Instrum.*, Vol. 58, No. 2, 1987.
136. M. Orme and E.P. Muntz. The manipulation of capillary stream breakup using amplitude-modulated disturbances: A pictorial and quantitative representation. *Phys. Fluids A 2*, Vol. 7, pp. 1124–1140, 1990.
137. K. Anders, N. Roth, and A. Frohn. Operation characteristics of vibrating-orifice generators: The coherence length. *Part. Part. Syst. Charact.*, Vol. 9, pp. 40–43, 1992.
138. E.P. Muntz and M. Dixon. Application to space operations of free-flying, controlled stream of liquids. *J. Spacecraft*, Vol. 23, No. 4, pp. 411–419, 1986.
139. G. Rosenstock. *Erzeugung schnell fliegender Tropfen für Tintendrucker mit Hilfe von Druckwellen.* PhD dissertation, Universität München, 1984.
140. D.B. Bogy and F.E. Talke. Experimental and theoretical study of wave propagation phenomena in drop-on-demand ink jet devices. *IBM J. Res. Develop.*, Vol. 28, No. 3, pp. 314–321, 1984.
141. R.L. Adams and J. Roy. A one-dimensional numerical model of a drop-on-demand ink jet. *J. Applied Mechanics*, Vol. 53, pp. 193–197, 1986.
142. T.W. Shield, D.B. Bogy, and F.E. Talke. A numerical comparison of one-dimensional fluid jet models applied to drop-on-demand printing. *J. Computational Physics*, Vol. 67, pp. 327–347, 1986.
143. E.H. Brandt. Levitation in physics. *Science*, Vol. 243, pp. 349–355, 1989.
144. J.R. Brock. The kinetics of ultrafine particles. In W.H. Marlow, editor, *Aerosol Microphysics I, Particle Interaction*, Topics in Current Physics. Springer-Verlag, Berlin, 1980.
145. F. Gelbhard and J.H. Seinfeld. Simulation of multicomponent aerosol dynamics. *J. Colloid and Interface Sci.*, Vol. 78, No. 2, pp. 485–501, 1980.
146. K.C. Young. The evolution of drop spectra due to condensation, coalescence and breakup. *J. Atmospheric Sciences*, Vol. 32, pp. 965–1411, 1975.
147. D.Y. Liu, K. Anders, and A. Frohn. Drag coefficients of single droplets moving in an infinite droplet chain on the axis of a tube. *Int. J. Multiphase Flow*, Vol. 14, pp. 217–232, 1988.
148. C.S. Connon and D. Dunn-Rankin. Droplet stream dynamics at ambient pressure. *Atomization and Sprays*, Vol. 6, pp. 485–497, 1996.
149. K. Anders, A. Frohn, A. Karl, and N. Roth. Flame propagation in planar droplet arrays and interaction phenomena between neighbouring droplet

streams. In *Proc. 26th Symp. (Int.) on Combustion*, pp. 1697–1703. The Combustion Institute, 1996.

150. J.M. Schneider, N.R. Lindblad, and C.D. Hendricks. An apparatus to study the collision and coalescence of liquid aerosols. *J. Colloid Sci.*, Vol. 20, pp. 610–616, 1965.
151. G.D. Kinzer and R. Gunn. The evaporation, temperature and thermal relaxation-time of freely falling waterdrops. *J. Meteorology*, Vol. 8, No. 2, pp. 71–83, 1951.
152. P.C. Nordine and R.M. Atkins. Aerodynamic levitation of laser-heated solids in gas jets. *Rev. Sci. Instrum.*, Vol. 53, No. 9, pp. 1456–1464, 1982.
153. J.F. Rex, A.E. Fuhs, and S.S. Penner. Interference effects during burning in air for stationary n-heptane, ethyl alcohol, and methyl alcohol droplets. *Jet Propulsion*, Vol. 26, pp. 179–187, 1956.
154. T. Kadoka, H. Hiroyasu, and A. Farazandehmehr. Soot formation by combustion of a fuel droplet in high pressure gaseous environments. *Combustion and Flame*, Vol. 29, pp. 67–75, 1977.
155. T. Niioka and J. Sato. Combustion and microexplosion behavior of miscible fuel droplets under high pressure. In *Proc. 21th Symp. (Int.) on Combustion*, pp. 625–631. The Combustion Institute, 1986.
156. R.S. Bradley, M.G. Evans, and R.W. Whytlaw-Gray. The rate of evaporation of droplets. Evaporation and diffusion coefficients, and vapour pressures of dibutyl phthalate and butyl stearate. *Proc. R. Soc. London, Ser. A*, Vol. 186, pp. 368–390, 1946.
157. R.S. Bradley. Rates of evaporation, IV. The rate of evaporation and vapour pressure of rhombic sulphur. *Proc. R. Soc. London, Ser. A*, Vol. 205, pp. 553–563, 1951.
158. P.J. Wyatt and D.T. Phillips. A new instrument for the study of individual aerosol particles. *J. Colloid and Interface Sci.*, Vol. 39, No. 1, pp. 125–135, 1972.
159. E.J. Davis and A.K. Ray. Single aerosol particle size and mass measurements using an electrodynamic balance. *J. Colloid and Interface Sci.*, Vol. 75, No. 2, pp. 566–576, 1980.
160. E.J. Davis, P. Ravindran, and A.K. Ray. Single aerosol particle studies. *Adv. Colloid and Interface Sci.*, Vol. 15, pp. 1–24, 1981.
161. L. Altwegg, M. Pope, S. Arnold, W.Y. Fowlkes, and M.A. El Hamamsy. Modified Millikan capacitor for photoemission studies. *Rev. Sci. Instrum.*, Vol. 53, No. 3, pp. 332–337, 1982.
162. P.F. Clancy, E.G. Lierke, R. Grossbach, and W.M. Heide. Electrostatic and acoustic instrumentation for material science processing in space. *Acta Astronautica*, Vol. 7, pp. 877–891, 1980.
163. S. Arnold. Determination of particle mass and charge by one electron differentials. *J. Aerosol Sci.*, Vol. 10, pp. 49–53, 1979.
164. B. Deepti and A.K. Ray. In situ measurement of photochemical reactions in microdroplets. *J. Aerosol Sci.*, Vol. 30, No. 3, pp. 279–288, 1999.
165. R.F. Wuerker, H. Shelton, and R.V. Langmuir. Electrodynamic containment of charged particles. *J. Appl. Phys.*, Vol. 30, No. 3, pp. 342–349, 1959.
166. S. Arnold and N. Hessel. Photoemission from single electrodynamically levitated microparticles. *Rev. Sci. Instrum.*, Vol. 56, No. 11, pp. 2066–2069, 1985.
167. S. Arnold and L.M. Folan. Fluorescence spectrometer for a single levitated microparticle. *Rev. Sci. Instrum.*, Vol. 57, No. 9, pp. 2250–2253, 1986.
168. S. Arnold and L.M. Folan. Spherical void electrodynamic levitator. *Rev. Sci. Instrum.*, Vol. 58, No. 9, pp. 1732–1735, 1987.

169. E.C. Okress and D.M. Wroughton. Electromagnetic levitation of solid and molten metals. *J. Appl. Phys.*, Vol. 23, No. 5, pp. 545–552, 1952.
170. T.B. Jones and J.P. Kraybill. Active feedback-controlled dielectrophoretic levitation. *J. Appl. Phys.*, Vol. 60, No. 4, pp. 1247–1252, 1986.
171. R.E. Rosenzweig. Magnetic fluids. *Scientific American*, Vol. 247, No. 4, pp. 124–132, 1982.
172. A. Ashkin. Acceleration and trapping of particles by radiation pressure. *Phys. Rev. Lett.*, Vol. 24, No. 4, pp. 156–159, 1970.
173. A. Ashkin and J.M. Dziedzic. Optical levitation of liquid drops by radiation pressure. *Science*, Vol. 187, pp. 1073–1075, 1975.
174. N. Roth, K. Anders, and A. Frohn. Size and evaporation rate measurements of optically levitated droplets. In *Proc. 3rd Int. Congr. on Optical Particle Sizing*, pp. 371–377, 1993.
175. A. Ashkin and J.M. Dziedzic. Optical levitation in high vacuum. *Appl. Phys. Lett.*, Vol. 29, No. 6, pp. 333–335, 1976.
176. K.F. Ren, G. Gréhan, and G. Gouesbet. Prediction of reverse radiation pressure by generalized Lorenz-Mie theory. *Applied Optics*, Vol. 35, No. 15, pp. 2702–2710, 1996.
177. G. Roosen and S. Slansky. Influence of the beam divergence on the exerted force on a sphere by a laser beam and required conditions for stable optical levitation. *Optics Communications*, Vol. 29, No. 3, pp. 341–346, 1979.
178. M.I. Angelova and B. Pouligny. Trapping and levitation of a dielectric sphere with off-centred Gaussian beams: I. Experimental. *Pure Appl. Opt.*, Vol. 2, pp. 261–276, 1993.
179. G. Martinot-Lagarde, B. Pouligny, M.I. Angelova, G. Gréhan, and G. Gouesbet. Trapping and levitation of a dielectric sphere with off-centred Gaussian beams: II. GLMT analysis. *Pure Appl. Opt.*, Vol. 4, pp. 571–585, 1995.
180. B.T. Unger and P.L. Marston. Optical levitation of bubbles in water by the radiation pressure of a laser beam: An acoustically quiet levitator. *J. Acoust. Soc. Am.*, Vol. 83, No. 3, pp. 970–975, 1988.
181. A. Ashkin. Applications of laser radiation pressure. *Science*, Vol. 210, No. 4474, pp. 1081–1088, 1980.
182. E. Leung, N. Jacobi, and T. Wang. Acoustic radiation force on a rigid sphere in a resonance chamber. *J. Acoust. Soc. Am.*, Vol. 70, No. 6, pp. 1762–1767, 1981.
183. Gor'kov. On the forces acting on a small particle in an acoustical field in an ideal fluid. *Soviet Physics-Doklady*, Vol. 6, No. 9, pp. 773–775, 1962.
184. L.V. King. On the acoustic radiation pressures on spheres. *Proc. R. Soc. London, Ser. A*, Vol. 147, pp. 212–240, 1934.
185. W.A. Oran, L.H. Berge, and H.W. Parker. Parametric study of an acoustic levitation system. *Rev. Science Instrum.*, Vol. 51, No. 5, pp. 626–631, 1980.
186. E. Eisner. Design of sonic amplitude transformers for high magnification. *J. Acoustical Soc. Am.*, Vol. 35, No. 9, pp. 1367–1377, 1963.
187. J.A.G. Juárez and G.R. Corral. Piezoelectric transducer for air-borne ultrasound. *Acoustica*, Vol. 29, pp. 234–239, 1973.
188. E.G. Lierke. Acoustic levitation - a comprehensive survey of principles and applications. *ACUSTICA·acta acustica*, Vol. 82, pp. 220–237, 1996.
189. E.H. Trinh and C.J. Hsu. Equilibrium shapes of acoustically levitated drops. *J. Acoust. Soc. Am.*, Vol. 79, No. 5, pp. 1335–1338, 1986.
190. E.H. Trinh. Compact acoustic levitation device for studies in fluid dynamics and material science in the laboratory and microgravity. *Rev. Science Instrum.*, Vol. 56, No. 11, pp. 2059–2065, 1985.

191. A.J. Marchese, F.L. Dryer, R.O. Colantonio, and V. Nayagam. Microgravity combustion of methanol and methanol/water droplets: Drop tower experiments and model predictions. In *Proc. 26th Symp. (Int.) on Combustion*, pp. 1209–1217. The Combustion Institute, 1996.
192. S. Kumagai, T. Sakai, and S. Okajima. Combustion of free fuel droplets in a freely falling chamber. In *Proc. 13th Symp. (Int.) on Combustion*, pp. 779–785. The Combustion Institute, 1971.
193. M.Y. Choi and K.-O. Lee. Investigation of sooting in microgravity droplet combustion. In *Proc. 26th Symp. (Int.) on Combustion*, pp. 1243–1249. The Combustion Institute, 1996.
194. D.F. Wang and B.D. Shaw. Droplet combustion in a simulated reduced-gravity environment. *Combustion Sci. Tech.*, Vol. 113-114, pp. 451–470, 1996.
195. D.L. Dietrich, J.B. Haggard Jr., F.L. Dryer, V. Nayagam, B.D. Shaw, and F.A. Williams. Droplet combustion experiments in Spacelab. In *Proc. 26th Symp. (Int.) on Combustion*, pp. 1201–1207. The Combustion Institute, 1996.
196. Y. Tian and R.E. Apfel. A novel multiple drop levitator for the study of drop arrays. *J. Aerosol Sci.*, Vol. 27, No. 5, pp. 721–737, 1996.
197. J.K.R. Weber, D.S. Hampton, D.R. Merkley, C.A. Rey, M.M. Zatarski, and P.C. Nordine. Aeroacoustic levitation: A method for containerless liquid-phase processing at high temperatures. *Rev. Sci. Instrum.*, Vol. 65, No. 2, pp. 456–465, 1994.
198. M. Seaver, A. Galloway, and T.J. Manuccia. Acoustic levitation in a free-jet wind tunnel. *Rev. Sci. Instrum.*, Vol. 60, No. 11, pp. 3452–3459, 1989.
199. G. König and A. Frohn. Visualization of uniformly disintegrated liquid jets by a new observation technique. In *Proc. 3rd Int. Symp. on Flow Visualization*, pp. 488–492, 1983.
200. L.E. Drain. *The Laser Doppler Technique*. John Wiley & Sons, New York, 1980.
201. F. Durst, A. Melling, and J.H. Whitelaw. *Principles and Practice of Laser-Doppler Anemometry*. Academic Press, London, 1981.
202. R. Schodl. Laser-two-focus velocimetry. In *Proc. Advanced Instrumentation for Aero Engines Components*, AGARD-CP-399, pp. 7.1–7.31, 1986.
203. W. Förster, R. Schodl, M. Beversdorff, A. Klemmer, and E. Rijmenants. Design and experimental verification of 3-d velocimeters based on the laser-2-focus technique. In *Proc. 5th Int. Symp. on Application of Laser Techniques to Fluid Mechanics*, pp. 8.4, 1990.
204. M. Raffel, C. Willert, and J. Kompenhans. *Particle Image Velocimetry: A Practical Guide*. Experimental Fluid Mechanics. Springer-Verlag, Berlin, 1998.
205. R. Ragucci, A. Cavaliere, and P. Massoli. Droplet sizing by laser light scattering exploiting intensity angular oscillation in the Mie regime. *Part. Part. Syst. Charact.*, Vol. 7, pp. 221–225, 1990.
206. N.A. Fomin. *Speckle Photography for Fluid Mechanics Measurements*. Experimental Fluid Mechanics. Springer-Verlag, Berlin, 1998.
207. K. Anders, N. Roth, and A. Frohn. Simultaneous measurements of droplet size and droplet velocity in monodisperse droplet streams. In *von Karman Institute for Fluid Dynamics, Lecture Series, Particle Image Displacement Velocimetry*, pp. 11–13, 1988.
208. K.F. Ren, D. Lebrun, C. Özkul, G. Kleitz, G. Gouesbet, and G. Gréhan. On the measurement of particles by imaging methods: Theoretical and experimental aspects. *Part. Part. Syst. Charact.*, Vol. 13, pp. 156–164, 1996.
209. H.-E. Albrecht, C. Tropea, M. Borys, and N. Damaschke. *Laser Doppler and Phase Doppler Measurement Techniques*. Experimental Fluid Mechanics. Springer-Verlag, Berlin, 2000.

210. P. Massoli, F. Beretta, and A. D'Alessio. Single droplet size, velocity, and optical characteristics by the polarization properties of scattered light. *Applied Optics*, Vol. 28, No. 6, pp. 1200–1205, 1989.
211. P. Massoli, F. Beretta, and A. D'Alessio. A new experimental technique for the determination of single droplets size, velocity, and optical properties inside sprays. *Chem. Eng. Comm.*, Vol. 75, pp. 171–180, 1989.
212. K. Anders. *Experimentelle und theoretische Untersuchung der Tropfenverdampfung für Knudsen-Zahlen im Übergangsbereich.* PhD dissertation, Universität Stuttgart, 1994. Verlag Shaker, Aachen.
213. K. Anders. Einsatz eines CCD-Bildsensors zur Bestimmung von Durchmessern und Verdampfungsraten von Flüssigkeitströpfchen im Größenbereich zwischen $10\,\mu$m und $300\,\mu$m. In *Proc. Sensor '85*, 1985.
214. K.H. Hesselbacher, K. Anders, and A. Frohn. Experimental investigation of Gaussian beam effects on the accuracy of a droplet sizing method. *Applied Optics*, Vol. 30, pp. 4930–4935, 1991.
215. H. Umhauer. Ermittlung von Partikelgrößenverteilungen hoher Konzentration mit Hilfe einer Streulichtmeßeinrichtung. *VDI-Berichte, Nr. 232*, pp. 101–105, 1975.
216. H. Umhauer. Particle size distribution analysis by scattered light measurement using an optically defined measuring volume. *J. Aerosol Sci.*, Vol. 14, No. 6, pp. 765–770, 1983.
217. H. Umhauer. Single particle light scattering size analysis: Delimitation of a measuring volume by purely optical means. In *Proc. 3th Int. Congr. on Optical Particle Sizing*, pp. 315–321, 1993.
218. F. Herrmann and A. Frohn. Größen- und Geschwindigkeitsmessung an Einzeltropfen durch Auswertung der Streulichtverteilung des Nahfelds. In *Proc. 1. Workshop über Sprays, Erfassung von Sprühvorgängen und Technologien der Fluidzerstäubung*, Erlangen, 1994. Lehrstuhl für Strömungsmechanik, LSTM.
219. C.F. Hess and C.P. Wood. The pulse displacement technique - a single particle counter with a size range larger than 1000:1. *Part. Part. Syst. Charact.*, Vol. 11, pp. 107–113, 1994.
220. C.P. Wood and C.F. Hess. A miniaturized instrument implementing pulse displacement to measure particles from $2\,\mu$m to $5000\,\mu$m. In *Proc. 7th Int. Symp. on Application of Laser Techniques to Fluid Mechanics*, pp. 24.6.1–24.6.8, 1994.
221. J.P.A.J. van Beeck and M.L. Riethmuller. Nonintrusive measurements of temperature and size of single falling raindrops. *Applied Optics*, Vol. 34, No. 10, pp. 1633–1639, 1995.
222. J.P.A.J. van Beeck and M.L. Riethmuller. Rainbow phenomena applied to the measurement of droplet size and velocity and to the detection of nonsphericity. *Applied Optics*, Vol. 35, No. 13, pp. 2259–2266, 1996.
223. O. Preining. Information theory applied to the acquisition of size distributions. *J. Aerosol Sci.*, Vol. 3, pp. 289–296, 1972.
224. P. Biswas, C.L. Jones, and R.C. Flagan. Distortion of size distributions by condensation and evaporation in aerosol instruments. *Aerosol Science and Technology*, Vol. 7, pp. 231–246, 1987.
225. A. Doyle, D.R. Moffett, and B. Vonnegut. Behavior of evaporating electrically charged droplets. *J. Colloid Sci.*, Vol. 19, pp. 136–143, 1964.
226. K.-H. Ahn and B.Y.H. Liu. Particle activation and droplet growth processes in condensation nuclei counter - I. Theoretical background. *J. Aerosol Sci.*, Vol. 21, No. 2, pp. 249–261, 1990.
227. B.B. Weiner. Particle and spray sizing using laser diffraction. *SPIE Optics in Quality Assurance II*, Vol. 170, 1979.

228. E.D. Hirleman. Modeling of multiple scattering effects in Fraunhofer diffraction particle size analysis. *Part. Part. Syst. Charact.*, Vol. 5, pp. 57–65, 1988.
229. E.D. Hirleman. Optimal scaling of the inverse Fraunhofer diffraction particle sizing problem: The linear system produced by quadrature. *Part. Charact.*, Vol. 4, pp. 128–133, 1987.
230. E.D. Hirleman and L.G. Dodge. Performance comparison of Malvern instruments laser diffraction drop size analysers. In *Proc. ICLASS 85 Conference*, pp. 8564/1–8564/14, 1985.
231. H. Mühlenweg and E.D. Hirleman. Laser diffraction spectroscopy: Influence of particle shape and a shape adaptation technique. *Part. Part. Syst. Charact.*, Vol. 15, pp. 163–169, 1998.
232. S. Wittig, H.-J. Feld, A. Müller, W. Samenfink, and A. Tremmel. Application of the dispersion-quotient-method under technical system conditions. In *Proc. 2nd. Int. Congr. on Optical Particle Sizing*, pp. 335–347, 1990.
233. K. Schaber, A. Schenkel, and R.A. Zahoransky. Einsatz und Bewertung eines Drei-Wellenlängen-Extinktionsverfahrens zur Charakterisierung von Aerosolen unter industriellen Bedingungen. *Technisches Messen*, Vol. 7/8, pp. 295–300, 1994.
234. L. Kai and P. Massoli. Scattering of electromagnetic-plane waves by radially inhomogeneous spheres: A finely stratified sphere model. *Applied Optics*, Vol. 33, No. 3, pp. 501–511, 1994.
235. L. Kai, P. Massoli, and A. D'Alessio. Some far-field scattering characteristics of radially inhomogeneous particles. *Part. Part. Syst. Charact.*, Vol. 2, pp. 385–390, 1994.
236. L. Kai and A. D'Alessio. Extinction efficiency of gradient-index microspheres. *Part. Part. Syst. Charact.*, Vol. 12, pp. 119–122, 1995.
237. K.A. Fuller. Scattering of light by coated spheres. *Optics Letters*, Vol. 18, No. 4, pp. 257–259, 1993.
238. K.A. Fuller. Scattering and absorption by inhomogeneous spheres and sphere aggregates. *SPIE*, Vol. 1892, 1993.
239. M. Schneider and E.D. Hirleman. Influence of internal refractive index gradients on size measurements of spherically symmetric particles by phase Doppler anemometry. *Applied Optics*, Vol. 33, No. 12, pp. 2379–2388, 1994.
240. K. Anders, N. Roth, and A. Frohn. Influence of refractive index gradients within droplets on rainbow position and implications for rainbow refractometry. *Part. Part. Syst. Charact.*, Vol. 13, pp. 125–129, 1996.
241. P.M. Aker, P.A. Moortgat, and J.-X. Zhang. MDSRS imaging: A spectrocopic tour through the diffuse part of the electric double layer. *SPIE*, Vol. 2547, pp. 110–116, 1995.
242. G. Pitcher, G. Wigley, and M. Saffman. Sensitivity of dropsize measurements by phase Doppler anemometry to refractive index changes in combustion fuel sprays. In *Proc. 5th Int. Symp. on Application of Laser Techniques to Fluid Mechanics*, pp. 14.4.1–14.4.9, 1990.
243. A. Naqwi, F. Durst, and X. Liu. Extended phase-Doppler system for characterization of multiphase flows. *Part. Part. Syst. Charact.*, Vol. 8, pp. 16–22, 1991.
244. G. Brenn, J. Domnick, F. Durst, C. Tropea, and T.-H. Xu. Investigation of polydisperse spray interaction using an extended phase-Doppler anemometer. In *Proc. 7th Int. Symp. on Application of Laser Techniques to Fluid Mechanics*, Vol. 1, pp. 21.1.1–21.1.8, 1994.
245. G. Brenn, C. Tropea, and T.-H. Xu. Investigations on accuracy and resolution of refractive index measurements with an extended phase-Doppler anemometer.

In *Proc. 8th Int. Symp. on Application of Laser Techniques to Fluid Mechanics*, Vol. 1, pp. 9.4.1–9.4.8, 1996.
246. F. Onofri, T. Girasole, G. Gréhan, G. Gouesbet, G. Brenn, J. Domnick, T.-H. Xu, and C. Tropea. Phase-Doppler anemometry with the dual burst technique for measurement of refractive index and absorption coefficient simultaneously with size and velocity. *Part. Part. Syst. Charact.*, Vol. 13, pp. 112–124, 1996.
247. P. Massoli, F. Beretta, A. D'Alessio, and M. Lazzaro. Temperature and size of single transparent droplets by light scattering in the forward and rainbow regions. *Applied Optics*, Vol. 32, No. 18, pp. 3295–3301, 1993.
248. J.K. Schaller, S. Wassenberg, and C.G. Stojanoff. A new method for temperature measurements of droplets. *SPIE*, Vol. 2122, pp. 186–194, 1994.
249. J.K. Schaller, S. Wassenberg, D.K. Fiedler, and C.G. Stojanoff. Refractive index determination as a tool for temperature measurement and process control: A new approach. *SPIE*, Vol. 2248, pp. 318–325, 1994.
250. N. Roth, K. Anders, and A. Frohn. Size insensitive rainbow refractometry: Theoretical aspects. In *Proc. 8th Int. Symp. on Application of Laser Techniques to Fluid Mechanics*, Vol. 1, pp. 9.2.1–9.2.6, 1996.
251. S.V. Sankar, D.M. Robart, and W.D. Bachalo. An advanced signal processor for improved accuracy in droplet temperature measurements. In *Proc. 8th Int. Symp. on Application of Laser Techniques to Fluid Mechanics*, Vol. 1, pp. 9.3.1–9.3.9, 1996.
252. J.C. Johnson. Measurement of the surface temperature of evaporating water drops. *J. Appl. Physics*, Vol. 21, pp. 22–23, 1950.
253. A.R. Hall. Experimental temperature gradients in burning drops. In *Proc. 7th Symp. (Int.) on Combustion*, pp. 399–406. The Combustion Institute, 1958.
254. N. Naudin. *Développement d'un système optronique infrarouge pour la mesure de température de gouttes en mouvement rapide et en combustion.* PhD thèse, L'école nationale supérieure de l'aéronautique et de l'espace, 1995.
255. C.D. Richards and R.F. Richards. Transient temperature measurements in a convectively cooled droplet. *Experiments in Fluids*, Vol. 25, pp. 392–400, 1996.
256. R. Vehring and G. Schweiger. Optical determination of the temperature of transparent microparticles. *Applied Spectroscopy*, Vol. 46, No. 1, pp. 25–27, 1992.
257. M. Seaver and J.R. Peele. Noncontact fluorescence thermometry of acoustically levitated waterdrops. *Applied Optics*, Vol. 29, No. 33, pp. 4956–4961, 1990.
258. M.R. Wells and L.A. Melton. Temperature measurements of falling droplets. *J. Heat Transfer*, Vol. 112, pp. 1008–1013, 1990.
259. Y. Tian and R.E. Holt, R.G. Apfel. A new method for measuring liquid surface tension with acoustic levitation. *Rev. Sci. Instrum.*, Vol. 66, No. 5, pp. 3349–3354, 1995.
260. S.V. Sankar, D.H. Buermann, and W.D. Bachalo. Application of rainbow thermometry to the study of fuel droplet heat-up and evaporation characteristics. *Transactions of the ASME, Journal of Engineering for Gas Turbines and Power*, Vol. 119, pp. 573–584, 1997.
261. P. Chylek. Resonance structure of Mie scattering: Distance between resonances. *J. Opt. Soc. Am. A*, Vol. 7, No. 9, pp. 1609–1613, 1990.
262. B.V. Ramarao and C. Tien. Role of Basset force on particle deposition in stagnation flow. *J. Aerosol Sci.*, Vol. 21, No. 5, pp. 597–611, 1990.
263. J.B. Edson and C.W. Fairall. Spray droplet modeling, 1, Lagrangian model simulation of the turbulent transport of evaporating droplets. *J. Geophys. Res.*, Vol. 99, pp. 25295–25311, 1994.

264. J.B. Edson, S. Anquetin, P.G. Mestayer, and J.F. Sini. Spray droplet modeling, 2, an iterative Eulerian-Lagrangian model of evaporating spray droplets. *J. Geophys. Res.*, Vol. 101, pp. 1279–1293, 1996.
265. J.Y. Poo and N. Ashgriz. Variation of drag coefficients in an interacting drop stream. *Experiments in Fluids*, Vol. 11, pp. 1–8, 1991.
266. I. Silverman and W.A. Sirignano. Multi-droplet interaction effects in dense sprays. *Int. J. Multiphase Flow*, Vol. 20, No. 1, pp. 99–116, 1994.
267. A.S. Sangani, D.Z. Zhang, and A. Prosperetti. The added mass, Basset, and viscous drag coefficients in nondilute bubbly liquids undergoing small-amplitude oscillatory motion. *Phys. Fluids*, Vol. 3, No. 12, pp. 2955–2970, 1991.
268. S.S. Sadhal, P.S. Ayyaswamy, and J.N. Chung. *Transport Phenomena with Drops and Bubbles.* Mechanical Engineering Series. Springer-Verlag, New York, 1997.
269. L. Talbot, R.K. Cheng, R.W. Schefer, and D.R. Willis. Thermophoresis of particles in a heated boundary layer. *J. Fluid Mech.*, Vol. 101, No. 4, pp. 737–758, 1980.
270. T. Kanki, S. Inchi, T. Miyazaki, and H. Ueda. On thermophoresis of relatively large aerosol particles suspended near a plate. *J. Colloid and Interface Sci.*, Vol. 107, No. 2, pp. 418–428, 1985.
271. Y. Sone and K. Aoki. Forces on a spherical particle in a slightly rarefied gas. In *Proc. 10th Int. Symp. on Rarefied Gas Dynamics*, pp. 417–433, 1976.
272. W. Schneider. Drag of droplets moving through their own vapor, part I - continuum flow. *PhysicoChem. Hydrodyn.*, Vol. 2, pp. 135–141, 1981.
273. W. Schneider. Drag of droplets moving through their own vapor, part II - free molecular flow. *PhysicoChem. Hydrodyn.*, Vol. 3, No. 2, pp. 119–126, 1082.
274. S.L. Lee. Particle drag in a dilute turbulent two-phase suspension flow. *Int. J. Multiphase Flow*, Vol. 13, No. 2, pp. 247–256, 1987.
275. R. Mei. Velocity fidelity of flow tracer particles. *Experiments in Fluids*, Vol. 22, pp. 1–13, 1996.
276. J.A. Mulholland, R.K. Srivastava, and J.O.L. Wendt. Influence of droplet spacing on drag coefficient in non-evaporating, monodisperse streams. *AIAA Journal*, Vol. 26, No. 10, pp. 1231–1237, 1988.
277. C.S. Connon and D. Dunn-Rankin. Flow behavior near an infinite droplet stream. *Experiments in Fluids*, Vol. 21, pp. 80–86, 1996.
278. L.J. Forney, D.B. van Dyke, and W.K. McGregor. Dynamics of particle-shock interactions: Part I: Similitude. *Aerosol Science and Technology*, Vol. 6, pp. 129–141, 1987.
279. L.J. Forney, A.E. Walker, and W.K. McGregor. Dynamics of particle-shock interactions: Part II: Effect of the Basset term. *Aerosol Science and Technology*, Vol. 6, pp. 143–152, 1987.
280. Th. Kraut. *Optische Untersuchungen an Tropfenkollektiven hoher Ordnung.* PhD dissertation, Universität Stuttgart, 1990.
281. I.P. Chung, D. Dunn-Rankin, R.F. Phalen, and M.J. Oldham. Low-cost wind tunnel for aerosol inhalation studies. *American Industrial Hygiene Association Journal*, Vol. 53, No. 4, pp. 232–236, 1992.
282. O. Witschger, K. Willeke, S.A. Grinshpun, V. Aizenberg, J. Smith, and P.A. Baron. Simplified method for testing personal inhalable aerosol samplers. *J. Aerosol Sci.*, Vol. 29, No. 7, pp. 855–874, 1998.
283. G. Ramachandran, A. Sreenath, and J.H. Vincent. Towards a new method for experimental determination of aerosol sampler aspiration efficiency in small wind tunnels. *J. Aerosol Sci.*, Vol. 29, No. 7, pp. 875–891, 1998.
284. G. Vali and E.J. Stansburg. Time-dependant characteristics of the heterogeneous nucleation of ice. *Can. J. Phys.*, Vol. 44, pp. 477–502, 1966.

285. *Ice Accretion Simulation*, AGARD-AR-344, 1997.
286. R. Gunn and G.D. Kinzer. The terminal velocity of fall for water droplets in stagnant air. *J. Meteorology*, Vol. 6, pp. 243–248, 1949.
287. W.R. Cotton and N.R. Gokhale. Collision, coalescence, and break-up of large water drops in a vertical wind tunnel. *J. Geophys. Res.*, Vol. 72, No. 16, pp. 4041–4049, 1967.
288. S.K. Mitra, J. Brinkmann, and H.R. Pruppacher. A wind tunnel study on the drop-to-particle conversion. *J. Aerosol Sci.*, Vol. 23, No. 3, pp. 245–256, 1992.
289. Y. Tian and R.E. Holt, R.G. Apfel. Deformation and location of an acoustically levitated drop. *J. Acoustic. Soc. Am.*, Vol. 93, No. 6, pp. 3096–3104, 1993.
290. A.L. Yarin, M. Pfaffenlehner, and C. Tropea. On the acoustic levitation of droplets. *J. Fluid Mech.*, Vol. 356, pp. 65–91, 1998.
291. W.T. Shi and R.E. Apfel. Deformation and position of acoustically levitated liquid drops. *J. Acoust. Soc. Am.*, Vol. 99, No. 4, pp. 1977–1984, 1998.
292. E.H. Trinh, P.L. Marston, and J.L. Robey. Acoustic measurement of the surface tension of levitated drops. *J. Colloid and Interface Sci.*, Vol. 124, No. 1, pp. 95–103, 1988.
293. R.E. Apfel. A novel technique for measuring the strength of liquids. *J. Acoust. Soc. Am.*, Vol. 49, No. 1, pp. 145–155, 1971.
294. J.T. Cronin and T.B. Brill. Acoustic levitation as a IR spectroscopy sampling technique. *Applied Spectroscopy*, Vol. 43, No. 2, pp. 253–257, 1989.
295. S. Temkin. Gasdynamic agglomeration of aerosols. I. Acoustic waves. *Phys. Fluids*, Vol. 6, No. 7, pp. 2294–2302, 1994.
296. G. Funcke. *Beeinflussung der Agglomeration feiner flüssiger Partikeln durch akustische Felder.* PhD dissertation, Universität Stuttgart, 1997. Verlag Shaker, Aachen.
297. P. Caperan, J. Somers, K. Richter, and S. Fourcaudot. Acoustic agglomeration of a glycol fog aerosol: Influence of particle concentration and intensity of the sound field at two frequencies. *J. Aerosol Sci.*, Vol. 26, No. 4, pp. 595–612, 1995.
298. M.V. Smoluchowski. Drei Vorträge über Diffusion, Brownsche Molekularbewegung und Koagulation von Kolloidteilchen. *Phys. Zeitschr.*, Vol. 8, pp. 557–599, 1916.
299. G. Rudinger, K.H. Chou, and D.T. Shaw. Motion of small particles in a field of oscillating shock waves. In *Proc. 13th Int. Symp. Shock Tubes and Waves*, pp. 664–672, 1981.
300. H. Örtel. *Stoßrohre.* Springer-Verlag, Wien, 1966.
301. J.K. Wright. *Shock Tubes.* John Wiley & Sons Inc., New York, 1961.
302. E.F. Green and J.P. Toennies. *Chemical Reactions in Shock Waves.* Academic Press, New York, 1964.
303. J.N. Bradley. *Shock Waves in Chemistry and Physics.* John Wiley & Sons Inc., New York, 1962.
304. F.E. Marble. Dynamics of dusty gases. *Annual Review of Fluid Mechanics*, Vol. 2, pp. 397–446, 1970.
305. R.S. Valentine, N.F. Sather, and W.J. Heideger. The motion of drops in viscous media. *Chem. Engng. Sci.*, Vol. 20, pp. 719–728, 1965.
306. F. Gelbard, L.A. Mondy, and S.E. Ohrt. A new method for determining hydrodynamic effects on the collision of two spheres. *J. Statistical Physics*, Vol. 62, No. 5/6, pp. 945–960, 1991.
307. Z.-T. Deng and S.-M. Jeng. Numerical simulation of droplet deformation in convective flows. *AIAA Journal*, Vol. 30, No. 5, pp. 1290–1296, 1992.

308. R.J. Haywood, M. Renksizbulut, and G.D. Raithby. Transient deformation and evaporation of droplets at intermediate Reynolds numbers. *Int. J. Heat Mass Transfer*, Vol. 37, No. 9, pp. 1401–1409, 1994.
309. W.G. Reinecke and G.D. Waldman. Shock layer shattering of cloud drops in reentry flight. *AIAA Paper, No. 75-152*, 1975.
310. A.A. Ranger and J.A. Nicholls. Aerodynamic shattering of drops. *AIAA Journal*, Vol. 7, No. 2, pp. 285–290, 1969.
311. A. Wierzba and K. Takayama. Shock tube study of stripping type droplet breakup. In *Proc. 16th Int. Symp. Shock Tubes and Waves*, pp. 335–341, 1987.
312. W.E. Krauss and B.M. Leadon. Deformation fragmentation of water drops due to shock wave impact. *AIAA Paper, No. 71-392*, 1971.
313. W. Klenk, N. Widdecke, and A. Frohn. Disintegration of monodisperse droplet streams by shock waves. In *Proc. 4th Asian Symp. on Visualization*, pp. 229–234, 1996.
314. N. Widdecke, W. Klenk, and A. Frohn. Impact of strong shock waves on monodisperse isopropanol droplet streams. In *Proc. 20th Int. Symp. on Shock Tubes and Waves*, pp. 89–94, 1995.
315. W.G. Reinecke. Drop breakup and liquid jet penetration. *AIAA Journal*, Vol. 16, No. 6, pp. 618–619, 1978.
316. C.W. Kauffman and J.A. Nicholls. Shock-wave ignition of liquid droplets. *AIAA Journal*, Vol. 9, No. 5, pp. 880–885, 1971.
317. K. Miyasaka and Y. Mizutani. Ignition of sprays by an incident shock. *Combustion and Flame*, Vol. 25, pp. 177–186, 1975.
318. S.L. Soo. *Fluid Dynamics of Multiphase Systems.* Blaisdell Publishing Company, Waltham, 1967.
319. A. Bürkholz. *Droplet Separation.* VCH Verlagsgesellschaft, Weinheim, 1989.
320. M. Rein. Phenomena of liquid drop impact on liquid and solid surfaces. *Fluid Dynamics Research*, Vol. 12, pp. 61–93, 1993.
321. M. Rein. Wave phenomena during droplet impact. In *Proc. IUTAM Symposium on Waves in Liquid/Gas and Liquid/Vapor Two-Phase Systems*, pp. 171–190, 1995.
322. J. Fukai, Y. Shiiba, T. Yamamoto, O. Miyatake, D. Poulikakos, C.M. Megaridis, and Z. Zhao. Wetting effects on the spreading of a liquid droplet colliding with a flat surface: Experiment and modeling. *American Industrial Hygiene Association Journal*, Vol. 7, No. 2, pp. 236–247, 1995.
323. P.G. de Gennes. Wetting: Statistics and dynamics. *Reviews of Modern Physics*, Vol. 57, part I, No. 3, pp. 827–863, 1985.
324. R. Rioboo, M. Marengo, and C. Tropea. Outcomes from a drop impact on solid surfaces. In *Proc. 15th Int. Conf. on Liquid Atomization and Spray Systems.* ILASS, 1999.
325. G.E. Cossali, A. Coghe, and M. Marengo. The impact of a single drop on a wetted solid surface. *Experiments in Fluids*, Vol. 22, pp. 463–472, 1997.
326. Z. Levin and P.V. Hobbs. Splashing of water drops on solid and wetted surfaces: Hydrodynamics and charge separation. *Philos. Trans. R. Soc. London, Ser. A*, Vol. 269, pp. 555–585, 1971.
327. E.R.G. Eckert and R.M. Drake Jr. *Analysis of Heat and Mass Transfer.* McGraw-Hill Book Company, New York, 1972.
328. I. Michiyoshi and K. Makino. Heat transfer characteristics of evaporation of a liquid droplet on heated surfaces. *Int. J. Heat Mass Transfer*, Vol. 21, No. 21, pp. 605–613, 1978.
329. K.J. Baumeister and F.F. Simon. Leidenfrost temperature - its correlation for liquid metals, cryogens, hydrocarbons and water. *J. Heat Transfer*, Vol. 95, No. 2, pp. 166–173, 1973.

330. J.J. Rizza. Triple-point evaporation of impacted droplets. *J. Spacecraft and Rockets*, Vol. 16, No. 4, pp. 210–213, 1979.
331. K. Makino and I. Michiyoshi. The behavior of a water droplet on heated surfaces. *Int. J. Heat Mass Transfer*, Vol. 27, No. 5, pp. 781–791, 1984.
332. A. Karl. *Untersuchung der Wechselwirkung von Tropfen mit Wänden oberhalb der Leidenfrost-Temperatur.* PhD dissertation, Universität Stuttgart, 1997. Verlag Shaker, Aachen.
333. D.J. Naber and P.V. Farrell. Hydrodynamics of droplet impingement on a heated surface. SAE Technical Paper Series 930919, SAE, 1993.
334. K. Anders, N. Roth, and A. Frohn. The velocity change of ethanol droplets during collision with a wall analysed by image processing. *Experiments in Fluids*, Vol. 15, pp. 91–96, 1993.
335. A. Karl, A. Anders, and A. Frohn. Measurements of interaction times during droplet wall collisions. In *Proc. 4th Triennal Int. Symp. on Fluid Control, Fluid Measurement, Fluid Mechanics, Visualization, Fluidics*, Vol. 2, pp. 731–735. CERT, ONERA, 1994.
336. A. Karl and A. Frohn. Experimental investigation of interaction processes between droplets and hot walls. *Phys. Fluids.* accepted for publication.
337. N. Roth, A. Karl, and A. Frohn. Observation of liquid-wall contact during droplet impact. In *Proc. 14th Int. Conf. on Liquid Atomization and Spray Systems*, pp. 103–107. ILASS, 1998.
338. A. Karl, J. Wolber, and A. Frohn. Interaction of droplet groups with hot walls above the Leidenfrost-temperature. In *Proc. 13th Int. Conf. on Liquid Atomization and Spray Systems*, pp. 458–464, 1997.
339. B. Lafaurie, C. Nardone, R. Scardovelli, S. Zaleski, and G. Zanetti. Fragmentation ain multiphase flows with SURFER. *J. Comput. Phys.*, Vol. 113, pp. 134–147, 1994.
340. C.W. Hirth and B.D. Nichols. Volume-of-fluid (VOF) method for the dynamics of free boundaries. *J. Comput. Phys.*, Vol. 39, No. 1, pp. 201–225, 1981.
341. M. Rieber. *Numerische Modellierung der Dynamik freier Grenzflächen in Zweiphasenströmungen.* PhD dissertation, Universität Stuttgart, 2000.
342. A. Karl, A. Anders, and A. Frohn. Disintegration of droplets colliding with hot walls. In A. Serizawa, T. Fukano, and J. Bataille, editors, *Advances in Multiphase Flow.* Elsevier Science B.V., Amsterdam, 1995.
343. H.E. Edgerton and R.J. Killian. *Moments of Vision - the Stroboscopic Revolution in Photography.* The MIT Press, Cambridge, 1979.
344. M. Rieber and A. Frohn. A numerical study on the mechanism of splashing. *Int. J. Heat and Fluid Flow*, 1999. accepted for publication.
345. P. Bhatnagar, E.P. Gross, and M.K. Krook. A model for collision processes in gases. I. Small amplitude processes in charged and neutral one-component systems. *Physical Review*, Vol. 94, No. 3, pp. 511–525, 1954.
346. X. Shan and H. Chen. Simulation of non-ideal gases and liquid-gas phase transitions by the lattice Boltzmann equation. *Physical Review E*, Vol. 49, pp. 2941–2948, 1994.
347. J.A. Somers and P.C. Rem. Analysis of surface tension in two-phase lattice gases. *Physica*, Vol. D47, pp. 39–46, 1991.
348. A. Karl, M. Rieber, M. Schelkle, K. Anders, and A. Frohn. Comparison of new numerical results for droplet wall interactions with experimental results. In *Proc. 2nd Int. Symp. on Numerical Methods for Multiphase Flows*, pp. 201–206. ASME Fluids Engineering Division, 1996.
349. P. Ebert, M. Kibler, A. Mainka, B. Tenberken, K. Baechmann, G. Frank, and J. Tschiersch. A field study of particle scavenging by raindrops of different sizes using monodisperse aerosol. *J. Aerosol Sci.*, Vol. 29, No. 1/2, pp. 173–186, 1998.

350. C.E. Abbott. A survey of waterdrop interaction experiments. *Reviews of Geophysics and Space Physics*, Vol. 15, No. 3, pp. 363–374, 1977.
351. V.A. Arkhipov, G.S. Ratanov, and V.F. Trofimov. Experimental investigation of the interaction of colliding droplets. *J. Applied Mechanics, Technology & Physics*, Vol. 2, No. 19, pp. 201–204, 1978.
352. P.R. Brazier-Smith, S.G. Jennings, and J. Latham. The interaction of falling water drops: Coalescence. *Proc. R. Soc., Ser. A*, Vol. 326, pp. 393–408, 1972.
353. N. Ashgriz and J.Y. Poo. Coalescence and separation in binary collisions of liquid drops. *J. Fluid Mech.*, Vol. 221, pp. 183–204, 1990.
354. P.R. Brazier-Smith, S.G. Jennings, and J. Latham. Accelerated rates of rain fall. *Nature*, Vol. 232, pp. 112–113, 1971.
355. Y.J. Jiang, A. Umemura, and C.K. Law. An experimental investigation on the collision behaviour on hydrocarbon droplets. *J. Fluid Mech.*, Vol. 234, pp. 171–190, 1992.
356. G. Brenn and A. Frohn. Collision and merging of two equal droplets of propanol. *Experiments in Fluids*, Vol. 7, pp. 441–446, 1989.
357. N.R. Lindblad. Effects of relative humidity and electric charge on the coalescence of curved water surfaces. *J. Colloid Sci.*, Vol. 19, pp. 729–743, 1964.
358. J. Qian and C.K. Law. Regimes of coalescence and separation in droplet collisions. *J. Fluid Mech.*, Vol. 331, pp. 59–80, 1997.
359. A.M. Podvysotsky and A.A. Shraiber. Coalescence and break-up of drops in two-phase flows. *Int. J. Multiphase Flow*, Vol. 10, No. 2, pp. 195–209, 1994.
360. G. Brenn and A. Frohn. Tropfenkollisionen. *Spektrum der Wissenschaft*, pp. 116–124, Dezember 1990.
361. M. Orme. Experiments on droplet collisions, bounce, coalescence and disruption. *Prog. Energy Combustion Sci.*, Vol. 23, pp. 65–79, 1997.
362. T.G.O. Berg, G.C. Fernish, and T.A. Gaukler. The mechanism of coalescence of liquid drops. *J. Atmospheric Sciences*, Vol. 20, pp. 153–158, 1963.
363. M.R.H. Nobari and G. Tryggvason. Numerical simulation of three-dimensional drop collisions. *AIAA Journal*, Vol. 34, No. 4, pp. 750–755, 1996.
364. M. Schelkle, M. Rieber, and A. Frohn. Numerische Simulation von Tropfenkollisionen. *Spektrum der Wissenschaft*, pp. 72–79, Januar 1999.
365. M. Schelkle. *Lattice-Boltzmann-Verfahren zur Simulation dreidimensionaler Zweiphasenströmungen mit freien Oberflächen.* PhD dissertation, Universität Stuttgart, 1997. Verlag Shaker, Aachen.
366. J.W. Ross, W.A. Miller, and G.C. Weatherby. Dynamic computer simulation of viscous flow sintering kinetics. *J. Appl. Phys.*, Vol. 52, No. 6, pp. 3884–3888, 1981.
367. P.S. Prokhorov. The effects of humidity deficit on coagulation processes and the coalescence of liquid drops. *Discussion of the Faraday Society*, Vol. 18, pp. 41–51, 1954.
368. Y. Hiram and A. Nir. A simulation of surface tension driven coalescence. *J. Colloid and Interface Sci.*, Vol. 95, No. 2, pp. 462–470, 1983.
369. H.A. Stone, B.J. Bentley, and L.G. Leal. An experimental study of transient effects in the breakup of viscous drops. *J. Fluid Mech.*, Vol. 173, pp. 131–158, 1986.
370. V.A. Arkhipov, G.S. Ratanov, and V.F. Trofimov. Experimental investigation of the interaction of colliding droplets. *J. Appl. Mech. Tech. Phys. USSR*, Vol. 2, pp. 73–77, 1978.
371. N. Roth, M. Rieber, and A. Frohn. High energy head-on collisions of droplets. In *Proc. 15th Int. Conf. on Liquid Atomization and Spray Systems.* ILASS, 1999.

372. J.P.A.J. van Beeck and M.L. Riethmuller. Détermination simultanée de la vitesse, de la température et du diamètre des gouttes par la méthode de l'arc-en-ciel. In *Proc. 5ème Congrès de Vélocimétrie Laser*, pp. E2–1 – E2–8, 1996.
373. C.K. Law. Recent advances in droplet vaporization and combustion. *Prog. Energy Combustion Sci.*, Vol. 8, pp. 171–201, 1982.
374. N. Roth, K. Anders, and A. Frohn. Determination of size, evaporation rate and freezing of water droplets using light scattering and radiation pressure. *Part. Part. Syst. Charact.*, Vol. 11, pp. 207–211, 1994.
375. W.-J. Yang, K.H. Guo, and T. Uemura. Theory and experiments on evaporating sessile drops of binary liquid mixtures. *Int. J. Heat Mass Transfer*, Vol. 32, No. 7, pp. 1197–1205, 1989.
376. A.L. Randolph, A. Makino, and C.K. Law. Liquid-phase diffusional resistance in multicomponent droplet gasification. In *Proc. 21th Symp. (Int.) on Combustion*, pp. 601–608. The Combustion Institute, 1986.
377. J.C. Yang and C.T. Avedisian. The combustion of unsupported heptane/hexadecane mixture droplets at low gravity. In *Proc. 22th Symp. (Int.) on Combustion*, pp. 2037–2044. The Combustion Institute, 1988.
378. M.M. Elkotb, S.L. Aly, and H.A. Elsalmawy. Evaporation characteristics of fuel and multifuel droplets. *Combustion and Flame*, Vol. 85, pp. 300–308, 1991.
379. N. Roth and A. Frohn. Measuring technique to investigate the evaporation process of droplets consisting of hydrocarbon mixtures. In *Proc. 13th Int. Conf. on Liquid Atomization and Spray Systems*, pp. 402–406. ILASS, 1997.
380. K. Anders, N. Roth, and A. Frohn. A new technique for investigating phase transition processes of optically levitated droplets consisting of water and sulfuric acid. *J. Geophys. Res. - Atm.*, Vol. 101, No. D14, pp. 19223–19229, 1995.
381. K. Anders, A. Frohn, G. Lavergne, and N. Roth. Mesures du diamètre, de la vitesse et de la température des gouttes par méthode optique. In *Proc. 3ème Congrès Francophone de Vélocimétrie Laser*, pp. 4.6.1–4.6.8, 1992.
382. A.J.K. Leiroz and R.H. Rangel. Numerical study of droplet-stream vaporization at zero Reynolds number. *Numerical Heat Transfer, Applications*, Vol. 27, No. 5, pp. 2053–2072, 1995.
383. K. Anders and A. Frohn. Experimental investigation of droplet evaporation in a wide Knudsen number range. In *Proc. of the 14th Int. Symp. on Rarefied Gas Dynamics*, pp. 975–982, 1884.
384. R. Kneer. *Grundlegende Untersuchungen zur Sprühstrahlausbreitung in hochbelasteten Brennräumen: Tropfenverdunstung und Sprühstrahlcharakterisierung.* PhD dissertation, Universität Karlsruhe, 1993. Verlag Shaker, Aachen.
385. G.A.E. Godsave. Studies of the combustion of drops in a fuel spray - the burning of single drops of fuel. In *Proc. 4th Symp. (Int.) on Combustion*, pp. 818–830. The Combustion Institute, 1952.
386. A.R. Hall and J. Diederichsen. An experimental study of the burning of single drops of fuel in air at pressures up to twenty atmospheres. In *Proc. 4th Symp. (Int.) on Combustion*, pp. 837–846. The Combustion Institute, 1952.
387. S. Okajima and S. Kumagai. Experimental studies on combustion of fuel droplets in flowing air under zero- and high-gravity conditions. In *Proc. 19th Symp. (Int.) on Combustion*, pp. 1021–1027. The Combustion Institute, 1982.
388. H. Hara and S. Kumagai. The effect of initial diameter on free droplet combustion with spherical flame. In *Proc. 25th Symp. (Int.) on Combustion*, pp. 423–430. The Combustion Institute, 1994.
389. J.F. Rex, A.E. Fuhs, and S.S. Penner. Interference effects during burning in air for stationary n-heptane, ethyl alcohol, and methyl alcohol droplets. *Jet Propulsion*, Vol. 26, pp. 179–187, 1956.

390. K. Miyasaka and C.K. Law. Combustion of stongly-interacting linear droplet arrays. In *Proc. 18th Symp. (Int.) on Combustion*, pp. 283–292. The Combustion Institute, 1981.
391. J.S. Tsai and A.M. Sterling. The combustion of linear droplet arrays. In *Proc. 23th Symp. (Int.) on Combustion*, pp. 1405–1411. The Combustion Institute, 1990.
392. M. Mikami, H. Kato, and M. Sato, J. Kono. Interactive combustion of two droplets in microgravity. In *Proc. 25th Symp. (Int.) on Combustion*, pp. 431–438. The Combustion Institute, 1994.
393. J.J. Sangiovanni and M. Labowsky. Burning times of linear fuel droplet arrays: A comparison of experiment and theory. *Combustion and Flame*, Vol. 47, pp. 15–30, 1982.
394. N. Roth, K. Anders, and A. Frohn. Size and velocity measurements to investigate interactions between burning fuel droplets. In *Proc. 5th Int. Symp. on Application of Laser Techniques to Fluid Mechanics*, pp. 31.4/1–31.4/6, 1990.
395. N. Roth, K. Anders, and A. Frohn. Experimental investigation of the reduction of burning rate due to a finite spacing between droplets. In R. S. Lee, J. H. Whitelaw, and T. S. Wung, editors, *Aerothermodynamics of Combustors*, pp. 175–183. Springer-Verlag, Berlin, 1992.
396. H.H. Chiu and T.M. Liu. Group combustion of liquid droplets. *Combustion Sci. Tech.*, Vol. 17, pp. 127–142, 1977.
397. S.M. Correa and M. Sichel. The group combustion of a spherical cloud of monodisperse fuel droplets. In *Proc. 19th Symp. (Int.) on Combustion*, pp. 981–991. The Combustion Institute, 1982.
398. W. Ryan, K. Annamalai, and J. Caton. Relation between group combustion and drop array studies. *Combustion and Flame*, Vol. 80, pp. 313–321, 1990.
399. F. Lacas, N. Darabiha, P. Versaevel, J.C. Rolon, and S. Candel. Influence of droplet number density on the structure of strained laminar spray flames. In *Proc. 24th Symp. (Int.) on Combustion*, pp. 1523–1529. The Combustion Institute, 1992.
400. H.H. Chiu, H.Y. Kim, and E.J. Croke. Internal group combustion of liquid droplets. In *Proc. 19th Symp. (Int.) on Combustion*, pp. 971–980. The Combustion Institute, 1982.
401. F. Akamatsu, Y. Mizutani, M. Katsuki, S. Tsushima, and Y.D. Cho. Measurement of local group combustion number of droplet clusters in a premixed spray stream. In *Proc. 26th Symp. (Int.) on Combustion*, pp. 1723–1729. The Combustion Institute, 1996.
402. N. Roth, K. Anders, and A. Frohn. Refractive-index measurements for the correction of particle sizing methods. *Applied Optics*, Vol. 30, No. 33, pp. 4960–4965, 1996.
403. S.V. Sankar, D.H. Buermann, K.M. Ibrahim, and W.D. Bachalo. Application of an integrated phase Doppler interferometer/rainbow thermometer/point diffraction interferometer for characterizing burning droplets. In *Proc. 25th Symp. (Int.) on Combustion*, pp. 413–421. The Combustion Institute, 1994.
404. J.Y. Zhu and D. Dunn-Rankin. Temperature characteristics of a combusting droplet stream. In *Proc. 24th Symp. (Int.) on Combustion*, pp. 1473–1481. The Combustion Institute, 1992.
405. R.H. Rangel and W.A. Sirignano. Unsteady flame propagation in a spray with transient droplet heating. In *Proc. 22th Symp. (Int.) on Combustion*, pp. 1931–1939. The Combustion Institute, 1988.
406. T.H. Lin and Y.Y. Sheu. Theory of laminar flame propagation in near-stoichiometric dilute sprays. *Combustion and Flame*, Vol. 84, pp. 333–342, 1991.

407. T.A. Brzustowski, A. Sobiesiak, and S. Wojcicki. Flame propagation along an array of liquid fuel droplets at zero gravity. In *Proc. 18th Symp. (Int.) on Combustion*, pp. 265–273. The Combustion Institute, 1981.
408. J.J. Sangiovanni and L.G. Dodge. Observations of flame structure in the combustion of monodispersed droplet streams. In *Proc. Colloq. on Turbulent-Combustion Interactions*, pp. 455–465, 1986.
409. M. Queiroz and S.-C. Yao. A parametric exploration on the chaotic behavior of flame propagation in planar sprays. In *Proc. Natural Heat Transfer Conference*, 1981.
410. Y. Mizutani and M. Ogasawara. Laminar flame propagation in droplet suspension of liquid fuel. *Int. J. Heat Mass Transfer*, Vol. 8, pp. 921–935, 1964.
411. S. Hayashi and S. Kumagai. Flame propagation in fuel droplet-vapor-air mixtures. In *Proc. 15th Symp. (Int.) on Combustion*, pp. 445–452. The Combustion Institute, 1974.
412. S. Hayashi, T. Ohtani, K. Iinuma, and S. Kumagai. Limiting factor of flame propagation in low-volatility fuel clouds. In *Proc. 18th Symp. (Int.) on Combustion*, pp. 361–367. The Combustion Institute, 1981.
413. D.R. Ballal and A.H. Lefebvre. Flame propagation in heterogeneous mixtures of fuel droplets, fuel vapor and air. In *Proc. 18th Symp. (Int.) on Combustion*, pp. 321–328. The Combustion Institute, 1981.
414. C.K. Law, A. Lee, and A.L. Randolph. Aerothermochemical studies of energetic liquid materials. 2. Combustion and microexplosion of droplets of organic azides. *Combustion and Flame*, Vol. 71, pp. 123–136, 1988.
415. J.C. Lasheras, L.T. Yap, and F.L. Dryer. Effect of the ambient pressure on the explosive burning of emulsified and multicomponent fuel droplets. In *Proc. 20th Symp. (Int.) on Combustion*, pp. 1761–1772. The Combustion Institute, 1984.
416. C.H. Wang, X.Q. Liu, and C.K. Law. Combustion and microexplosion of freely falling multicomponent droplets. *Combustion and Flame*, Vol. 56, pp. 175–197, 1984.
417. A.L. Randolph and C.K. Law. Time-resolved gasification and sooting characteristics of droplets of alcohol/oil blends and water/oil emulsions. In *Proc. 21th Symp. (Int.) on Combustion*, pp. 1125–1131. The Combustion Institute, 1986.
418. M. Mattiello, L. Cosmai, L. Pistone, F. Beretta, and P. Massoli. Experimental evidence for microexplosions in water/fuel oil emulsion flames inferred by laser light scattering. In *Proc. 24th Symp. (Int.) on Combustion*, pp. 1573–1578. The Combustion Institute, 1992.
419. M. Tsue, T. Kadota, H. Hamaya, and H. Yamasaki. Combustion behavior of an emulsified fuel droplet under high pressure conditions. In *Proc. 3rd Asian-Pacific Int. Symp. on Combustion and Energy Utilization*, pp. 378–383, 1995.
420. M. Tsue, T. Kadota, D. Segawa, and H. Yamasaki. Statistical analysis on onset of microexplosion for an emulsion droplet. In *Proc. 26th Symp. (Int.) on Combustion*, pp. 1629–1635. The Combustion Institute, 1996.
421. H. Appleman. The formation of exhaust condensation trails by jet aircraft. *Bulletin American Meteorological Society*, Vol. 34, No. 1, pp. 14–20, 1953.
422. R.G. Knollenberg. Measurements of the growth of the ice budget in a persisting contrail. *J. Atmospheric Sciences*, Vol. 29, pp. 1367–1374, 1972.
423. B. Kärcher, T. Peter, U.M. Biermann, and U. Schumann. The initial composition of jet condensation trails. *J. Atmospheric Sciences*, Vol. 53, No. 21, pp. 3066–3083, 1996.
424. A.J. Heymsfield and C.M.R. Platt. A parameterization of the particle size spectrum of ice clouds in terms of the ambient temperature and the ice water content. *J. Atmospheric Sciences*, Vol. 41, No. 5, pp. 846–855, 1984.

425. A.J. Heymsfield and R.M. Sabin. Cirrus crystal nucleation by homogeneous freezing of solution droplets. *J. Atmosperic Sciences*, Vol. 46, No. 14, pp. 2252–2264, 1989.
426. R.L. Pitter and H.R. Pruppacher. A wind tunnel investigation of freezing of small water drops falling at terminal velocity in air. *Quart. J. Roy. Met. Soc.*, Vol. 99, pp. 540–550, 1973.
427. B. Krämer, M. Schwell, O. Hübner, H. Vortisch, T. Leisner, E. Rühl, H. Baumgärtel, and L. Wöste. Homogeneous ice nucleation observed in single levitated microdroplets. *Ber. Bunsenges. Phys. Chem.*, Vol. 100, No. 11, pp. 1911–1914, 1996.
428. F. Feuillebois, A. Lasek, P. Creismeas, F. Pigeonneau, and A. Szaniawski. Freezing of a subcooled liquid droplet. *J. Colloid and Interface Sci.*, Vol. 169, pp. 90–102, 1995.
429. W.P. Arnott, Y.Y. Dong, and J. Hallett. Extinction efficiency in the infrared ($2 - 18\,\mu$m) of laboratory ice clouds: Observations of scattering minima in the Christiansen bands of ice. *Applied Optics*, Vol. 34, No. 3, pp. 541–551, 1995.
430. A. Macke. Scattering of light by polyhedral ice crystals. *Applied Optics*, Vol. 32, No. 15, pp. 2780–2788, 1993.
431. V. Freudenthaler, F. Homburg, and H. Jäger. Spatial and optical parameters of contrails in the vortex and dispersion regime determined by means of a ground-based scanning lidar. In *Proc. Int. Colloq. on Impact of Aircraft Emissions upon the Atmosphere*, pp. 159–166. Onera, 1996.
432. K. Sassen. Depolarization of laser light backscattered by artificial clouds. *J. Applied Meteorology*, Vol. 13, pp. 923–933, 1974.
433. N. Roth and A. Frohn. Size and polarization behaviour of optically levitated frozen water droplets. *Atmospheric Environment*, Vol. 32, No. 18, pp. 3139–3143, 1998.
434. A. Umemura and Y. Shimada. Characteristics of supercritical droplet gasification. In *Proc. 26th Symp. (Int.) on Combustion*, pp. 1621–1628. The Combustion Institute, 1996.
435. G. Bauer, G. Lavergne, and A. Frohn. Behaviour of ethane droplets injected near the critical state. In *Proc. 1st French-German Seminar, Research on Liquid Rocket Propulsion, Combustion Phenomena Studies*, 1995.
436. T. Kadota, H. Hiroyasu, and A. Farazandehmer. Soot formation by combustion of a fuel droplet in high pressure gaseous environment. *Combustion and Flame*, Vol. 29, pp. 67–75, 1977.
437. J. Magill, F. Capone, R. Beukers, P. Werner, and R.W. Ohse. Pulsed laser heating of acoustically levitated microspheres under pressure. *High Temperatures - High Pressures*, Vol. 19, pp. 461–471, 1987.
438. P.R. Wieber. Calculated temperature histories of vaporizing droplets to the critical point. *AIAA Journal*, Vol. 1, No. 12, pp. 2764–2770, 1963.
439. J.A. Manrique and G.L. Borman. Calculations of steady state droplet vaporization at high ambient pressures. *Int. J. Heat Mass Transfer*, Vol. 12, pp. 1081–1095, 1969.
440. R.L. Matlosz, S. Leipziger, and T.P. Torda. Investigation of liquid drop evaporation in a high temperature and high pressure environment. *Int. J. Heat Mass Transfer*, Vol. 15, pp. 831–852, 1972.
441. J.-P. Delplanque and W.A. Sirignano. Numerical study of the transient vaporization of an oxygen droplet at sub- and super-critical conditions. *Int. J. Heat Mass Transfer*, Vol. 36, pp. 303–314, 1993.
442. D.G. Talley, R.D. Woodward, T.L. Kaltz, L.N. Long, and M.M. Micci. Experimental and numerical studies of transcritical LOX droplets. In *Proc. 14th Int. Conf. on Liquid Atomization and Spray Systems*, pp. 615–620. ILASS, 1998.

443. U. Meier, M. Decker, A. Schik, and W. Stricker. Mixing properties of cryogenic jets at high pressures studied by Raman imaging. In *Proc. 14th Int. Conf. on Liquid Atomization and Spray Systems*, pp. 609–614. ILASS, 1998.
444. A. Kowalewicz. *Combustion Systems of High-Speed Piston I. C. Engines*. Elsevier, Amsterdam, 1984.
445. B.M. Kim, Y.J. Jang, and J.K. Kim. Burning rate of a pool fire with downward-directed sprays. *Fire Safety Journal*, Vol. 27, pp. 37–48, 1996.
446. T.R. Marker, C.P. Sarkos, and R.G. Hill. Water spray system development and evaluation for enhanced postcrash fire survivability and in-flight protection in cargo compartments. In *Proc. Aircraft Fire Safety*, AGARD-CP-587, pp. 12.1–12.12, 1997.
447. S. Tinker, M. di Marzo, P. Tartarini, S. Chandra, and Y.M. Quiao. Dropwise evaporative cooling: Effect of dissolved gases and effect of surfactants. In *Proc. Int. Conf. on Fire Research and Engineering*, pp. 91–96, 1995.
448. C.C. Chu and M.L. Corradini. One-dimensional transient fluid model for fuel/coolant interaction analysis. *Nucl. Sci. and Engineering*, Vol. 101, pp. 48–71, 1989.
449. T.G. Theofanous. The study of steam explosions in nuclear systems. *Nuclear Engineering and Design*, Vol. 155, pp. 1–26, 1995.
450. C.C. Chu, J.J. Sienicki, B.W. Spencer, W. Frid, and G. Löwenhielm. Ex-vessel melt-coolant interactions in deep waterpool: Studies and accident management for Swedish BWRs. *Nuclear Engineering and Design*, Vol. 155, pp. 159–213, 1995.
451. K. Nikander. Nebulization of suspensions. *J. Aerosol Med.*, Vol. 10, No. 1, pp. 89, 1997.
452. L. Laube and G.W. Benedict. Nebulized insulin. *J. Aerosol Med.*, Vol. 10, No. 1, pp. 90, 1997.
453. Th. Voshaar and D. Köhler. Heparin aerosols. *J. Aerosol Med.*, Vol. 10, No. 1, pp. 90, 1997.
454. J.F. Lewis. Surfactant aerosols. *J. Aerosol Med.*, Vol. 10, No. 1, pp. 91, 1997.
455. R. Jäger-Waldau. Feasibility of drug delivery to the respiratory tract by a mechanical spray pump. *J. Aerosol Med.*, Vol. 7, No. 2, pp. 147–154, 1994.
456. M. Knoch, E. Wunderlich, and S. Geldner. A nebulizer system for highly reproducible aerosol delivery. *J. Aerosol Med.*, Vol. 7, No. 3, pp. 229–237, 1994.
457. O.N.M. McCallion, K.M.G. Taylor, M. Thomas, and A.J. Taylor. Ultrasonic nebulization of fluids with different viscosities and surface tensions. *J. Aerosol Med.*, Vol. 8, No. 3, pp. 281–284, 1995.
458. D.L. Johnson, N.A. Esmen, and M. Polikandritou-Lambros. Spinning disk aerosol generators as a low energy alternative to medical air-jet nebulizers. *J. Aerosol Med.*, Vol. 9, No. 2, pp. 249–261, 1996.
459. W.-J. Li, M. Perzl, R. Ferron, G.A. Batycky, J. Heyder, and D.A. Edwards. The macrotransport properties of aerosol particles in the human oral-pharyngeal region. *J. Aerosol Sci.*, Vol. 29, No. 8, pp. 995–1010, 1998.
460. I. Balásházy, W. Hofmann, and T. Heistracher. Computation of local enhancement factors for the quantification of particle deposition patterns in airway bifurcations. *J. Aerosol Sci.*, Vol. 30, No. 2, pp. 185–203, 1999.
461. I. P. Chung, R. F. Phalen, and D. Dunn-Rankin. Predicted aerosol aspiration efficiency for infants, children, and adults. *Appl. Occup. Environ. Hyg.*, Vol. 8, No. 7, pp. 639–644, 1993.
462. J. Lacey. Special issue: Sampling and rapid assay of bioaerosols. *J. Aerosol Sci.*, Vol. 28, No. 3, pp. 349–538, 1997.
463. K. Masters. *Spray Drying*. George Godwin Limited, London, 2nd edition, 1976.

464. E. Weidner, Z. Knez, and Z. Novak. PGSS (Particles from gas saturated solutions) - a new process for powder generation. In *Proc. 3rd Int. Symp. Supercritical Fluids*, pp. 229–234, 1994.
465. Y.A. Levendis and R.C. Flagan. Combustion of uniformly sized glassy carbon particles. *Combustion Sci. and Tech.*, Vol. 53, pp. 266–271, 1987.
466. J.J. Wu, H.V. Nguyen, and R.C. Flagan. A method for the synthesis of submicron particles. *Langmuir*, Vol. 3, No. 2, pp. 266–271, 1987.
467. E.L. Dreizin. Uniform solid and hollow metal spheres; formation in a pulsed micro-arc and applications. *Proc. Mat. Res. Soc.*, Vol. 372, pp. 263–268, 1995.
468. O. Andersen, S. Hansmann, and K. Bauckhage. Production of fine particles from melts of metals or highly viscous fluids by ultrasonic standing wave atomization. *Part. Part. Syst. Charact.*, Vol. 13, pp. 217–223, 1996.
469. A.R. Frost. Rotary atomization in ligament formation mode. *J. agric. Engng. Res.*, Vol. 26, pp. 63–78, 1981.
470. A. Hewitt. Droplet size and agricultural spraying, part I: Atomization, spray transport, description, drift, and droplet size measurement techniques. *Atomization and Sprays*, Vol. 7, No. 3, pp. 236–244, 1997.
471. M.E. Teske and A.J. Bilanin. Drop size scaling analysis of non-Newtonian fluids. *Atomization and Sprays*, Vol. 4, No. 4, pp. 473–483, 1994.
472. R.W. Dexter. Measurement of extensional viscosity of polymer solutions and its effects on atomization from a spray nozzle. *Atomization and Sprays*, Vol. 6, No. 2, pp. 167–191, 1996.
473. A.H.M. Bresser and W. Salomons. International overview and assessment. In D.C. Adriano and W. Salomons, editors, *Acidic Precipitation*, Vol. 5 of *Advances in Environmental Science*. Springer-Verlag, New York, 1990.
474. R.R. Rogers and M.K. Yau. *A Short Course in Cloud Physics*. International Series in Natural Philosophy. Pergamon Press, Oxford, 3rd edition, 1991.
475. Q. Han. Near-global survey of effective droplet radii in liquid water clouds using ISCCP data. *J. Climate and Applied Meteorology*, Vol. 7, pp. 465–497, 1994.
476. U. Schumann, A. Chlond, A. Ebel, B. Kärcher, H. Pak, H. Schlager, A. Schmitt, and P. Wendling, editors. *Pollutants from air traffic - results of atmospheric research 1992-1997*. Mitteilung 97-04. Deutsches Zentrum für Luft- und Raumfahrt e.V., DLR, 1997.
477. H. Appleman. The formation of exhaust condensation trails by jet air-craft. *Bull. Am. Meteor. Soc.*, Vol. 34, No. 1, pp. 14–20, 1953.
478. R. Busen and U. Schumann. Visible contrail formation from fuels with different sulfur contents. *Geophys. Res. Lett.*, Vol. 22, No. 11, pp. 1357–1360, 1995.
479. J.K. Luers. Wing contamination: Threat to safe flight. *Astronautics and Aeronautics*, Vol. 21, No. 11, pp. 54–59, 1983.
480. L.J. Forney. Droplet impaction on a supersonic wedge: Consideration of similitude. *AIAA Journal*, Vol. 28, No. 4, pp. 650–654, 1990.
481. R. Ashenden, W. Lindberg, and J. Marwitz. Two-dimensional NACA 23012 airfoil performance degradation by supercooled cloud, drizzle, and raindrop icing. *AIAA Paper, No. 96-0870*, 1996.
482. D.R. Miller, H.E. Addy, and R.F. Ide. A study of large droplet ice accretions in the NASA Lewis IRT at near-freezing conditions. *AIAA Paper, No. 96-0934*, 1996.
483. D.C. Jenkins. Disintegration of raindrops by shockwaves ahead of conical bodies. *Philos. Trans. R. Soc. London, Ser. A*, Vol. 260, pp. 153–160, 1966.
484. E.P. Muntz and M. Orme. Characteristics, control, and uses of liquid streams in space. *AIAA Journal*, Vol. 25, No. 5, pp. 746–756, 1987.

485. A.T. Mattick and A. Hertzberg. Liquid droplet radiators for heat rejection in space. *J. Energy*, Vol. 5, pp. 387–393, 1981.

486. A.T. Mattick and A. Hertzberg. The liquid droplet radiator-ultralightweight heat rejection system for efficient energy conversion in space. *Acta Astronautica*, Vol. 9, No. 3, pp. 165–172, 1982.

487. K. Hickman and D.J. Trevoy. Evaporation from liquid surfaces in vacuum. *Vacuum*, Vol. 2, No. 1, pp. 3–18, 1952.

488. H.K. Cammenga. Evaporation mechanisms of liquids. *Current Topics in Material Science*, Vol. 5, pp. 1111–1115, 1980.

489. E.P. Muntz, S.-S. Qian, and M. Dixon. The injection of liquid into high vacuum for R.G.D. and space applications. In *Proc. 14th Int. Symp. Rarefied Gas Dynamics*, Vol. 2, pp. 919–935, 1984.

490. S.F. Simpson, J.R. Kincaid, and F.J. Holler. Microdroplet mixing for rapid reaction kinetics with Raman spectrometric detection. *Anal. Chem.*, Vol. 55, No. 8, pp. 1420–1422, 1983.

491. E. Urlaub, M. Lankers, I. Hartmann, J. Popp, M. Trunk, and W. Kiefer. Applications of the optical trapping technique to analyze chemical reactions in single emulsion particles. *Fresenius J. Anal. Chem.*, Vol. 355, pp. 329–331, 1996.

492. M. Trunk, J. Popp, M. Lankers, and W. Kiefer. Microchemistry: Time dependence of an acid-base reaction in a single optically levitated microdroplet. *Chemical Physics Letters*, Vol. 264, pp. 233–237, 1997.

493. J. Musick, J. Popp, M. Trunk, and W. Kiefer. Polymerization and copolymerization reaction observed in optically levitated aerosol particles. *J. Aerosol Sci.*, Vol. 27, Supplement 1, pp. S561–S562, 1996.

494. A. Lawley. Atomization of specialty powders. *J. Metals*, Vol. 13, pp. 13–18, 1981.

495. E.J. Lavernia, J.D. Ayers, and T.S. Srivatsan. Rapid solification processing with specific application to aluminium alloys. *Int. Materials Reviews*, Vol. 32, No. 1, pp. 1–44, 1992.

496. W.T. Pimbley. Drop formation from a liquid jet: A linear one-dimensional analysis considered as a boundary value problem. *IBM J. Res. Develop.*, Vol. 20, pp. 148–156, 1976.

497. W.T. Pimbley and H.C. Lee. Satellite droplet formation in a liquid jet. *IBM J. Res. Develop.*, Vol. 21, pp. 21–30, 1977.

498. A. Prosperettti and M.S. Plesset. Vapour-bubble growth in superheated liquid. *J. Fluid Mech.*, Vol. 85, part 2, pp. 349–368, 1978.

499. S. Fujikawa and T. Akamatsu. Effects of the non-equilibrium condensation of vapour on the pressure wave produced by the collapse of a bubble in a liquid. *J. Fluid Mech.*, Vol. 97, part 3, pp. 481–512, 1980.

500. D.B. Bogy and F.E. Talke. Experimental and theoretical study of wave propagation phenomena in drop-on-demand ink jet devices. *IBM J. Res. Develop.*, Vol. 28, No. 3, pp. 314–320, 1984.

501. J.E. Fromm. Numerical calculation of the fluid dynamics of drop-on-demand jets. *IBM J. Res. Develop.*, Vol. 28, No. 3, pp. 322–333, 1984.

502. A.A. Allen, J.D. Meyer, and W.R. Knight. Thermodynamics and hydrodynamics of thermal ink jets. *Hewlett-Packard Journal*, Vol. 35, No. 5, pp. 21–27, 1985.

503. A. Asai, T. Hara, and I. Endo. One-dimensional model of bubble growth and liquid flow in bubble jet printers. *Japanese J. Applied Physics*, Vol. 26, No. 10, pp. 1794–1801, 1987.

504. J.F. Olivier. Initial stages of ink jet drop impaction, spreading and wetting on paper. *Tappi Journal*, Vol. 67, No. 10, pp. 90–94, 1984.

505. W. Runge. *Berrechnungsmodell thermischer Tintenschreibwerke.* Fortschr.-Ber. VDI Reihe 1, Nr.219. VDI-Verlag, Düsseldorf, 1993.
506. S. Aden, J.H. Bohórquez, D.M. Collins, M.D. Crook, A. García, and U.E. Hess. The third-generation HP thermal InkJet printhead. *Hewlett-Packard Journal*, Vol. 45, No. 1, pp. 41–45, 1994.
507. W.R. Cox, T. Chen, D.W. Ussery, D.J. Hayes, and R.F. Hoenigman. Microjet printing of an amorphic microlens arrays. *SPIE*, Vol. 2687, pp. 89–98, 1996.
508. C.A. Chen and J.-H. Chun. Development of a droplet-based manufacturing process for free-form fabrication. *Annals of the CIRP*, Vol. 46, No. 1, pp. 131–134, 1997.
509. F. Gao and A.A. Sonin. Precise deposition of molten microdrops: The physics of digital microfabrication. *Proc. R. Soc. London, Ser. A*, Vol. 444, pp. 533–554, 1994.
510. St. Schiaffino and A.A. Sonin. Motion and arrest of a molten contact line on a cold surface: An experimental study. *Phys. Fluids*, Vol. 9, No. 8, pp. 2217–2226, 1997.
511. St. Schiaffino and A.A. Sonin. On the theory for the arrest of an advancing molten contact line on a cold solid of the same material. *Phys.Fluids*, Vol. 9, No. 8, pp. 2227–2233, 1997.
512. M.E. Orme, C. Huang, and J. Courter. Precision droplet-based manufacturing and material synthesis: Fluid dynamics and thermal control issues. *Atomization and Sprays*, Vol. 6, No. 3, pp. 305–329, 1996.
513. J.-H. Chun and C.H. Passow. Droplet based manufacturing. *Annals of the CIRP*, Vol. 42, No. 1, pp. 235–238, 1993.
514. P.F. Blazdell, J.R.G. Evans, M.J. Edirisinghe, P. Shaw, and M.J. Binstead. The computer aided manufacture of ceramics using multilayer jet printing. *J. Material Science Letters*, Vol. 14, pp. 1562–1565, 1995.
515. S.T. Vuong and S.S. Sadhal. Growth and translation of a liquid-vapour compound drop in a second liquid. *J. Fluid Mech.*, Vol. 209, part 1, pp. 617–637, 1989.
516. S.T. Vuong and S.S. Sadhal. Growth and translation of a liquid-vapour compound drop in a second liquid. *J. Fluid Mech.*, Vol. 209, part 2, pp. 639–660, 1989.
517. H. Hashimoto and S. Kawano. Mass transfer around a moving encapsulated liquid drop. *Int. J. Multiphase Flow*, Vol. 19, No. 1, pp. 213–218, 1993.
518. J.D. Dziezak. Microencapsulation and encapsulated ingredients. *Ford Technology*, Vol. 42, pp. 136–151, 1988.
519. N.P. Rav, N. Tymiak, J. Blum, A. Neuman, H.J. Lee, S.L. Girshik, P.H. McMurray, and J. Heberlein. Hypersonic plasma particle deposition of nanostructured silicon and silicon carbide. *J. Aerosol Sci.*, Vol. 29, No. 516, pp. 707–720, 1998.
520. A. Gomez, D. Bingham, L. de Juan, and K. Tang. Production of protein nanoparticles by electrospray drying. *J. Aerosol Sci.*, Vol. 29, No. 516, pp. 561–574, 1998.
521. R.W. Siegel. Synthesis and properties of nanophase materials. *Materials Science and Engineering*, Vol. A 168, No. 2, pp. 189–197, 1993.
522. P.J. Antaki. Liquid vaporization and combustion from slurry fuel droplets. In *Thermal Treatment Hazardous Wastes*, Vol. 1 of *Encyclopedia of Environmental Control Technology.* Gulf Publishing Company, Houston, 1989.
523. T. Sakai and M. Saito. Single-droplet combustion of coal slurry fuels. *Combustion and Flame*, Vol. 51, pp. 141–154, 1983.

524. D.J. Maloney and J.F. Spann. Evaporation, agglomeration, and explosive boiling characteristics of coal-water fuels under intense heating conditions. In *Proc. 22th Symp. (Int.) on Combustion*, pp. 1999–2008. The Combustion Institute, 1988.
525. C.K. Law, H.K. Law, and C.H. Lee. Combustion characteristics of coal/oil and coal/oil/water mixtures. *Energy*, Vol. 4, pp. 329–339, 1979.
526. K. Miyasaka and C.K. Law. Combustion and agglomeration of coal oil mixtures in furnace environments. *Combustion Sci. Tech.*, Vol. 24, pp. 71–82, 1980.
527. P.M. Walsh, M. Zhang, W.F. Farmayan, and J.M. Beér. Ignition and combustion of coal-water slurry in a confined turbulent diffusion flame. In *Proc. 20th Symp. (Int.) on Combustion*, pp. 1401–1407. The Combustion Institute, 1984.
528. R. Yavuz, A. Williams, and Küçükbayrac. Combustion behavior of low rank coal water slurries. In *Proc. 13th Annual Int. Pittsburgh Coal Conf.*, Vol. 1, pp. 500–505, 1996.
529. W.A. Sirignano and R. Bhatia. Metal slurry droplet and spray combustion. *Progress in Astronautics and Aeronautics, Advances in Combustion Science*, Vol. 173, pp. 117–130, 1996.
530. F. Okasha and M. Miccio. Prediction of coal-water dispersion in a fluidized bed combustor. In *Proc. 26th Symp. (Int.) on Combustion*, pp. 3277–3285. The Combustion Institute, 1996.
531. K. Takeno, K. Tokuda, and T. Ichinose. Fundamental experiment on the combustion of coal-water mixture and modeling of the process. In *Proc. 26th Symp. (Int.) on Combustion*, pp. 3223–3230. The Combustion Institute, 1996.
532. P. Becher. *Emulsions: Theory and Practice.* Reinhold Publishing Corporation, New York, 2nd edition, 1965.
533. S.-H. Chen and R. Rajagopalan, editors. *Micellar Solutions and Microemulsions.* Springer-Verlag, New York, 1965.
534. W.H. Marlow. Introduction: The domains of aerosol physics. In W.H. Marlow, editor, *Aerosol Microphysics I, Particle Interaction*, Topics in Current Physics. Springer-Verlag, Berlin, 1980.
535. N.A. Fuchs. *The Mechanics of Aerosols.* Pergamon Press, Oxford, 1964.
536. C.N. Davies. *Aerosol Science.* Academic Press, London, 1966.
537. R.P. Dron. The wettability of road aggregates with doped bituminous binders. In R.F. Gould, editor, *Contact Angle, Wettability and Adhesion*, Advances in Chemistry Series 43. American Chemical Society, Washington D.C., 1964.

Index

Printing: Mercedes-Druck, Berlin
Binding: Stein + Lehmann, Berlin

Lightning Source UK Ltd.
Milton Keynes UK
20 November 2009

146442UK00007B/144/A

9 783540 658870